PENGUIN BOOKS

THE NEW PENGUIN DICTIONARY OF GEOLOGY

Philip Kearey was born in July 1948 and educated at the University of Durham, where he gained his Ph.D. in geophysics in 1973. He began his career as Senior Research Assistant in the Department of Geological Sciences at Durham and he is now Senior Lecturer in Geophysics at Bristol University. He is both Fellow and Chartered Geologist of the Geological Society of London. He has published numerous papers in various scientific journals and is the editor of *The Encyclopedia of the Solid Earth Sciences* (1993). He co-wrote *An Introduction to Geophysical Exploration* (1984; revised 1991) with M. Brooks, and with F. J. Vine he has published *Global Tectonics* (1990).

PHILIP KEAREY

THE NEW PENGUIN DICTIONARY OF GEOLOGY

PENGUIN BOOKS

PENGUIN BOOKS

Published by the Penguin Group
Penguin Books Ltd, 27 Wrights Lane, London W8 5TZ, England
Penguin Books USA Inc., 375 Hudson Street, New York, New York 10014, USA
Penguin Books Australia Ltd, Ringwood, Victoria, Australia
Penguin Books Canada Ltd, 10 Alcorn Avenue, Toronto, Ontario, Canada M4V 3B2
Penguin Books (NZ) Ltd, 182–190 Wairau Road, Auckland 10, New Zealand

Penguin Books Ltd, Registered Offices: Harmondsworth, Middlesex, England

First published 1996
1 3 5 7 9 10 8 6 4 2

Typeset, from data supplied, by Datix International Limited, Bungay, Suffolk
Printed in England by Clays Ltd, St Ives plc
Set in 8.75/10pt Monophoto Bembo

To Jane

INTRODUCTION

This dictionary contains 7606 entries which cover definitions of terms relevant to Geology. Because of the overlap which exists between the many branches of the Earth Sciences, for example between Geomorphology and Sedimentology or between Palaeontology and Biology, some of the entries might be considered to be at the borders of Geology, but they are included because of their importance to the subject.

For reasons of economy of space, I have concentrated on producing concise definitions of the entries rather than longer treatments. If further information should be required about any particular entry, it should be sought from the 348 references, arranged by topic, provided in the Bibliography, which lists modern works covering all aspects of Geology.

Also for reasons of space, I have omitted descriptions of individual geologists. Similarly, it would have been impossible to include definitions of, for example, every mineral ever described, every small local glacial event or every local stratigraphy. I hope that the reader will agree that the most important have been selected.

Entries and synonyms are distinguished by bold type, whilst those words which appear in the entries and which are entries themselves appear in italic type. Cross-references to related terms are labelled 'See also' and references to contrasting terms are labelled 'cf.' Abbreviations used in the text are as follows: Ma = million years, Ga = billion (10^9) years, L. = Lower, M. = Middle, U. = Upper, ~ = approximately.

Spelling in most cases follows the conventions used in Great Britain. Some, but not all, alternative spellings are given in American usage. Alternatives have not always been given for words containing once-ligatured diphthongs, e.g. 'ae'. Thus 'aeolian' appears but not 'eolian'. I hope that this will not seriously inconvenience readers.

All stratigraphic ages are given according to Harland, W. B., Armstrong, R. L., Cox, A. V., Craig, L. E., Smith, A. G. and Smith, D. G. (1990) *A Geologic Timescale 1990*. Cambridge University Press, Cambridge.

I would welcome suggestions for the inclusion of terms which do not appear in the present dictionary.

PHILIP KEAREY
Bristol, January 1995

Era	Sub-era / Period / Sub-period	Epoch	Stage	Age Ma	Stage abbrev.	Intervals Ma	
Cenozoic (Cz)	Quaternary or Pleistogene	Holocene		0.01	Hol	0.01	
		Pleistocene		1.64	Ple	1.63	
	Tertiary (TT), Neogene (Ng)	Pliocene 2	Piacenzian	3.4	Pia	3.6	22
		Pli 1	Zanclian	5.2	Zan		
		Miocene 3	Messinian	6.7	Mes	5.2	
			Tortonian	10.4	Tor		
		Miocene 2	Serravallian	14.2	Srv	5.9	
			Langhian	16.3	Lan		
		Miocene 1	Burdigalian	21.5	Bur	7.0	
		Mio	Aquitanian	23.3	Aqt		
	Paleogene (Pg)	Oligocene 2	Chattian	29.3	Cht	6.0	42
		Oli 1	Rupelian	35.4	Rup	6.1	
		Eocene 3	Priabonian	38.6	Prb	3.2	
		Eocene 2	Bartonian	42.1	Brt	11.4	
			Lutetian	50.0	Lut		
		Eoc 1	Ypresian	56.5	Ypr	6.5	
		Paleocene 2	Thanetian	60.5	Tha	4.0	
		Pal 1	Danian	65.0	Dan	4.5	
Mesozoic (Mz)	Cretaceous (K)	Gulf (Gul), Senonian (Sen)	Maastrichtian	74.0	Maa	9.0	81
			Campanian	83.0	Cmp	9.0	
			Santonian	86.5	San	3.6	
			Coniacian	88.5	Con	1.9	
		Gallic (Gal)	Turonian	90.4	Tur	1.9	
			Cenomanian	97.0	Cen	6.6	
		K_1	Albian	112.0	Alb	15.0	
			Aptian	124.5	Apt	12.5	
			Barremian	131.8	Brm	7.3	
		Neocomian (Neo)	Hauterivian	135.0	Hau	3.2	
			Valanginian	140.7	Vlg	5.7	
			Berriasian	145.6	Ber	4.9	
	Jurassic (J)	Malm (J_3, Mlm)	Tithonian	152.1	Tth	6.5	62
			Kimmeridgian	154.7	Kim	2.6	
			Oxfordian	157.1	Oxf	2.4	
		Dogger (J_2, Dog)	Callovian	161.3	Clv	4.2	
			Bathonian	166.1	Bth	4.8	
			Bajocian	173.5	Baj	7.4	
			Aalenian	178.0	Aal	4.5	
		Lias (J_1, Lia)	Toarcian	187.0	Toa	9.0	
			Pliensbachian	194.5	Plb	7.5	
			Sinemurian	203.5	Sin	9.0	
			Hettangian	208.0	Het	4.5	
	Triassic (Tr)	Tr_3	Rhaetian	209.5	Rht	1.5	37
			Norian	223.4	Nor	13.9	
			Carnian	235.0	Crn	11.6	
		Tr_2	Ladinian	239.5	Lad	4.5	
			Anisian	241.1	Ans	1.6	
		Scythian (Tr_1, Scy)	Spathian	241.9	Spa	0.8	
			Nammalian	243.4	Nml	1.5	
			Griesbachian	245.0	Gri	1.6	
Paleozoic	Permian (P)	Zechstein (Zec)	Changxingian	247.5	Chx	2.5	45
			Longtanian	250.0	Lgt	2.5	
			Capitanian	252.5	Cap	2.5	
			Wordian	255.0	Wor	2.5	
			Ufimian	256.1	Ufi	1.1	
		Rotliegendes (Rot)	Kungurian	259.7	Kun	3.6	
			Artinskian	268.8	Art	9.1	
			Sakmarian	281.5	Sak	12.7	
			Asselian	290.0	Ass	8.5	
	Carboniferous, Pennsylvanian (C_2)	Gzelian (Gze)	Noginskian	293.6	Nog	3.6	33
			Klazminskian	295.1	Kla	1.5	
		Kasimovian (Kas)	Dorogomilovskian	298.3	Dor	3.2	
			Chamovnicheskian	299.9	Chv	1.6	
			Krevyakinskian	303.0	Kre	3.1	
		Moscovian (Mos)	Myachkovskian	305.0	Mya	2.0	
			Podolskian	307.1	Pod	2.1	
			Kashirskian	309.2	Ksk	2.1	
			Vereiskian	311.3	Vrk	2.1	
		Bashkirian (Bsh)	Melekesskian	313.4	Mel	2.1	
			Cheremshanskian	318.3	Che	4.9	
			Yeadonian	320.6	Yea	2.3	
			Marsdenian	321.5	Mrd	0.9	
			Kinderscoutian	322.8	Kin	1.3	
	C_1	Serpukhovian	Alportian	325.6	Alp	2.8	
			Chokierian		Cho	2.7	40

Era	Sub-era / Period / Sub-per	Epoch	Stage	Age Ma/Ga	Stage abbrev.	Intervals Ma	
Paleozoic (Pz)	Carboniferous (C), C_2	Bashkirian	Marsdenian	321.5	Mrd	11.5	33
			Kinderscoutian	322.8	Kin		
	Mississippian (C_1)	Serpukhovian (Spk)	Alportian	325.6	Alp	10	40
			Chokierian	328.3	Cho		
			Arnsbergian	331.1	Arn		
			Pendleian	332.9	Pnd		
		Visean (Vis)	Brigantian	336.0	Bri	17	
			Asbian	339.4	Asb		
			Holkerian	342.8	Hlk		
			Arundian	345.0	Aru		
			Chadian	349.5	Chd		
		Tournaisian (Tou)	Ivorian	353.8	Ivo	13	
			Hastarian	362.5	Has		
	Devonian (D)	D_3	Famennian	367.0	Fam	15	46
			Frasnian	377.4	Frs		
		D_2	Givetian	380.8	Giv	9	
			Eifelian	386.0	Eif		
		D_1	Emsian	390.4	Ems	22	
			Pragian	396.3	Pra		
			Lochkovian	408.5	Lok		
	Silurian (S), S_4	Pridoli (Prd)		410.7	Prd	2	31
	S_3	Ludlow (Lud)	Ludfordian	415.1	Ldf	13	
			Gorstian	424.0	Gor		
	S_2	Wenlock (Wen)	Gleedonian	425.4	Gle	6.5	
			Whitwellian	426.1	Whi		
			Sheinwoodian	430.4	She		
	S_1	Llandovery (Lly)	Telychian	432.6	Tel	8.5	
			Aeronian	436.9	Aer		
			Rhuddanian	439.0	Rhu		
	Ordovician (O), Bala (Bal)	Ashgill (Ash)	Hirnantian	439.5	Hir	4	71
			Rawtheyan	440.1	Raw		
			Cautleyan	440.6	Cau		
			Pusgillian	443.1	Pus		
		Caradoc (Crd)	Onnian	444.0	Onn	21	
			Actonian	444.5	Act		
			Marshbrookian	447.1	Mrb		
			Longvillian	449.7	Lon		
			Soudleyan	457.5	Sou		
			Harnagian	462.3	Har		
			Costonian	463.9	Cos		
	Dyfed (Dfd)	Llandeilo (Llo)	Late	465.4	Llo_3	4.5	
			Mid	467.0	Llo_2		
			Early	468.6	Llo_1		
		Llanvirn (Lln)	Late	472.7	Lln_2	7.5	
			Early	476.1	Lln_1		
	Canadian (Cnd)	Arenig		493.0	Arg	17	
		Tremadoc		510.0	Tre	17	
	Cambrian (€)	Merioneth (Mer)	Dolgellian	514.1	Dol	7	60
			Maentwrogian	517.2	Mnt		
		St David's (StD)	Menevian	530.2	Men	19	
			Solvan	536.0	Sol		
		Caerfai (Crf)	Lenian	554	Len	18	
			Atdabanian	560	Atb	16	
			Tommotian	570	Tom		
Sinian (Z)	Vendian (V)	Ediacara (Edi)	Poundian	580	Pou	20	40
			Wonokan	590	Won		
		Varanger (Var)	Mortensnes	600	Mor	20	
			Smalfjord	610	Sma		
	Sturtian			0.80	Stu		190
Riphean (Rif)			Karatau	1.05	Kar		250
			Yurmatin	1.35	Yur		300
			Burzyan	1.65	Buz		300
Animikean				2.2	Ani		550
Huronian				2.45	Hur		250
Randian				2.8	Ran		350
Swazian				3.5	Swz		700
Isuan				3.8	Isu		300
Hadean (Hde)			Early Imbrian	3.85	Imb		50
			Nectarian	3.95	Nec		100
			Basin Groups 1–9	4.15	BG1–9		200
			Cryptic	4.56	Cry		410

Table 1 **The Stratigraphic Column.** From Harland, W.B., Armstrong, R.L., Cox, A.V., Craig, L.E., Smith, A.G. and Smith, D.G. (1990) *A Geologic Timescale 1990*. Cambridge University Press, Cambridge.

Quantity	SI name	SI Symbol	c.g.s. equivalent	Imperial (U.S.A.) equivalent
Mass	kilogram	kg	10^3 g	2.205 lb
Time	second	s	s	s
Length	metre	m	10^2 cm	39.37 in 3.281 ft
Acceleration	metre s^{-2}	m s^{-2}	10^2 cm s^{-2} = 10^2 gal	39.37 in s^{-2}
Gravity	gravity unit	gu = μm s^{-2}	10^{-1} milligal (mgal)	3.937×10^{-5} in s^{-2}
Density	megagram m^{-3}	Mg m^{-3}	g cm^{-3}	3.613×10^{-2} lb in^{-3} 62.421 lb ft^{-3}
Force	newton	N	10^5 dyne	0.2248 lb (force)
Pressure	pascal	Pa = N m^{-2}	10 dyne cm^{-2} = 10^{-5} bar	1.45×10^{-4} lb in^{-2}
Energy	joule	J	10^7 erg	0.7375 ft lb
Power	watt	W = J s^{-1}	10^7 erg s^{-1}	0.7375 ft lb s^{-1} 1.341×10^{-3} hp
Temperature	T	°C*	°C	(1.8T + 32) °F
Current	ampere	A	A	A
Potential	volt	V	V	V
Resistance	ohm	Ω = V A^{-1}	Ω	Ω
Resistivity	ohm m	Ω m	10^2 Ω cm	3.281 ohm ft
Conductance	siemen	S = $Ω^{-1}$	mho	mho
Conductivity	siemen m^{-1}	S m^{-1}	10^{-2} mho cm^{-1}	0.3048 mho ft^{-1}
Dielectric constant	dimensionless			
Magnetic flux	weber	Wb = V s	10^8 maxwell	
Magnetic flux density (B)	tesla	T = Wb m^{-2}	10^4 gauss (G)	
Magnetic anomaly	nanotesla	nT = 10^{-9} T	gamma (γ) = 10^{-5} G	
Magnetizing field (H)	ampere m^{-1}	A m^{-1}	$4\pi 10^{-3}$ oersted (Oe)	
Inductance	henry	H = Wb A^{-1}	10^9 emu (electromagnetic unit)	
Permeability of vacuum (μ_0)	henry m^{-1}	$4\pi 10^{-7}$ H m^{-1}	1	
Susceptibility	dimensionless	k	4π emu	
Magnetic pole strength	ampere m	A m	10 emu	
Magnetic moment	ampere m^2	A m^2	10^3 emu	
Magnetization (J)	ampere m^{-1}	A m^{-1}	10^{-3} emu cm^{-3}	

* Strictly, SI temperatures should be stated in Kelvin (K = 273.15 + °C). In this book, however, temperatures are given in the more familiar Centigrade (Celsius) scale.

Table 2 **S.I., C.g.s. and Imperial (Customary U.S.) Units and Conversion Factors.**

A

Å Symbol for *Ångström.*

A-subduction See *Amperer subduction.*

aa A solidified *lava flow* with a very rough or clinkery surface.

Aalenian The lowest *stage* of the M. *Jurassic* (*Dogger*), 178.0–173.5 Ma.

AAS See *atomic absorption spectroscopy.*

Ab Method of indicating *plagioclase feldspar* composition as a percentage of *albite,* e.g. Ab_{15} indicates a composition of 15% *albite*, 85% *anorthite.*

abandonment The voluntary surrender of legal rights or title to a mining claim.

Abbé refractometer An instrument for determining the *refractive index* of *minerals* and liquids.

ablation The removal of detritus by wind action.

abnormal pressure (overpressure, geopressure) The pressure in a formation exceeding the *hydrostatic pressure* of a water column of *density* 1.114 Mg m^{-3}. Important in controlling the *migration* of fluids.

abrasion Mechanical *erosion* by debris-charged wind, water or ice, which also removes the eroded material. See also *ablation.*

abrasion platform Any horizontal surface cut into a slope.

abrasive Any material suitable for grinding, polishing, scouring or cutting, e.g. *diamond, corundum, sand, pumice.*

absarokite A *porphyritic basalt* with a small amount of *orthoclase* in the *groundmass.*

absolute age. The age of a rock or *formation* with respect to the present. Determined by *radiometric dating* methods. cf. *relative age.*

absolute permeability The ability of a rock at 100% saturation to transmit a particular fluid. cf. *effective permeability.*

absolute plate motion The motion of a *plate* with respect to the interior of the Earth. Absolute motions can be determined by making use of the *hotspot reference frame*. cf. *relative plate motion.*

absolute temperature Temperature measured on the Kelvin scale, with respect to absolute zero (−273.15°C).

Abukuma-type metamorphism *Metamorphism* at a high *geothermal gradient* and low pressure, characterized by the presence of *andalusite* and *sillimanite* in *pelitic* rocks.

abutment load The weight transferred to adjacent solid rock in a deep excavation.

abyss A very deep oceanic depression.

abyssal plain (basin plain, basin floor) A flat, generally smooth, sediment-covered, deep ocean floor.

ac joint See *cross joint*.

Acadian orogeny An *orogeny* affecting the northern parts of the *Appalachian fold belt* in the *Devonian*, corresponding to the *Caledonian orogeny* of Europe. See also *Taconic orogeny*.

acanthite (Ag_2S) An *ore mineral* of *silver*.

Acanthodiformes An order of subclass *Acanthodii*, class *Osteichthyes*, superclass *Pisces*; small, shark-like bony *fish*. Range U. *Silurian*–L. *Permian*.

Acanthodii A subclass of class *Osteichthyes*, superclass *Pisces*; primitive *fish* characterized by bony spines along the front edges of their fins and no bony internal skeleton. Range *Silurian*–*Permian*.

accelerated erosion The increased rate of *erosion* resulting from human activity.

accelerator mass spectrometry dating (AMS dating, accelerator radiocarbon dating) A *radiocarbon dating* method in which a *mass spectrometer* is used to detect ^{14}C atoms. Makes use of smaller quantities of material than required for conventional *radiocarbon dating* and extends its range to beyond 50 000 years.

accelerator radiocarbon dating See *accelerator mass spectrometry dating*.

accessory lithic A *clast* in a *pyroclastic* rock formed of material torn from a *volcanic vent*'s walls during a *volcanic eruption*.

accessory mineral A *mineral* comprising less than ~10% of a rock which is insignificant to nomenclature or classification. cf. *essential mineral*.

accessory plate A plate of a specially cut *mineral* for use in a polarizing microscope, used in determining the character of a *mineral* in *thin section*.

accidental lithic A *clast* in a *pyroclastic* rock plucked from the ground during the transport of *tephra*.

accommodation structure A small *structure* which allows a bed to fill all available space created during *deformation*.

accordion fold See *chevron fold*.

accretion The process by which inanimate objects increase in size by the addition of material to their surfaces, e.g. accretion of continents, in which a *craton* increases in size by the welding to it of *lithosphere* brought into juxtaposition by *subduction*.

accretion vein A *mineral vein* in which more than one phase of *fracture* and infilling has occurred.

accretionary prism (accretionary wedge, subduction complex) A pile of sediments, characterized by repeated *thrust faults*, which accumulates on the oceanward side of a *subduction zone* and grows by the *offscraping* of sediment from the top of the downgoing *plate* by the leading edge of the overriding *plate*.

accretive plate margin See *constructive plate margin*.

ACF diagram Triangular graph whose axes are Al_2O_3, CaO and FeO + MgO, used to illustrate chemical variation in a suite of rocks.

achnelith (Pelé's hair) Hair-like volcanic *glass* formed by *lava* exuding through a small orifice and blown by the wind.

achondrite A class of *stony meteorite* with no *chondrules*.

achroite A white, potassium-rich variety of *tourmaline*.

acicular Needle-like.

acid clay A clay which releases hydrogen ions when in water suspension, e.g. *Fuller's earth*.

acid rain Rainwater with a *pH* of less than 4 arising from the *dissolution* of gases produced naturally or from industrial processes.

acid rock See *acidic rock*.

acidic rock (acid rock) An *igneous rock* with >10% free *silica*.

aclinic line See *magnetic equator*.

acme zone (epibole, peak zone) A *biozone* in which a particular species or genus is at its maximum abundance.

acmite See *aegirine*.

acoustic basement The boundary between overlying sediments and underlying *igneous* or *metamorphic rocks*, characterized by strong *seismic reflections* on a *seismogram*.

acoustic impedance The product of *seismic velocity* and bulk *density*. The acoustic impedance contrast across a boundary determines the proportions of seismic energy transmitted and reflected at the boundary.

acoustic log A *geophysical borehole log* in which measurements are taken which utilize the properties of acoustic wave propagation.

acritarchs A diverse group of micro-organisms with hollow, organic-walled vesicles, 5–500 μm in diameter. Probably produced by several groups of *protists* and useful in *biostratigraphy*. Range *Precambrian–Recent*.

actinides The elements with atomic numbers 89–104, with properties similar to actinium.

actinolite ($Ca_2(Mg,Fe)_5Si_8O_{22}(OH)_2$) A green *amphibole*.

Actinopterygii A subclass of class *Osteichthyes*, superclass *Pisces*; the ray-fin *fishes*. Range *Devonian–Recent*.

activation analysis A technique of identifying stable isotopes by irradiating a sample with neutrons, charged particles or gamma rays. The induced radioactivity allows identification of isotopes from their characteristic radiation.

active continental margin A *continental margin* that is also a *subduction zone*. cf. *passive continental margin*.

Actonian A *stage* of the *Ordovician*, 444.5–444.0 Ma.

actualism See *uniformitarianism*.

adamantine lustre A brilliant, sparkling *lustre* arising from a *mineral*'s being transparent and having a high *refractive index*, e.g. *diamond*, *cerussite*.

adamellite A coarse-grained, *acid igneous rock* with *plagioclase* comprising more than two thirds of the *feldspar* present.

Adams-Williamson equation An equation describing the relationship between *seismic velocity*, *gravity* and the *adiabatic* change in *density* within the Earth.

adaptive divergence See *adaptive radiation*.

adaptive radiation (adaptive divergence) An evolutionary trend by which organisms increase in *diversity* as they adapt to occupy a large number of ecological environments.

addition rule See *Weiss zone law*.

Adelaidean orogeny An *orogeny* affecting S Australia from late *Proterozoic–Ordovician*.

adhesion A process in which dry *sand* blown onto a damp surface is held by the surface tension of water rising between the grains by capillary action.

adhesion lamination (quasi-planar adhesion stratification) A flat to low angle *lamination* with a crinkly appearance and very good *sorting*. Forms in well-sorted *sand* by *adhesion* resulting from a strong wind acting on a slightly damp surface.

adhesion plane bed The bedform corresponding to *adhesion lamination*.

adhesion pseudo cross-lamination The cross-laminae forming in *tabular* sets when climbing *adhesion ripples* deposit well-sorted *sand*.

adhesion ripple (aeolian microridge, anti-ripplet) A millimetric scale, straight or sinuous ridge forming orthogonal to a unidirectional wind by *adhesion*. The *ripples* have steep upwind sides and shallow dipping *lee sides*, and migrate downwind.

adhesion structure A sedimentary *structure* formed by *adhesion*.

adhesion wart A submillimetric to millimetric, irregular protuberance believed to form by *adhesion* under strong, frequently shifting winds.

adiabatic Describing the relationship between pressure and volume as a substance expands or compresses without emitting or absorbing heat.

adinole An *argillaceous* rock which has undergone *albitization* during *contact metamorphism*.

adit A subhorizontal tunnel driven into a hillside.

admissible section A geological section in which the observed *structures* maintain the same style below the surface. A *balanced section* can be constructed only from such a section.

admission The substitution of a *trace element* for a *major element* with similar ionic radius during *magma crystallization*.

adobe A clay similar to *loess*.

adoral On the same side of the body as the mouth.

adsorption Adhesion to the surface of a material.

adularescence The milky-bluish sheen shown by *moonstone*.

adularia ($KAlSi_3O_8$) A colourless, *translucent* variety of *potash feldspar*.

advection The lateral transport of fluid, such as occurs in a *convection current*.

adventive cone See *parasitic cone*.

aegirine (acmite) ($NaFe^{3+}Si_2O_6$) A sodic *pyroxene* commonly occurring as a late product in the *crystallization* of alkaline *magma*.

aegirine-augite A *pyroxene* intermediate in composition between *aegirine* and *augite*.

aegirite An obsolete name for *aegirine.*

aenigmatite (cossyrite) ($Na_2Fe^{2+}{}_5Ti O_2(Si_2O_6)_3$) A rare, titanium-bearing silicate.

aeolian bounding surface An *erosion* surface truncating *strata* within wind-deposited sediments.

aeolian micro-ridge See *adhesion ripple.*

aeolian placer A *placer deposit* commonly formed by the reworking of a *beach placer* by the wind.

aeolian plane-bed A *slip-faced* bedform formed on a subhorizontal surface in an *aeolian sand sheet.*

aeolian sand sheet An extensive area of wind-deposited *sand*, with only rare *slip-faced* bedforms, commonly found around *dune fields.*

aeolian stratification The *bedding* and *lamination* formed as wind-blown sediment, usually of *sand* grade, accumulates on a dry surface. See also *adhesion structure.*

aeolianite A cemented, wind-blown sediment. The *clasts* may be grains of *quartz*, calcium carbonate, *gypsum*, etc. and the cement commonly calcium carbonate, although other water-soluble *minerals* such as *gypsum* have been described. The cement originates as grains within the *sand* mobilized by infiltrating rainwater or from *groundwater* which subsequently evaporates and *re-precipitates* the *mineral.* Commonly found along coasts within 40° of the equator.

aerial photography Successive photographs in the visible and very near infrared bands taken by a downward-pointing camera mounted on an aircraft. Three-dimensional topography can be studied by observing sequential overlapping pairs of photographs through a *stereoscope*, which allows geological *structure*, *geomorphology*, vegetation etc. to be assessed.

aerodynamic ripple A *wind ripple* similar in size to an *impact ripple*, but more sinuous and lying parallel to the wind. May be caused by wind vortices close to the surface.

aerolite See *stony meteorite.*

Aeronian A *stage* of the *Silurian*, 436.9–432.6 Ma.

aerosol A suspension or dispersion of solid particles in a gas.

AFM diagram A triangular graph with axes **A**l_2O_3, **F**eO and **M**gO, used to represent the composition of metamorphosed *pelites.*

AFMAG **A**udio **F**requency **Mag**netic field method. An *electromagnetic induction method* for the location of electrically-conducting bodies in the subsurface which makes use of natural variations in the *geomagnetic field* in the audiofrequency range known as *sferics.* Used both on land and in airborne versions.

aftershock An *earthquake* in the same area as, and following, the 'main' *earthquake.* Probably caused by the transfer of *stress* to *faults* in the proximity of the *focus.* The number of aftershocks depends on the *magnitude* of the 'main' *earthquake* and decreases exponentially with time. cf. *foreshock.*

agalmatolite (pagodite) A compact variety of *pyrophyllite.*

agate Concentric layers of *chalcedony* with different colours and *porosity*.

age A fourth order geological time unit.

age of the Earth The oldest rocks yet found provide radiometric ages of 3.89 Ga, while *meteorites* and Moon rocks give ages of 4.66 Ga. U-Pb studies suggest that the *mantle* became a closed system at the latter date, implying that the Earth formed at the same time as the other planets of the Solar system.

agglomerate A *breccia* or *conglomerate* formed during volcanic activity.

agglutinate Welded splatter, commonly of *basaltic* composition, deposited ballistically in *strombolian* or *fire fountain volcanic eruptions*. See also *welded tuff*.

aggradation The process of building up a surface with sediment deposited by wind or water.

aggregate 1. A mass of rock fragments and/or *mineral* grains. 2. Any granular solid material used alone, e.g. as *ballast*, or mixed with a binding material, e.g. concrete.

aggressivity A measure of the capacity of water to dissolve calcium carbonate. An important parameter in the development of *karst*.

agmatite *Migmatite* which appears to form a network of *veins* in the *host rock*.

Agnatha A class of superclass *Pisces*; *fish* with no biting jaws and a sucker-like mouth, including the hagfish and lampreys. Range *Ordovician–Recent*.

Agnostida An order of subphylum *Trilobita*, phylum *Arthropoda*; small, eyeless *trilobites* with a similar cephalon and pygidium, two thoracic segments and no facial sutures. Range L. *Cambrian–Ordovician*.

agonic line The line joining locations of zero magnetic *declination*.

agrichnia A *trace fossil* indicative of farming activity.

ahermatypic coral A *coral* with no symbiotic *algae*, usually solitary, non-*reef* forming and capable of living in deeper water than *hermatypic corals*.

Aijiashan (Neichiashan) An *Ordovician* succession in China covering part of the *Arenig*, the *Llanvirn*, *Llandeilo*, *Costonian* and part of the *Harnagian*.

Ainusian A *Cretaceous* succession in the far east of the former USSR equivalent to the *Cenomanian*.

air wave Seismic energy transmitted through the atmosphere from a *shot*.

airborne geophysical survey A *magnetic*, *electromagnetic*, *radiometric* or *gravity survey* undertaken alone or in concert from a fixed-wing aeroplane or helicopter. More cost-effective than a ground survey but with limited *depth of penetration*.

airgun A mechanical marine *seismic source* in which a burst of high-pressure air is suddenly released into the water, forming a bubble which oscillates in a similar fashion to an explosion.

Airy hypothesis A mechanism of *isostatic compensation* in which surface topography is balanced by variation in the thickness of a *crust* of constant *density*. Mountain ranges would be underlain by crustal roots and ocean basins by anti-roots, i.e. elevated *Moho*. cf. *Pratt hypothesis*.

Aistopoda An order of subclass *Lepospondyli*, class *Amphibia*; small, snake-like *amphibians*. Range L. *Carboniferous*–early *Permian*.

Akchagylian See *Apsheronian*.

åkermanite ($Ca_2MgSi_2O_7$) A *feldspathoid* of the *melilite* group.

A'KF diagram A triangular graph with axes **A'** ($Al_2O_3 + Fe_2O_3$)-($Na_2O + K_2O + CaO$), **K** (K_2O) and **F** ($FeO + MgO + MnO$) used to show the composition of a *metamorphic rock*.

aklé A complex *dune* with interlocking, crescentic, generally parallel ridges forming orthogonal to the wind direction in areas of high *sand* supply.

alabandite (MnS) A rare *ore mineral* of manganese.

alabaster See *gypsum*.

alas A *thermokarst* feature consisting of a steep-sided, flat-floored depression where local melting of *permafrost* has occurred.

alaskite A bimineralic, *leucocratic*, *granitic* rock composed of *quartz* and *alkali feldspar*.

Albertian A *Cambrian* succession in the E USA covering part of the *Lenian* and the *St. David's*.

albertite A pitch-black, solid *bitumen* of the *asphaltite* group.

Albian A *stage* of the *Cretaceous*, 112.0–97.0 Ma.

albite ($NaAlSi_3O_8$) The sodic end-member of the *plagioclase feldspars*, An_{0-10}.

albite twinning *Twinning* which forms zebra-like stripes, seen in *plagioclase feldspars*.

albitite A variety of *syenite* composed almost wholly of *albite* and produced by soda *fenitization*.

albitization The *metasomatic replacement* of an existing *mineral*, usually another *feldspar*, by *albite* as sodium ions are introduced into the rock.

alcove An arcuate, steep-sided valley on the side of a rock *outcrop* produced by water *erosion*.

alcrete A *duricrust* of indurated *bauxite* formed when aluminium sesquioxides accumulate in the zone of *weathering*.

Alcyonaria See *Octocorallia*.

Alexandrian An *Ordovician/Silurian* succession in North America covering part of the upper *Ordovician*, the *Rhuddanian* and part of the *Aeronian*.

alexandrite A *gem* variety of *chrysoberyl*.

algae Non-taxonomic name for simple plants which did not differentiate into root, stem and leaves.

algal mat See *stromatolite*.

algal ridge A morphological feature found on some *reef flats*.

alginite (bituminite) A *coal maceral* of the *exinite* group made up of coalified *algal* remains, characteristic of *boghead coal*.

Algoma-type iron formation A type of *banded iron formation* developed as oxide, carbonate and sulphide facies. Generally centimetres to hundreds of metres thick and less than a few kilometres in length. Characteristic of *Archaean greenstone belts* and probably formed in an *ocean trench* with a volcanic source of iron.

Algoman orogeny See *Kenoran orogeny*.

aliasing A phenomenon in sampling a waveform when frequencies greater than half the sampling interval (the *Nyquist frequency*) cause distortion of the low frequency part of the signal. Avoided by making use of an *anti-alias filter*.

aliquot A small sample of a material *assayed* to estimate the properties of the whole.

alkali basalt (olivine basalt) *Basalt* containing *normative nepheline* and *olivine*.

alkali feldspar A general term for *feldspar* of the *albite-anorthite* series.

alkali flat See *playa lake*.

alkali granite A *granite* containing alkali *amphibole* or *pyroxene*.

alkali lake A lake in an arid region rich in dissolved sodium and potassium carbonate, sodium chloride and other alkaline salts.

alkali-lime index The SiO_2 value at which $CaO = Na_2O + K_2O$ on a graph of these quantities against SiO_2 for a suite of related *igneous rocks*.

alkali metal Lithium, sodium, potassium, rubidium, caesium and francium.

alkaline earth metals Calcium, strontium, barium and radium.

alkaline rock An *igneous rock* with a high proportion of alkalis ($Na_2O + K_2O$) relative to *silica*, i.e. *silica-undersaturated*, containing *normative nepheline* or *leucite* and characterized by *alkali feldspar*, *feldspathoids*, alkali-rich *pyroxenes* and *amphiboles*, *phlogopite* and Zr, Nb, Rb, Ba, Ti and P enrichment.

allanite (orthite) ($(Ca,Ce)_3(Fe^{2+},Fe^{3+})Al_2O(SiO_4)(Si_2O_7)(OH)$) A *sorosilicate* found as an *accessory mineral* in many *igneous rocks*.

Alleghenian orogeny An *orogeny* affecting the central and southern parts of the *Appalachian fold belt* in the *Carboniferous–Permian*, corresponding to the *Variscan orogeny* of Europe.

alleghanyite ($Mn_5(SiO_4)_2(OH)_2$) A hydrated manganese silicate of the *humite* group, the manganese analogue of *chondrodite*.

allemontite (AsSb) A rare *vein mineral* comprising a *solid solution* of arsenic and antimony.

Aller A *Permian* succession in NW Europe covering the lower part of the *Capitanian*.

allochem Abbreviation of **allochemical** constituent. An organized *aggregate* of *calcite*, once used synonymously with grain or particle.

allochemical metamorphism *Metamorphism* in which there is removal or addition of material and the chemical composition of the rock is altered.

allochthon (heterochthon) A large structural unit, e.g. a *suspect terrane*, originating at a long distance from its present position. cf. *autochthon*, *parautochthon*.

allocyclic Referring to controls on sedimentary accumulation external to a sedimentary system, such as climate, tectonic activity, sea level change and source area geology. cf. *autocyclic*.

allodapic Descriptive of sediment deposited by *mass flow*, particularly *limestone*.

allogenic Descriptive of material originating from existing rock and transported to its present location. cf. *authigenic*.

allogenic stream A stream flowing through an area where it gains no *discharge*, occurring where *discharge* is derived from further up the *catchment* or where *stream flow* is supplemented by *groundwater*.

allopatric speciation The evolutionary divergence of geographically separated populations to form distinct species.

allophane ($Al_2Si_2O_5(OH)_2$) A white, amorphous *clay mineral* found along *fractures*.

Allotheria A subclass of class *Mammalia* whose only order is the *Multituberculata*.

allotriomorphic texture A *texture* of an *igneous rock* in which crystals exhibit a form related to surrounding, previously crystallized *minerals*.

allotropy The existence of an element in two or more forms.

alluvial architecture The three-dimensional distribution of sedimentary *facies* in an *alluvial deposit*, governed by *autocyclic* and *allocyclic* controls.

alluvial deposit An accumulation of sediment deposited from running fresh water in a channel or on an alluvial, coastal or *deltaic* plain and comprising *gravel*, *sand*, *mud*, *coal* and chemical precipitates. The alluvial architecture is controlled by channel type, vegetation cover, channel density, source area geology, climate, *tectonic* setting and surface *deformation*.

alluvial fan An *alluvial deposit* with a semi-conical, downstream-broadening shape formed where the topographic gradient reduces and the transporting capacity is diminished as the width of flow increases, such as along mountain fronts, *fault* scarps, valley sides and *glacier* margins.

alluvial placer An *alluvial deposit* containing economic *minerals*.

alluviation The deposition of *alluvium*.

alluvium Sediment transported and deposited by running fresh water.

almandine ($Fe_2Al_2Si_3O_{12}$) A *garnet*. Appreciable amounts of Mg and Mn generally replace Fe^{2+}.

almandine spinel See *ruby spinel*.

Alminian An *Eocene* succession in the former USSR equivalent to the *Priabonian*.

alnöito An *alkali basalt*.

alpha quartz (α-quartz, low quartz) The *quartz polymorph* stable below 573°C, common in all types of rock. cf. *beta quartz*.

Alpides The mountain belt stretching from the Alps to the Himalaya.

Alpine-Himalayan orogeny An *orogeny* affecting Europe and Asia in *Triassic–Miocene* times caused by the closure of the *Tethys* ocean.

Alportian The highest *stage* of the *Carboniferous*, 325.6–322.8 Ma.

alstonite ($CaBa(CO_3)_2$) An *orthorhombic carbonate mineral*.

altaite (PbTe) A very rare *ore mineral* of lead.

alteration Any chemical or mineralogical change in a rock caused by chemical or physical action.

alteration halo A rim of *minerals* formed in the *wall rock* round a *vein* or *ore* as the result of *hydrothermal alteration*. *Mineral* zoning in the halo indicates the changing nature of the *hydrothermal solution* during its passage.

alternation frequency map A *subsurface facies map* which allows distinction between a sequence of many alternating thin units and fewer thick units within the same lithology.

altiplanation terrace See *cryoplanation*.

Altonian A *Miocene* succession in New Zealand covering the upper part of the *Burdigalian*.

alum A series of double sulphate *isomorphs* with potash alum, approximating the formula ($KAl(SO_4)_2.12H_2O$).

alum shale *Shale* containing *alum*, usually formed by the *weathering* of *pyrite*-bearing *shale*.

alumstone See *alunite*.

alunite (alumstone, löwigite) ($KAl_3(SO_4)_2(OH)_6$) A *mineral* used in the production of *alum*.

alunitization The development of new *alunite* by *hydrothermal alteration*.

alveoles (honeycombs, stone lattice) Centimetric-scale small depressions formed by *weathering*, making up a dense network of holes on gently-sloping rock surfaces, possibly as a result of salt crystallization or other chemical or biological processes.

alvikite A medium- to fine-grained *carbonatite*.

amalgam (AgHg) A naturally occurring *silver-mercury solid solution*.

amazonite A green variety of *microcline*.

amazonstone A semi-precious *gem*-quality variety of *amazonite*.

amber A *mineral* formed from fossilized resin.

amber ice Ice containing dispersed, fine-grained sediment with an *amber*-like appearance.

amblygonite ($LiAlFPO_4$) A rare *mineral* found in *granite pegmatites*.

Amblypoda An order of infraclass *Eutheria*, subclass *Theria*, class *Mammalia*; large *mammals* with broad-surfaced, low-crowned cheek teeth. Range U. *Palaeocene*–M. *Oligocene*.

amesite A variety of *septechlorite* rich in magnesium and aluminium.

amethyst A purple variety of *quartz*.

Amgan A *Cambrian* succession in Siberia covering parts of the *Solvan* and *Menevian*.

Aminikan An *era* of the *Precambrian*, 2200–1650 Ma.

Ammonoidea An order of class *Cephalopoda*, phylum *Mollusca*; *molluscs* with a coiled shell containing folded internal compartment walls. Range *Devonian*–U. *Cretaceous*.

Amniota Vertebrate animals (*reptiles*, *mammals* and *birds*) laying an egg covered by a tough shell and containing all water, food and facilities necessary for the complete development of the embryo. cf. *Anamniota*.

Amonton's Law $\tau = \mu\sigma$, where τ = *shear stress*, μ = *coefficient of friction*, σ = *normal stress*. Relates to frictional sliding along a *fault plane*.

amorphous Descriptive of a *mineral* with no regular arrangement of atoms, i.e. non-crystalline.

amosite (brown asbestos) An *asbestiform* variety of the *amphibole cummingtonite*.

Amperer subduction (A-subduction) *Subduction* during *continental collision* in which the *crust* of the overriding *plate* becomes detached from the *mantle* part of the *lithosphere* so that the *crust* of the downgoing *plate* directly underthrusts that of the overriding *plate*.

Amphibia/amphibians A class of *vertebrates* capable of free existence on dry land but requiring water for egg laying and the maturing of larval forms prior to metamorphosis to the adult state. Range *Devonian–Recent*.

amphiboles *Silicate minerals* with an internal structure consisting of a double chain of linked silicate tetrahedra and cations occupying sites between oxygen ions at the edges of the chains. General chemistry $A_{0-1}B_2C_5D_8O_{22}(OH)_2$, where A = vacant or Na,K, B = Na,Ca,Mg or Fe^{2+}, C = Mg,Fe^{2+},Fe^{3+} or Al, D = Si or Al and some OH may be replaced by F,Cl or O^{2-}. They exhibit *vitreous lustre* and perfect *cleavage*. They occur commonly in *igneous* and *metamorphic rocks*.

amphibolite A non-*foliated, metamorphic rock* composed mainly of *amphibole*, formed by the *regional metamorphism* of *basic igneous rocks*.

amphibolite facies (epidote-amphibolite facies) A *metamorphic facies* formed under moderate to high temperature and pressure.

amphibolitization See *uralitization*.

amphidromic point A nodal point in the sea where no vertical tidal movement occurs, and tidal currents are most rapid.

Amphineura A class of phylum *Mollusca*; *molluscs* whose surface is covered by seven or eight calcareous plates. Range *Cambrian–Recent*.

amphoteric With both acidic and alkaline properties.

AMS dating See *accelerator mass spectrometry dating*.

amygdale (amygdule) A *vesicle* in *lava* or *pyroclastic rock* filled with low-temperature *minerals*, e.g. *calcite*, *quartz*.

amygdaloidal Containing *amygdales*.

amygdule See *amygdale*.

An A method of indicating *plagioclase feldspar* composition as a percentage of *anorthite*, e.g. An_{15} indicates a composition of 15% *anorthite*, 85% *albite*.

anabranch A river channel pattern in which the width of islands is more than three times the river width at average *discharge*.

anaerobic (anoxic, azoic) Descriptive of an environment in which the concentration of dissolved oxygen is too low to support *metazoan* life. Corresponds to an oxygen concentration less than 0.1 ml l^{-1}.

anagenesis An evolutionary change in a single species lineage.

analcime (analcite) ($NaAlSi_2O_6.H_2O$) A *zeolite* or *feldspathoid*.

analcite See *analcime*.

analcitite An *igneous rock* in which abundant *plagioclase* crystals have been replaced by *analcite* during late stage *deuteric alteration* in which *magmatic* fluids penetrated *cleavages* and *fractures* in the *plagioclase*.

Anamniota Vertebrate animals (*fish* and *amphibians*) which produce an egg without a tough shell. cf. *Amniota*.

Anapsida/anapsids A subclass of class *Reptilia*; *reptiles* lacking apertures in the skull behind the eyes. Range *Triassic–Recent*.

anastomosing Descriptive of a pattern characterized by branching and rejoining sinuous routes, e.g. an anastomosing river.

anatase (TiO_2) A relatively rare titanium *oxide mineral*.

anatectite See *anatexite*.

anatexis (palingenesis) *Partial melting* and fusion producing a melt of smaller proportion than the original rock, leaving an unmelted refractory residuum. The melt composition depends on phase relationships in the solid and the temperature and pressure conditions of melting. cf. *syntexis*.

anatexite (anatectite) A rock formed by *anatexis*.

anauxite ($Al_2(SiO_7)(OH)_4$) A *clay mineral*.

andalusite (Al_2SiO_5) A *nesosilicate* important in *metamorphic rocks*.

Andean mountain belt An *orogenic belt* forming in response to the steady-state *subduction* of oceanic *lithosphere* beneath *continental lithosphere*, typified by the South American Andes. The recognition of *suspect terranes* in the Andes now casts doubt on this simple process. cf. *Himalayan mountain belt*.

Anderken An *Ordovician* succession in Kazakhstan covering part of the *Soudleyan*, *Longvillian* and *Marshbrookian*.

Andernach lava See *Mayen lava*.

andesine A *plagioclase feldspar* (An_{30-50}).

andesite A *volcanic rock* of *intermediate* composition characteristic of the *calc-alkaline basalt-andesite-dacite-rhyolite* association. Commonly *porphyritic* with *phenocrysts* of zoned sodic *plagioclase*, *biotite*, *hornblende* and *pyroxene* in a *groundmass* of the same *minerals* and more sodic *plagioclase* and *quartz*. The extrusive equivalent of *diorite*, grading into *latite* with increasing *alkali feldspar* and into *dacite* with increasing *alkali feldspar* and *quartz*. Typical of the *subduction zone* environment.

andesite line A line separating *island arc* and continental *andesitic* rocks from the *basaltic* oceanic region.

andradite ($Ca_3Fe_2^{3+}Si_3O_{12}$) A *garnet*. Fe is frequently partially replaced by Al. Typically formed by the *metamorphism* of impure calcareous rocks and found in *metasomatic skarns*.

anelasticity Time-dependent *elasticity*.

Angaraland A large continent in the northern hemisphere which existed in *Carboniferous* and *Permian* times, comprising the former USSR east of the Urals.

Angiospermopsida/angiosperms A class of division *Trachaeophyta*, kingdom *Plantae*; flowering *land plants*. Range *Cretaceous–Recent*.

angle of contact The angle between a liquid surface and a solid with which it is in contact.

angle of draw The angle between the end of a subsurface working and the point on the surface to which its associated subsidence extends.

angle of internal friction (angle of shearing resistance) The angle relating the *shear stress* required for sliding (e.g. along a *fault*) to the *normal stress* on a plane. The inverse tangent of the angle defines the *coefficient of friction*.

angle of repose The maximum angle of slope that can be maintained by an accumulation of material.

angle of shearing resistance See *angle of internal friction*.

anglesite ($PbSO_4$) A *secondary mineral* formed by *weathering* of lead *ores*.

Ångström (Å) 10^{-10}m.

angular shear strain The *shear strain* produced in an original right angle during *strain*, defined as the tangent of the change in angle.

angular velocity The rate of change of an angle between two lines. Used to quantify the rate of rotation of one *plate* with respect to another, as angular velocity is not dependent on the distance from the *pole of rotation*. cf. *tangential velocity*.

angularity (roundness) The ratio between the average radius of circles drawn within the corners and edges of a *clast* and the radius of the largest circle which could be drawn within the *clast*; thus a measure of the sharpness of the corners of a *clast*.

anhedral Descriptive of a *mineral* with no crystal *form* developed.

anhydrite ($CaSO_4$) An *evaporite mineral*.

Anika A *Silurian* succession in the Mirnyy Creek area of NE Siberia equivalent to the *Telychian*.

anilite (Cu_7S_4) A rare sulphide *ore mineral*.

Anisian The lower *stage* of the M. *Triassic*, 241.1–239.5 Ma.

anisotropic fabric A *fabric* in a rock which shows *preferred orientation*, e.g. clay flakes in a *mudrock* which orient as they settle from *suspension* and on *compaction*.

anisotropy The presence of a *preferred orientation*.

ankaramite A *porphyritic*, *phenocryst*-rich *alkali olivine basalt* with more *pyroxene* than *olivine* and a *groundmass* of *augite/titanaugite microlites*, *labradorite* and *biotite*. Occurs in association with *basanite* and *alkali basalt*.

ankerite ($Ca(Fe,Mg)(CO_3)_2$) A *carbonate mineral* of the *dolomite* group.

annabergite (nickel bloom) ($Ni_3(AsO_4)_2.8H_2O$) A green *secondary mineral* of nickel.

annealing The movement of grain boundaries and grain growth in a polycrystalline rock by which the rock undergoes *recovery* from *deformation*. Driven by elevated temperature or occurring slowly over a long period.

Annelida Segmented worms with

well-defined heads. Range *Precambrian–Recent.*

annite ($K_2Fe_6Si_6Al_2O_{20}(OH)_4$) The ferrous iron end-member of the *biotite micas.*

annular drainage A *drainage* pattern in which *subsequent streams* follow paths approximating portions of concentric circles. Often occurs around dissected *domes* or *basins.*

anomalous lead Lead with an isotopic ratio which gives an incorrect radiometric age. cf. *common lead.* See also *B-type lead, J-type lead.*

anorogenic Unrelated to *tectonism.*

anorthite ($CaAl_2Si_2O_8$) The most calcium rich *plagioclase feldspar,* An_{70-90}.

anorthoclase ($(Na,K)AlSi_3O_8$) An *alkali feldspar.*

anorthosite A rock consisting of >90% *plagioclase feldspar,* virtually restricted to rocks of *Precambrian* age.

anoxia The state of being *anoxic.*

anoxic See *anaerobic.*

anoxic event A geographically widespread, stratigraphically limited interval of organic-rich deposition, whose *carbon* assists in the removal of oxygen, e.g. in the *Turonian.*

antecedent drainage A drainage system maintaining its general direction by cutting through a localized uplifted area at a rate greater than the *uplift.*

antecedent platform theory The theory that *coral-algal reefs* and *atolls* grow upwards after colonizing submarine banks which have built up to a suitable depth, as *corals* can survive only in relatively shallow water. No longer widely accepted, as an explanation based on a combination of subsidence and *glacial control theory* is now preferred.

anteconsequent stream A stream flowing as a consequence of an early *uplift* but *antecedent* to later periods of *uplift.* cf. *consequent, inconsequent, insequent, obsequent, subsequent* and *resequent streams.*

anthophyllite ($(Mg,Fe)_7Si_8O_{22}(OH)_2$) A white *amphibole.*

Anthozoa A class of phylum *Cnidaria* which includes the *corals.* Range *Ordovician–Recent.*

anthracite A high *rank coal* with a high *carbon* content and a very low volatile content.

anthracitization The process of conversion of *brown coal* to *anthracite.*

Anthracosauria An order of subclass *Labyrinthodontia,* class *Amphibia;* possible ancestors of the *Amphibia* of *Carboniferous* age.

anthraxolite A hard black *asphalt* of the *asphaltite* group found in *veins* and as masses in *sedimentary rocks,* such as *oil shale.*

anthraxylon Vitreous *coal* constituents derived from woody tissue.

anthropogeomorphology The study of man as a *geomorphological* agent.

anti-alias filter A *filter* applied to a signal so as to avoid *aliasing.*

anti-ripplet See *adhesion ripple.*

Antiarchii An order of class *Placodermi,* superclass *Pisces;* small, grotesque, normally freshwater, placoderm *fish*. Range M. *Devonian*–top *Devonian*.

anticlinal trap An *oil trap* consisting of a domed *structure* dipping away from a central high point. May be generated by *halokinesis*.

anticline A *fold*, *closing* in any direction, in which the older rocks occupy the core.

anticlinorium A large composite *anticline* made up of smaller *folds*. cf. *synclinorium*.

anticrack faulting See *transformational faulting*.

antidune A bedform with a symmetrical shape whose crest is orthogonal to the flow direction, commonly found in fast, shallow flows. Rarely identified in ancient deposits.

antiferromagnetism Rock magnetism in which the magnetic lattices possess magnetizations which are equal and in opposite directions so that the rock has a zero external *magnetic field*.

antiform A *fold closing* upwards for which no information is available on the *younging* direction.

antiformal stack An *imbricate structure* in which different amounts of *thrust displacements* relative to *fault* separations result in a *duplex* dipping towards both *hinterland* and *foreland*.

antigorite ($Mg_3Si_2O_5(OH)_4$) A *phyllosilicate* of the *serpentine* group.

antimonite See *stibnite*.

antimony (Sb) A rare *vein mineral*.

antimony glance See *stibnite*.

antiperthite An intergrowth of *alkali feldspars* in which the sodium-rich phase predominates. cf. *perthite*.

antitaxial growth A *vein* fill growing from the centre towards the walls. cf. *syntaxial growth*.

antithetic In a direction opposite to the prevailing sense of *vergence* or asymmetry, applied to *structures* or *fabrics*. cf. *synthetic*.

Antler orogeny An *orogeny* of *Devonian* age in NW America.

antlerite ($Cu_3SO_4(OH)_4$) A *secondary mineral* formed by the *alteration* of copper *ores*.

Anura An order of subclass *Lissamphibia*, class *Amphibia*; the frogs and toads. Range L. *Jurassic–Recent*.

apalhraun An Icelandic flow of *aa*.

apatite ($Ca_5(PO_4)_3(F,Cl,OH)$) A *mineral* important as source of phosphate for fertilizer.

apex An American mining law term denoting the outcrop of a *vein* reaching the surface or the shallowest point of a subsurface *vein*.

apex law An American mining law whereby the holder of mining rights to the outcrop or *apex* of a *vein* may mine it down-*dip* beyond the vertical projection of his *claim*.

aphanitic texture (aphyric texture) A *texture* of an *igneous rock* comprising a *microcrystalline groundmass* with no *phenocrysts*. Forms by rapid crystallization of a melt close to its *liquidus* temperature with no large suspended crystals.

Aphebian The lower part of the *Proterozoic* of Canada, ~ 2560–1800 Ma.

aphotic zone The level in a water

body where light is absent, which occurs at depths >200 m in the oceans. cf. *diphotic zone*, *photic zone*.

aphyric texture See *aphanitic texture*.

API gravity (API scale) The American Petroleum Institute standard for expressing the specific gravity of oil: API gravity = 141.5/(*density*)–131.5.

API scale See *API gravity*.

Aplacophora A subclass of class *Amphineura*, phylum *Mollusca*; *molluscs* whose dorsal surface is covered by seven or eight non-calcareous plates, and thus with no fossil record.

aplite A fine- to medium-grained *igneous rock* occurring as thin (<20 cm) *veins* within coarser-grained *plutonic* rocks. Has a distinctive *subhedral-anhedral fabric* with a sugary *texture*. Forms by the squeezing out of residual *granitic magma* into *fractures* or *joints* formed at a late stage of cooling of the intrusion. Commonly associated with *pegmatite*. Without qualification aplite refers to *granitic* composition, but it can range in composition from *gabbroic* to *granitic*.

Apoda An order of subclass *Lissamphibia*, class *Amphibia*; legless, snake-like *amphibians*. Range L. *Jurassic–Recent*.

apophyllite ($KCa_4(Si_4O_{10})_2F.8H_2O$) A *phyllosilicate*.

apophysis A *vein* or protuberance connected to a larger intrusive body. Used to describe centimetric-scale branches of *dykes* and *sills*.

Appalachian fold belt A *fold belt* extending along the east coast of North America, affected by the *Acadian orogeny*, *Alleghenian orogeny* and *Taconian orogeny* during the *Palaeozoic*.

apparent cohesion The cohesion of grains by the surface tension of *pore fluid*.

apparent conductivity The inverse of *apparent resistivity*.

apparent dip The angle of inclination of a plane to the horizontal measured in a plane not orthogonal to the *strike* of the plane. Always less than the *true dip*.

apparent polar wander The pattern traced by *palaeomagnetic poles* of sequential age of a single *lithospheric* block plotted assuming the block remained stationary. cf. *true polar wander*.

apparent resistivity A complex weighted mean of the *electrical resistivities* in the subsurface which varies with the location and separation of the measuring electrodes.

apparent velocity The velocity of *seismic waves* measured from a segment of the graph of *arrival time* against *range*. If the refractor is dipping or undulating, this does not represent a true velocity.

appinite A heterogeneous group of medium- to coarse-grained *hornblende*- and *biotite*-rich *igneous rocks* forming small intrusions, varying from *subalkaline* to *alkaline* in composition and containing prominent *hornblende phenocrysts*.

applied geomorphology The field of *geomorphology* which applies *geomorphological* knowledge and techniques to the solution of environmental problems and the support of *environmental engineering* projects.

applied geophysics (geophysical exploration) Methods of investigating the subsurface by taking measurements at or near the Earth's surface. Includes *gravity, magnetic, electrical, electromagnetic, seismic* and *radiometric surveying.*

apron A *fan* of unconsolidated sediment at the base of a mountain or *glacier.*

Apsheronian (Akchagylian) A *Pliocene* succession on the Russian Platform covering the upper part of the *Piacenzian.*

Aptian A *stage* of the *Cretaceous,* 124.5–112.0 Ma.

aptychus A calcareous plate which represents a lower jaw plate of a *Mesozoic ammonite.*

aquamarine A green/blue, *gem*-quality variety of *beryl.*

aquiclude A formation with a low *permeability,* important in controlling flow in adjacent overlying and underlying permeable formations. cf. *aquifuge, aquitard.*

aquifer A geological unit containing sufficient saturated *permeable* rock to yield significant amounts of water.

aquifer loss The water lost by flow from the *aquifer* to the *well* when *drawdown* occurs.

aquifer test A test in which known quantities of water are added to or withdrawn from a *well* and subsequent changes in *head* measured.

aquifuge A geological unit with no connected pores which neither absorbs nor transmits water. cf. *aquiclude, aquitard.*

Aquitanian The lowest *stage* of the *Miocene* 23.3–21.5 Ma.

aquitard A formation allowing the throughflow of water at a much slower rate than an *aquifer.* cf. *aquiclude, aquifuge.*

Arachnida A class of subphylum *Chelicerata,* phylum *Arthropoda*; the spiders and scorpions. Range *Silurian–Recent.*

Araeoscelidia An order of subclass *Eurapsida,* class *Reptilia*; primitive, extinct *reptiles.* Range U. *Carboniferous*–M. *Permian.*

aragonite ($CaCO_3$) A *polymorph* of calcium carbonate. A major skeletal component of many modern invertebrates and so a major constituent of modern carbonate accumulations. Changes by *neomorphism* to *calcite* with increasing age.

aragonite compensation depth See *carbonate compensation depth.*

Araneida An order of class *Arachnida,* subphylum *Chelicerata,* phylum *Arthropoda*; the spiders. Range ?*Carboniferous–Recent.*

Aratauran A *Jurassic* succession in New Zealand equivalent to the *Hettangian* and *Sinemurian.*

arborescent See *dendritic.*

archae- Ancient.

Archaean (Archaeozoic, Azoic, Cryptozoic) The older *eon* of the *Precambrian* ranging from the formation of the Earth at ~4600 Ma to 2500 Ma.

Archaeocyatha Small, cup-like, calcareous, marine, *benthonic metazoans* probably related to the *Porifera.* Range *Cambrian.*

archaeomagnetism *Palaeomagnetism* applied to archaeological phenomena.

Archaeornithes A subclass of *Aves*

whose only representative is Archaeopteryx, the first *bird*, of M. *Kimmeridgian* age.

Archaeozoic See *Archaean*.

archetype The hypothetical ancestral form of an organism in which all basic characteristics are established without specialization.

Archie's Formula $\rho = a\rho_w\varphi^{-m}$ An empirical law relating the *electrical resistivity* of a porous rock ρ to its *porosity* φ and the resistivity of the *pore fluid* ρ_w, where a and m are constants.

archipelago A group of islands.

Archosauria/archosaurs A subclass of class *Reptilia* including the *crocodiles*, *dinosaurs* and *pterosaurs*. Range L. *Triassic–Recent*.

arenaceous Sandy or *sand*-like.

Arenig An *epoch* of the *Ordovician*, 493.0–476.1 Ma.

arenite 1. *Sandstone*. 2. A *sandstone* whose *matrix* makes up less than 15% of the rock by volume.

arête A narrow, steep-sided ridge formed during the convergence of *cirques*.

arfvedsonite ($Na_3Fe^{2+}{}_4Fe^{3+}Si_8O_{22}(OH)_2$) A deep green *amphibole*.

argentiferous Containing *silver*.

argentite (silver glance) (Ag_2S) A high temperature form of the *ore mineral acanthite*.

argentopyrite ($AgFe_2S_3$) A sulphide of the *wurtzite group*.

argillaceous Descriptive of a detrital *sedimentary rock* with particles < 4 mm.

argillite A hard, slightly metamorphosed, *argillaceous* rock.

argillization A type of *wall rock alteration* forming *clay minerals* in a *host rock* containing mineralization. The important subtypes are *advanced argillic alteration* and *intermediate argillic alteration*.

argon-argon dating A technique of *radiometric dating* based on the irradiation of a potassium-bearing sample with thermal and fast neutrons to produce ^{39}Ar. Measurement of the ratio of ^{40}Ar to ^{39}Ar then allows the age to be calculated.

argyrodite (canfieldite) (Ag_8GeS_6) A grey germanium-*silver* double sulphide found in *vein* deposits associated with *argentite* and other silver *minerals*.

arkose A *sandstone* with $< 15\%$ *matrix* by volume containing a *feldspar*:*quartz* ratio of at least 25% and a *feldspar*:rock fragment ratio of at least 50% by volume.

arkosic arenite A *sandstone* comprising $> 25\%$ *feldspar*, with the *feldspar* content exceeding that of rock fragments and $< 15\%$ mud *matrix* (material < 30 μm in diameter).

arkosic wacke (feldspathic greywacke, feldspathic wacke) A *sandstone* comprising $> 5\%$ of *sand* grade particles, with the *feldspar* content exceeding that of rock fragments and $> 15\%$ by volume of mud *matrix* (material < 30 μm in diameter).

armalcolite ($(Fe,Mg)Ti_2O_5$) An *oxide mineral* with the same structure as *pseudobrookite*, first found on the Moon, subsequently in *meteorites* and on Earth. Indicative of a highly reducing environment.

Armorican orogeny See *Hercynian orogeny*.

armoured mud ball (clay ball, mud ball, mud pebble, pudding ball) An approximately spherical, centimetric-sized lump of cohesive sediment that has been gouged from a stream bed/bank. Often found in *badlands* and along *ephemeral streams*, and also in tidal channels and on *beaches*.

armoured surface *Armouring* found in an equilibrium river channel, periodically disrupted during floods. cf. *paved bed*.

armouring Heterogeneous river bed material forming a thin layer of coarse grains one to two grains thick that inhibits transportation of underlying finer material.

Arnold An *Oligocene* succession in New Zealand comprising the *Bortonian* and *Kaiatan*.

Arnsbergian A *stage* of the *Carboniferous*, 331.1–328.3 Ma.

aromatics (C_nH_{2n-6}) Benzene-based compounds which occur in many *crude oils* up to a concentration of ~10%.

Arowhanan A *Cretaceous* succession in New Zealand covering part of the *Cenomanian*.

arrival time (onset time) The travel time of a *seismic wave* from source to detector.

arroyo (coulée, dry wash, wash) A steep-sided, *ephemeral stream* channel found in deserts, in which several metres of poorly-sorted sediment from occasional *runoff* may accumulate.

arsenical pyrites See *arsenopyrite*.

arsenopyrite (arsenical pyrites) (FeAsS) An arsenic *mineral* common in metallic sulphide *ores*.

arterite A *migmatite* with *vein*-like *granitic* intrusions into the *country rock*.

artesian aquifer (confined aquifer) An *aquifer* in which water is under sufficient pressure to drive it to the surface when penetrated by a *well*.

artesian basin A *basin*-shaped area of *artesian wells*.

artesian discharge The discharge from a *well* penetrating an *artesian aquifer*.

artesian head The hydrostatic head of an *artesian aquifer*.

artesian pressure The hydrostatic pressure of an *artesian aquifer* at the surface.

artesian spring A *spring* fed from an *artesian aquifer*.

artesian well A *well* fed from an *artesian aquifer*.

Arthrodira An order of class *Placodermi*, superclass *Pisces*, *fish* with bony plated bodies and heavily armoured heads. Range *Devonian*.

Arthropleurida A class of superclass *Myriapoda*, phylum *Uniramia*; terrestrial, herbivorous uniramians with multisegmented limbs with a large, lobed plate at their base. Range *Devonian*–U. *Carboniferous*.

Arthropoda/arthropods A diverse invertebrate phylum characterized by an exoskeleton and jointed appendages. Bilaterally symmetrical, with the body usually organized into specialized divisions, e.g. head, thorax, abdomen. Range L. *Cambrian*–*Recent*.

Articulata 1. A class of phylum *Brachiopoda* in which the *valves* are held together solely by muscles. Range L. *Cambrian–Recent*. 2. A subclass of class *Crinoidea*, subphylum *Pelmatozoa*; stemmed or stemless *crinoids* with simple cups. Range L. *Triassic–Recent*.

artificial recharge The replenishment of an *aquifer* by pumping water into it from surface supplies or other *aquifers*.

Artinskian A *stage* of the *Permian*, 268.8–259.7 Ma.

Artiodactyla/artiodactyls An order of infraclass *Eutheria*, subclass *Theria*, class *Mammalia*; hoofed *mammals* including cattle, sheep and other even-toed herbivores. Range *Eocene–Recent*.

Arundian A *stage* of the *Carboniferous*, 345.0–342.8 Ma.

asbestiform With an extremely fibrous form.

asbestos A group of *industrial minerals* with *asbestiform habit*, including *chrysotile*, *crocidolite* and *amosite*.

Asbian A *stage* of the *Carboniferous*, 339.4–336.0 Ma.

asbolane (asbolite, earthy cobalt) A type of *wad* containing cobalt.

asbolite See *asbolane*.

aseismic Free from *earthquakes*.

aseismic ridge An oceanic ridge lacking regular *earthquake* activity, possibly formed by the passage of the *lithosphere* over a near-stationary *hotspot* in the *mantle*, e.g. Emperor-Hawaiian Islands ridge.

aseismic slip Movement along a *fault plane* unaccompanied by *earthquake* activity, generally at rates of less than 0.1 m s^{-1}. cf. *seismic slip*.

ash 1. A fine-grained volcanic material. 2. An incombustible inorganic residue remaining after the burning of *coal*.

ash flow (pyroclastic flow) A concentrated dispersion of hot, juvenile, volcanic fragments (*ash, pumice, scoria*) in a gas which moves in response to *gravity*. Forms deposits ranging in volume from 0.01 km^3 to more than 1000 km^3.

ash-flow field (ash-flow plain, ignimbrite plain, ignimbrite plateau) A plain of flat-lying, gently-dipping *ignimbrites* erupted from numerous, closely-spaced *calderas*, often swamping existing topography and forming subhorizontal *plateaux*, sometimes covering areas of up to 20 000 km^2.

ash-flow plain See *ash-flow field*.

Ashgill The oldest *epoch* of the *Ordovician*, 443.1–439.0 Ma.

ashlar A block of *building stone* with straight edges.

asparagus stone *Apatite* of a yellow-green colour resembling asparagus.

aspect ratio The ratio of width to height of an object, shape, etc.

asperity ploughing A mechanism of *slickenside* formation in which a *fault* surface is abraded by a small protuberance (asperity) to form a *lineation* known as a *scratch, groove, tool track* or *gutter*.

asphalt (bitumen, pitch, tar) A solid member of the *petroleum* group.

asphalt-based crude oil (black oil) *Crude oil* dominantly composed of naphthenic compounds, comprising some 15% of world supplies.

asphaltene An *aggregate* of high-molecular weight *hydrocarbons* occurring in *asphalt.*

asphaltic pyrobitumen A black, structureless *asphalt* insoluble in carbon disulphide and generally with <5% oxygen.

asphaltite A group name for the organic compounds *albertite, anthraxolite, grahamite, impsonite, nigrite* and *uintaite.*

assay The amount of metal or metals in an *ore.*

assay boundary (assay limit, cut off limit) The boundary of economic *mineral* concentration of an *orebody.*

assay limit See *assay boundary.*

assay plan A map showing the variation in *grade* and distribution of metals in and around an *orebody.*

Asselian The lowest *stage* of the *Permian,* 290.0–281.5 Ma.

assemblage zone (coenozone) A *biostratigraphic zone* based on the temporal range of a group of species.

assimilation (contamination) The chemical and/or physical incorporation of rock into an unrelated *magma.*

associated gas Gas produced during the formation of *petroleum* and accompanying oil in an *oilfield.* cf. *non-associated gas.*

asterism A property of some crystals which appear to contain a star-like shape, arising from the reflection of light from oriented inclusions.

asteroid A small, rocky or metallic interplanetary body.

Asteroidea/asteroids A class of subphylum *Eleutherozoa,* phylum *Echinodermata* including starfish and brittle stars. Range L. *Ordovician–Recent.*

asthenosphere A mechanically weak layer of the *mantle* immediately beneath the *lithosphere,* corresponding to the depth range within the Earth where the melting temperature is most closely approached. The top is near the surface beneath *oceanic ridges,* 120–180 km deep under old ocean basins and at least 250 km deep, if present at all, beneath the continents.

astraeoid A condition of massive *coral* in which the *corallite* walls become thin or disappear but the septae remain.

Astrapotheria An order of infraclass *Eutheria,* subclass *Theria,* class *Mammalia;* rhinoceros-sized South American ungulates with long bodies, short legs and protruding lower incisors. Range U. *Palaeocene–Miocene.*

astrobleme A terrestrial *crater* over 10 km in diameter formed by the impact of an extraterrestrial body.

astrophyllite ($(K,Na)_3(Fe,Mn)_7(Ti,Zr)_2Si_8(O,OH)_{31}$) A rare titanium-bearing silicate.

asymmetric valley A valley with sides of unequal slope. Possibly the result of the geological *structure* as in *uniclinal shifting.* They are common in *periglacial* regions where they form as a result of differences in slope aspect and thus solar radiation received.

asymmetrical fold A *fold* which is not mirror symmetrical about the *axial* *plane*, caused by differing *limb* length or thickness.

atacamite ($CuCl(OH)_3$) A *secondary mineral* of copper.

Atdabanian A *stage* of the *Cambrian*, 560–554 Ma.

Atlantic-type coast (transverse coast, discordant coast) A coastline formed where the topography and geological *structure* are orthogonal to it. cf. *Pacific-type coast*.

Atokan A *Carboniferous* succession in the USA covering the lower part of the *Moscovian*.

atoll An irregular, subcircular, annular *coral-algal reef* surrounding a central *lagoon* in oceanic waters.

atollon A small *atoll* lying on the flank of a larger one.

atomic absorption spectroscopy (AAS) A method of *chemical analysis* in which a solution of a sample is passed into a flame which atomizes it. The amount of light absorbed from a source of the wavelength of a particular element focused on the flame gives a measure of that element's concentration.

atomic orbital A descriptor of the spatial distribution of an electron in an atom.

Atrypida An order of class *Articulata*, phylum *Brachiopoda*; *brachiopods* with biconvex, impunctate shells with a short, rounded hinge and pedicle foramen or notch. Range M. *Ordovician*–U. *Devonian*.

attached groundwater The water retained on the pore walls in rocks above the *water table*.

attapulgite ($2MgO_3SiO_2.4H_2O–Al_2O_35SiO_2.6H_2O$) One of the *minerals* making up the *palygorskite* group of *clay minerals*. The constituent of *Fuller's earth*.

attenuation The progressive reduction in amplitude of waves with increasing time or distance.

Atterberg limit The result of tests on cohesive soil that characterize changes from solid to plastic to liquid states.

attrital-anthraxylous coal A lustrous *coal* with the ratio of *anthraxylon* to *attritus* in the range 1:1 to 1:3.

attrital coal A US Geological Survey term used in the field description of *coal* in terms of its *fusain*, *vitrain* and *attritus* constituents, the latter being described according to its *lustre*.

attritus The microconstituents of *coal* including spores, cuticles, resins and granular opaque matter.

aubrite An *enstatite*-rich, calcium-poor *achondrite*.

augen gneiss A *gneiss* with a planar or linear *shape fabric* consisting of eye-like lensoid shapes, often resulting from the *deformation* of *porphyritic* coarse-grained *igneous rocks* or by the growth of *porphyroblasts*.

augen structure (rib structure) An eye-like marking on the surface of a *joint*.

augite ($(Ca,Na)(Mg,Fe,Al)(Si,Al)_2O_6$) The most common *pyroxene*, dark green to black in colour.

aulacogen An inactive *rift valley*, partially or completely filled with sediment, formed by the failure of one arm of a *rift-rift-rift triple junction* during *continental splitting*.

aureole See *metamorphic aureole.*

auric See *auriferous.*

aurichalcite ($(Zn,Cu)_3(CO_3)_2(OH)_3$) A rare *mineral* formed by the *alteration* of *base metal ores.*

auriferous (auric) Containing *gold.*

aurostibite ($AuSb_2$) *Gold*-antimony *ore.*

Austin A *Cretaceous* succession on the Gulf Coast of the USA equivalent to the *Coniacian* and *Santonian.*

austral Of southern hemisphere origin. cf. *boreal.*

authigenic Originating in place. Usually applied to *diagenetic minerals* forming in sediments after deposition.. cf. *allogenic.*

autobrecciated lava A viscous *lava flow* whose congealed crust has been fragmented by continued movements of molten *lava* in the interior. The smooth-faced blocks of crust weld together if sufficiently hot, but otherwise remain entrained within the mobile *lava.*

autochthon A large structural unit, e.g. a *nappe,* which originated close to its present position. cf. *allochthon, parautochthon.*

autoclastic Descriptive of a rock which has fragmented *in situ.*

autocorrelation function A function describing the *correlation* of a waveform with itself in order to measure similarity and periodicity along it.

autocyclic Referring to controls on sedimentation inherent to a sedimentary system, such as channel type, sediment discharge, *diagenesis.* cf. *allocyclic.*

autointrusion The injection of *magma* into fissures of its earlier crystallized rock.

autolith (cognate inclusion) A *xenolith* of early-formed material from the parental *magma* of an *igneous rock.*

autometamorphism The process whereby *metamorphic* changes take place by the action of residual fluids as an *igneous body* cools.

autometasomatism The *metasomatism* of newly-crystallized *igneous rock* by its residual fluids.

automorphic See *euhedral.*

autosuspension The condition in a fluid, approached only in *turbidity currents,* in which the hydraulic forces of motion provide sufficient energy to maintain a suspended load in the flow.

autunite ($Ca(UO_2)_2(PO_4)_2.10\text{–}12\ H_2O$) A uranium *ore mineral.*

auxiliary plane The *nodal plane* of an *earthquake* orthogonal to the *fault plane.* See *focal mechanism solution.*

avalanche A sudden, rapid movement of disaggregated ice, snow, earth or rock down a slope.

Avalonian orogeny An *orogeny* affecting the region from Georgia, USA, to Newfoundland, Canada, from *Cambrian–Ordovician* times.

aven See *pothole.*

aventurescence The visual phenomenon exhibited by *aventurine* which arises from the presence of small inclusions of coloured *minerals*.

aventurine A *gem* variety of *quartz* or, rarely, *oligoclase*, containing coloured inclusions, such as *haematite* (red) or chrome *mica* (green).

average velocity For *groundwater*, the volume of water passing through a given area of rock in unit time divided by its *porosity*.

Aves (birds) A class of feathered vertebrates with an egg typical of the *Amniota*. Possibly evolved from an *archosaur* and first appearing in the fossil record as Archaeopteryx in the U. *Jurassic*. Rare as *fossils* because of the delicacy of their bones.

avulsion The process by which a channel changes course by a sudden diversion. The periodicity of avulsion is controlled by the rate of topographic change, fluctuations in discharge and occurrence triggering events, e.g. *earthquakes*.

awaruite (Ni_3Fe) A rare alloy of *native metals* found in *peridotites*.

axial dipole (axial geocentric dipole) A model in which the time-averaged *geomagnetic field* is approximated by that of a single magnet aligned along the Earth's spin axis.

axial flattening (axial strain) *Strain* in which one *principal strain* direction is much smaller than the similarly-sized other two.

axial geocentric dipole See *axial dipole*.

axial modulus The ratio of longitudinal *stress* to longitudinal *strain* when there is no lateral *strain*.

axial plane The plane containing two crystallographic axes.

axial plane cleavage A set of *cleavage* planes subparallel to the *axial plane* of a *fold* and related to the formation of the *fold*.

axial plane foliation A set of *foliation* planes subparallel to the *axial plane* of a *fold* and related to the formation of the *fold*.

axial rift (median valley) A valley, 2–3 km deep and 30–50 km wide, found on slow-spreading *oceanic ridges* such as the Mid-Atlantic Ridge.

axial strain See *axial flattening*.

axially-symmetric extension *Extension* in one *principal strain* direction and equal *shortening* in all orthogonal directions. See *strain ellipsoid*.

axially-symmetric shortening *Shortening* in one *principal strain* direction and equal *extension* in all orthogonal directions. See *strain ellipsoid*.

axinite ($(Ca,Fe,Mn)_3Al_2(BO_3)_3(Si_4O_{12})(OH))$ A *cyclosilicate* found in *vugs* in *granites* or their *contact metamorphic* zones.

axiolitic structure An intergrowth of elongate fibres of *alkali feldspar* with *cristobalite* which have grown outwards from the sides of a linear *fracture* in *rhyolitic glass* by solid state growth during *devitrification*, with the *fracture* acting as the nucleus.

axis of rotation The axis passing through the centre of the Earth about which the motion of a *plate* can be described by a single angle of rotation. See *Euler's Theorem*.

axonometry The measurement of crystal axes.

azimuth The horizontal angle of a direction measured clockwise from north.

Azoic See *Archaean.*

azoic See *anaerobic.*

azurite ($Cu(CO_3)_2(OH)_2$) A blue, *secondary mineral* of copper.

B

***β*-quartz** See *beta quartz*.

B-type lead (Bleiberg-type lead) An *anomalous lead* whose isotopic ratios give an age older than its *host rocks*, possibly because the lead was derived from lead remobilized from an older deposit.

back The roof of an underground working.

back-arc basin (back-arc sea, marginal basin/sea) An isolated marine *basin* behind a *subduction zone*, formed either by *back-arc spreading* or by a step-back in the location of *underthrusting* adding oceanic *lithosphere* to the leading edge of the overriding *plate*. Up to three generations may be developed behind a single *subduction zone*.

back-arc sea See *back-arc basin*.

back-arc spreading The process whereby an *island arc* is rifted longitudinally and the two sections moved laterally apart to form a *back-arc basin* between them.

backfill The refill introduced into mine or quarry workings to support the worked area, ameliorate subsidence or dispose of waste materials.

backfold (retrocharriage) A *fold* with a *vergence* in the opposite sense to the majority of *folds* in a region.

backlimb The limb of an *asymmetric fold* with the lower *dip*. cf. *forelimb*.

backreef facies A sedimentary facies developing on the landward side of a *coral-algal reef* after growth and *cementation*. cf. *forereef facies*.

backset bed A *cross-bed* inclined against the flow direction.

backshore The area of a *beach* between normal high *tide* level and the highest point reached by marine action such as storm waves.

backswamp An area of waterlogged land adjacent to a river, containing fine-grained sediment and organic matter, common where *aggradation* of *fluvial* deposits along a river channel formed *levées* which are overtopped during peak *discharge*.

backthrust A low-angle *reverse fault* with a *vergence* different from the majority of *reverse faults* in the area.

backwash The *gravity*-fed return flow of water after a wave breaks on a *beach*. An important factor in determining *beach* gradient. cf. *swash*.

backwearing (parallel retreat) *Erosional* hillslope retreat without change in slope morphology. Occurs where the rate of *erosion* on different sections of the slope is proportional to the dip. cf. *downwearing*.

baddeleyite (ZrO_2) A zircon *mineral*, a by-product of working certain *carbonatites*.

badland A landscape produced by the extensive incision and *erosion* of weakly *cohesive* rocks consisting of deep gullies and ravines separated by steep ridges, small *mesas* and *buttes*. Usually devoid of vegetation which has been stripped by *erosion*.

bafflestone An *autochthonous* carbonate rock whose original components were bound organically during deposition, the organisms forming baffles to trap finer *matrix* material. cf. *bindstone*.

Bagenov A *Jurassic* succession in W Siberia equivalent to the *Tithonian*.

bajada A low-lying area of confluent *pediment* slopes and *alluvial fans* at the base of mountains around a desert.

Bajocian A *stage* of the *Jurassic*, 173.5–166.1 Ma.

Bakhchisaraian A *Palaeocene/Eocene* succession in the former USSR covering parts of the *Thanetian* and *Ypresian*.

Bala The youngest sub-*period* of the *Ordovician*, 463.9–439.0 Ma.

balanced section (restored section) A *cross-section* in which folded and/or faulted *strata* are arranged into their form prior to *deformation*. Provided certain factors attain, the balanced section can be used to check the veracity of the initial, deformed section, which should be an *admissible section*.

balas ruby A red, *gem* variety of *spinel*.

ball clay (pipe clay) A fine-grained *clay* comprising up to 70% *kaolinite* plus *illite, quartz, montmorillonite, chlorite* and 2–3% carbonaceous material, important to the ceramics industry.

ball-and-pillow structure A *structure* caused by the *wet sediment deformation* of interbedded *sand* and *mud*, characterized by globular protrusions and isolated pillows of *sandstone* which form by differential settling of unconsolidated *sand* into *mud*.

ballas A variety of *industrial diamond* in the form of very hard, dense, globular aggregates of minute, concentric *diamond* crystals.

ballast Crushed rock used for road beds or on railway tracks.

ballistic ripple (impact ripple) The most common form of *wind ripple* which runs straight and orthogonal to the wind direction with wavelengths from <10 mm to 20 m and heights from a few mm to 1 m.

ballstone A *sedimentary rock* comprising subspherical *nodular* masses, usually calcareous, in an *argillaceous matrix*.

Baltica (Baltoscandia) A continent made up of modern NW Europe which existed from *Cambrian* to early *Devonian* times.

Baltoscandia See *Baltica*.

Banan A *Triassic* succession in China equivalent to the *Carnian*.

band silicate See *inosilicate*.

band theory An approach to the calculation of the electronic structure of solids in which atoms arranged in a lattice have many neighbours so that the *atomic orbitals* interact to form a band of molecular orbitals whose energies lie in certain energy ranges. The proximity of these bands governs the facility with which electrons could be excited into an empty orbital, where its high mobility would allow a high *thermal* and *electrical conductivity*.

banded coal (humic coal) A *coal* of heterogeneous origin banded at a centimetric scale by organic layers of varying appearance and diverse origin.

banded iron formation (BIF, cherty iron formation, itabirite, jaspillite) An iron-rich sediment with layers of *chert* or *silica* and iron *minerals* (commonly *haematite*) 5–30 mm thick and laminated at millimetric or submillimetric scale. Virtually all are of *Precambrian* age and of *Algoma-* or *Superior-type*. Constitutes a very considerable economic *resource*.

bank erosion The *erosion* of a river bank by the detachment of particles or rotational *failure* following undercutting.

banket An early *Proterozoic gold*-uranium-bearing *quartz pebble conglomerate* of *placer* origin of the Witwatersrand Goldfield, South Africa.

bankfull discharge The maximum *discharge* possible in a river channel without overlapping the banks.

bannisterite ($(K,Na,Ca)Mn_{10}Al_2Si_{15}O_{44}.10H_2O$) A dark brown, *translucent*, hydrated manganese silicate.

Banquereau A *Cretaceous/Palaeocene* succession on the Scotian shelf of Canada covering parts of the *Maastrichtian* and *Danian*.

bar 1. The non-*SI* unit of pressure, equivalent to 10 *pascals*. 2. A linear deposit of *sand/gravel*, generally parallel to subparallel to a coastline or river channel.

bar finger sand An elongate deposit at a high angle to the coast caused by the *progradation* of distributary channel mouth systems in which the channels are fixed by cohesive muds of the *delta* front and interdistributory bays.

Barabin A *Jurassic* succession in W Siberia covering part of the *Callovian* and the *Oxfordian*.

barchan An isolated, crescentic *dune*, 0.3–30 m in height, with a gentle windward slope and a *slip-face* bounded by two arms extending downwind. Capable of movement with little change of shape at rates of 5–10 m a^{-1}.

barite (baryte) ($BaSO_4$) An *industrial mineral* important because of its high *density*.

Barker Index of Crystals A once popular method of identifying crystals by measurement of *interfacial angles*.

barkevikite ($(Na,K)Ca_2(Fe,Mg,Mn)_5(Si,Al)_8O_{22}(OH,F)_2$) An obsolete name for alumino-ferro-*hornblende*.

barranca A steep-sided gully formed by *erosion*, similar to a *donga*.

barred basin A partially restricted *basin* in which free water circulation is impeded by a barrier. Often *anoxic* or the site of *evaporite* deposition.

barrel The traditional unit for expressing a volume of oil, equivalent to 35 Imperial gallons or 42 US gallons.

Barremian A *stage* of the *Cretaceous*, 131.8–124.5 Ma.

barrier beach (bay bar) A *coastal bar* permanently exposed above sea level.

barrier boundary A hydraulic boundary of an *aquifer*, particularly one preventing expansion of a *cone of depression*.

barrier island An island, mainly sandy, elongate parallel to the shore and separated from it by a marsh or *lagoon.*

barrier reef A coastal *coral-algal reef*, separated from the shore by a *lagoon.*

Barrovian metamorphism *Regional metamorphism*, first recognized in Scotland, in which zones of increasing *metamorphic grade* are characterized by the appearance of a suite of *index minerals*; *chlorite, biotite, garnet, staurolite, kyanite, sillimanite.*

Barrovian zone A zone of *Barrovian metamorphism.*

Bartonian A *stage* of the *Eocene*, 42.1–38.6 Ma.

baryte See *barite.*

barytocalcite ($BaCa(CO_3)_2$) A calcium-barium double carbonate found in lead *veins.*

basal sapping The process of undercutting the base of a slope by *erosion* along a *spring*-line, *salt weathering* or *glacial* action.

basal tar mat A band of *heavy oil* and *tar* at the oil-water interface of an *oilfield* which can plug pores and inhibit production by the prevention of natural *water drive* from maintaining *reservoir* pressure.

basalt An *aphanitic mafic igneous rock* comprising *plagioclase feldspar* more calcic than An_{50} and *pyroxene*, perhaps with *nepheline, olivine* or *quartz* and with *accessory* iron-titanium oxide. Originates from the *partial melting* of *mantle peridotite* and makes up layer 2 of the *oceanic crust.*

basaltic hornblende See *oxy-hornblende.*

basaltic layer An outmoded term for the lower *continental crust.* cf. *granitic layer.*

basanite A *silica-undersaturated alkali olivine basalt* containing *olivine, clinopyroxene* and *plagioclase feldspar* with $>10\%$ *feldspathoids* in the form of *nepheline* or *leucite.* Found in association with other *alkaline igneous rocks.*

base level The lower limit of subaerial *erosional* activity, usually defined by the level of running water, but with sea level as the general base level for continents.

base metal A mining industry term for copper, lead, tin or zinc.

base station A reference location used during a geophysical survey.

base surge A cloud of turbulent solids, possibly including water vapour, which travels rapidly along the ground from a large explosion or *volcanic eruption* and eventually deposits laminated and *cross-bedded tephra.* Common in *phreatomagmatic eruptions.*

baseflow The subsurface *runoff* made up of *throughflow* and/or *groundwater.*

baselap A term employed in *seismic stratigraphy* to describe the termination of a sequence along its lower boundary.

basement The surface beneath which no *sedimentary rock* is found.

Bashkirian An *epoch* of the *Carboniferous*, 322.8–311.3 Ma.

basic rock A *quartz*-free *igneous rock* containing calcic *feldspars* and 45–55% *silica*, often with *pyroxene* and *olivine.*

basin 1. A *synformal structure* with a subcircular *outcrop.* 2. A topographic depression containing, or capable of receiving, sediment.

basin-and-range Terrain in which *normal faults* create interspersed block mountains and *basins*, frequently containing lakes.

basin floor See *abyssal plain.*

Basin Groups 1–9 An *era* of the *Precambrian*, 4150–3950 Ma.

basin plain See *abyssal plain.*

bassanite (hemihydrate) ($2CaSO_4.H_2O$) An *evaporite mineral* which may be an intermediate stage in the formation of *anhydrite* by the dehydration of *gypsum.*

bastard coal (batt) 1. Thin layers of *coal* found in *shales* immediately above a *coal seam.* 2. Any *coal* with a high *ash* content.

bastard ganister A *sandstone* similar to *ganister* but containing more interstitial material and often with incomplete secondary *silicification.*

bastard rock An impure *sandstone* with thin lenticular partings of *shale* or *coal.*

bastite (schillerspar) A variety of *serpentine.*

bastnaesite ($(Ce,La)CO_3(F,OH)$) A fluorocarbonate of lanthanum and cerium, an *ore mineral* of the *rare earth elements.*

batch melting model A *partial melting* model in which the melt remains at the melting site, in equilibrium with the solid, before migrating as a single batch of *magma.*

Bath Stone A soft *Jurassic freestone* of honey coloured, *oolitic limestone* used as a *building stone.*

batholith A large *composite intrusion* with a surface area >100 km² made up of multiple *plutons* of *gabbroic* to *granitic* composition.

Bathonian A *stage* of the *Jurassic*, 166.1–161.3 Ma.

bathyal Descriptive of a sea depth of 200–4000 m.

batt 1. See *bastard coal.* 2. A British term for a hardened *clay* other than *fireclay.* 3. A British term for compact, black, fissile, carbonaceous *shale* often intercalated with thin layers of *coal* or *ironstone.*

battery ore A manganese *ore* of *grade* suitable for the production of MnO_2 for use in dry cell batteries.

bauxite An earthy rock composed almost wholly of aluminium hydroxide, often formed by the intense chemical *weathering* of existing rocks in the tropics under high rainfall. The principal *ore* of aluminium.

bauxitization The process of intense tropical *weathering* of rock in which all soluble components and *iron oxide* are removed to leave *bauxite.*

bay bar See *barrier beach.*

bayou A swampy creek or backwater.

bc joint See *longitudinal joint.*

beach cusp A regular *lunate* feature, 1–60 m in width, occurring in the upper *swash zone* of a *sand* or *gravel beach* and probably formed by *wave* action.

beach The site of accumulation of sediment deposited by *waves* and currents around a sea or lake margin.

beach deposit The unconsolidated sediment of a *beach*, dominated by *swash-*

produced sedimentary *structures*.

beach placer deposit A *placer deposit* found on ancient and modern *beaches*, especially well developed by storm conditions and *longshore drift*. Important *minerals* include *cassiterite*, *diamond*, *gold*, *ilmenite*, *magnetite*, *monazite*, *rutile*, *xenotime* and *zircon*.

beach ridge A linear accumulation of wave-deposited sediment on a *beach*.

beachrock A *beach deposit* lithified by *cementation* by *minerals* (normally *aragonite*) which form as sea water evaporates.

bearing capacity The maximum load per unit area that a surface can support in safety without *shear failure*.

Becke line The bright line at the margin of a grain visible when viewed in *thin section* which can be used to determine the *refractive index* relative to another grain or the mounting medium.

becquerel The *SI* unit of radioactivity, the activity in which there is a decay of 1 nucleus s^{-1}.

becquerelite ($CaU_6O_{19}.11H_2O$) A hydrated uranium oxide, formed by the *alteration* of *uraninite*.

bed 1. The smallest unit used in *lithostratigraphy*. See also *formation*, *group*, *member*, *supergroup*. 2. See *bedding*.

bed roughness The frictional resistance to flow in a sedimentary system.

bedding (bed, stratum) A centimetric- to metric-scale layer of *sedimentary rock* bounded above and below by *bedding planes*.

bedding-cleavage intersection The intersection of *bedding* and *cleavage* which produces a *lineation* often parallel, or nearly parallel, to a related *fold axis*.

bedding mullion A *mullion structure* consisting of a polished or striated *bedding plane*.

bedding plane A distinct surface separating two *beds* which marks a break in the continuity of sedimentation caused, for example, by a period of *erosion* or cessation of sediment supply.

bedding plane cleavage A *cleavage* parallel to the *bedding*.

bedding plane schistosity The *preferred orientation* of planar *minerals* parallel to *bedding* in a *metamorphic rock*.

bedding plane slip The relative *displacement* of two *beds* during *flexural slip folding*. May produce *slickenside striations* on the *bedding plane*.

bedding plane thrust A *thrust fault* parallel to the *bedding*.

bedform The shape of the surface of a bed of granular sediment produced by fluid flow over it.

bedform theory A group of hypotheses attempting explanation of the physical reasons for the initiation, development, stability and characteristics of natural *bedforms*.

bedload The material transported along the floor of a flowing fluid by rolling, sliding or *saltation*.

bedrock Unweathered rock beneath unconsolidated material.

beekite A variety of *chalcedony*.

beforsite A *carbonatite* in which the dominant *carbonate mineral* is *dolomite*.

beidellite ($(Ca,Na)_{0.3}Al_2(OH)_2(Al,Si)_4O_{10}.4H_2O$) A *clay mineral* of the *montmorillonite* group.

Belbekian See *Bodrakian.*

Belemnitida/belemnites An order of subclass *Coleoidea*, class *Cephalopoda*, phylum *Mollusca*; *molluscs* characterized by a bullet-shaped internal shell. Range *Carboniferous–Eocene.*

bell A conical *nodule* or *concretion* in the roof of a *coal* mine which may fall without warning.

Bell gravimeter An axisymmetric marine *gravimeter* with improved accuracy over beam-based instruments.

bell hole The cavity left after the fall of a *bell.*

bell-metal ore See *stannite.*

bell pit An obsolete, shallow mining method in which material extraction led to a bell-shaped excavation.

Bellerophontida An order of class *Monoplacophora*, phylum *Mollusca*; non-septate, planispirally-coiled *molluscs*. Range *Cambrian–Triassic.*

belt of no erosion The area of a hillslope, extending away from the *interfluve*, where surface *runoff* is incapable of eroding the soil surface. Generally on bare soil and often in *badland* areas.

bench mark A site of known elevation above *sea level datum.*

Bendigonian An *Ordovician* succession in Australia covering the lower part of the early *Arenig.*

benefication The separation of desired *minerals* from *gangue* during exploitation of a *mineral* deposit, by which the *mineral* is concentrated prior to refining.

Benioff-Wadati zone (Benioff zone) The active seismic zone in a *subduction zone*, originally mapped as a single plane of *earthquake foci* dipping beneath the overriding *plate*. Recently recognized as two planes, the less pronounced upper zone delineating the top of the downgoing slab, the main lower zone 10–20 km below, in which *earthquakes* originate from internal *deformation* of the slab.

Benioff zone See *Benioff-Wadati zone.*

benitoite ($BaTiSi_3O_9$) A rare *cyclosilicate* found in *serpentine.*

benmoreite An alkali *lava* intermediate in composition between *mugearite* and *trachyte.*

benthonic Living on the sea floor.

benthos Organisms living on the seafloor.

bentonite A clay mainly made up of *montmorillonite* formed by the *alteration* of volcanic *ash.*

bentonite clays See *smectite clays.*

beraunite ($Fe^{2+}Fe^{3+}{}_5(OH)_5(H_2O)_4(PO_4)_4.2H_2O$) A hydrated iron phosphate found in deposits of iron *ore.*

bergschrund A *crevasse* at the head of a *glacier* separating mobile and stationary ice, which develops when a *glacier* moves downslope.

Bering land bridge The land connection of *Beringia* during the *Cenozoic*, which was the only route available into North America for *mammals.*

Beringia An area of the Bering Strait and adjacent Siberia and Alaska which was intermittently dry land during the *Mesozoic* and *Cenozoic* and constituted a migration route.

berm 1. A triangular feature orthogonal to the shore with a subhorizontal top and a more steeply dipping seaward surface found on certain *beaches*. 2. A narrow, man-made shelf on a slope of an *opencast mine* or quarry.

Bernoulli equation An equation describing the conservation of energy in a steady flow of an ideal, frictionless, incompressible fluid: $p/\rho + gz + v^2/2$ is constant along any *streamline*, where p = fluid pressure, ρ = fluid density, g = *gravity*, z = vertical height above a datum, v = fluid velocity.

Berriasian The lowest *stage* of the *Cretaceous*, 145.6–140.7 Ma.

berthierine A green variety of *chamosite*.

Bertrand lens An auxiliary lens of the petrological microscope between the eyepiece and objective used to view *interference figures*.

bertrandite ($Be_4Si_2O_7(OH)_2$) A hydrated beryllium silicate found in *pegmatites* and exploited as a source of beryllium.

beryl ($Be_3Al_2(Si_6O_{18})$) A *cyclosilicate* found in *pegmatites*. A source of beryllium; coloured varieties may be valued as *gemstones*, see *aquamarine*, *emerald*.

beryllonite ($NaBePO_4$) A rare, colourless to yellow *gem*.

bessemer ore A valuable iron *ore* with a phosphorus content $< 0.045\%$.

Besshi-type deposit (Kieslager deposit) A *Proterozoic* and *Palaeozoic*, volcanic-associated, *massive sulphide deposit*. Commonly a copper-zinc *ore* with or without *gold* and *silver* and probably originating within a *rift* in a deep *back-arc basin* accompanied by *basaltic* volcanism.

beta (β) pole diagram The representation on a *stereogram* of the *great circle* traces of successive positions of a folded surface, which intersect at the β pole or axis. cf. *pi (π) diagram*.

beta quartz (β-quartz, high quartz) A *polymorph* of *quartz* stable in the range 573–867°C. cf. *alpha quartz*.

betafite ($(Ca,Fe,U)_{2-x}(Nb,Ti,Ta)_2O_6(OH,F)_{1-z}$) A hydrated uranium niobate, tantalate and titanate of the *pyrochlore* group found in *pegmatites*.

biaxial stress (plane stress) A *stress* system in which the maximum *principal stress* is greater than the intermediate and the minimum *principal stress* is zero.

bieberite (cobalt vitriol) ($CoSO_4.7H_2O$) A hydrated cobalt sulphate found as *stalactites* and encrustations in workings containing other cobalt *minerals*.

BIF See *banded iron formation*.

bifurcation ratio A quantitative measure of the rate at which a stream network bifurcates.

billabong Australian term for a permanent/semi-permanent waterbody on a *floodplain* isolated by a change in mainstream location, e.g. an *oxbow* lake.

bimodal Descriptive of a frequency distribution which contains two modes, i.e. two values significantly more common than the rest.

binary system A chemical system of only two components.

binding coal See *caking coal.*

bindstone An *autochthonous* carbonate rock whose original components were bound organically during deposition, with the organisms binding finer *matrix* material together. cf. *bafflestone.*

Bingham plastic A fluid which requires an initial *yield stress* to be overcome before it moves, e.g. some *lava flows.*

biocenosis (biocoenosis, life assemblage) An assemblage of *fossils* which lived together before death and fossilization. cf. *thanatocenosis.*

biochron The length of time represented by a *biozone.*

bioclast A grain in a *limestone* consisting of the skeletal material of organisms.

bioclastic limestone A *limestone* in which the predominant grains are *bioclasts.*

biocoenosis See *biocenosis.*

biodegradation The decomposition of *crude oil* by *aerobic* bacteria carried into the *reservoir* by *meteoric water.*

bioerosion trace A *trace fossil* consisting of drillings or raspings on a hard substrate.

biofacies A body of *sedimentary rock* characterized by specific and distinctive biological characteristics. cf. *lithofacies.*

biogenic Descriptive of material produced by organisms or their activities.

biogenic sedimentary structure See *trace fossil.*

bioherm A discrete lens-shaped mass of organic origin, e.g. a shell bank.

biohorizon 1. An interface between *strata* at which a *biostratigraphic* change has occurred. 2. A *biostratigraphic marker bed.*

biokarst (phytokarst) *Karst* resulting from biological action, generally small-scale and especially common in the coastal zone, e.g. rock surfaces bored and abraded by marine organisms.

biolithite A *limestone* in which there is evidence for the binding of grains by organic processes.

biomicrite A *bioclastic limestone* with a *matrix* of *micrite.*

biomineralization The *calcification* of organic tissue.

biophile Descriptive of an element typically found in organisms.

biopyribole See *pyribole.*

biosome A sediment mass formed under uniform biological conditions.

biostasy The state of stability of a landscape when conditions favour *weathering* and soil formation rather than *erosion* and *denudation.* cf. *rhexistasy.*

biostratigraphy The subdivision and correlation of sequences of *strata* using *fossils.*

biostratinomy The study of the *post-mortem* history of an organism before burial.

biostrome A *bioherm* consisting of a bedded accumulation of skeletal material.

biotite ($K(Mg,Fe)_3(AlSi_3O_{10})(OH)_2$) A brown/green *phyllosilicate mica*.

biotitization A type of *wall rock alteration* caused by the percolation of *hydrothermal fluids*. It gives rise to new, often fine-grained *biotite* which may replace and *pseudomorph* primary *biotite* as well as growing elsewhere in the *host rock*. Often the main feature of *potassic alteration*.

biotope 1. An area of uniform ecology. 2. An environment in which a certain assemblage of organisms lives or lived. cf. *ecotope*.

bioturbation The disruption of sediments mainly by the burrowing activities of organisms, e.g. by feeding.

biozone See *zone*.

bird 1. See *Aves*. 2. An instrument towed behind an aeroplane containing a package of instruments.

birdseyes See *fenestrae*.

birefringence The property shown by a *mineral* possessing two *refractive indices*, so that a double image is produced through it. Interference colours are caused when viewed in polarized light.

birnessite ($Na_4Mn_{14}O_{27}.9H_2O$) A manganese *mineral* found in *manganese nodules*.

BIRPS **B**ritish **I**nstitutions **R**eflection **P**rofiling **S**yndicate. The consortium responsible for deep *seismic reflection* studies in the UK and elsewhere.

Birrimian orogeny An *orogeny* affecting *greenstone belts* in W Africa during the *Proterozoic*.

bischofite ($MgCl_2.6H_2O$) A hydrated magnesium chloride found in *evaporite* deposits.

bismite (BiO_3) *Monoclinic* bismuth trioxide. cf. *sillénite*.

bismuth A rare *native element* found in certain *vein* deposits.

bismuth glance See *bismuthinite*.

bismuthinite (bismuth glance) (Bi_2S_3) A bismuth *ore mineral*.

bismutite ($(BiO)_2CO_3$) Bismuth carbonate, found as a *secondary mineral* in the oxidized zones of *veins* and *pegmatites* containing *primary* bismuth *minerals*.

bisulcate Descriptive of a shell with two grooves or depressions.

bitter lake A lake rich in sulphates and carbonates.

bitumen See *asphalt*.

bituminite See *alginite*.

bituminous brown coal See *pitch coal*.

bituminous coal The normal type of *coal*, intermediate in *rank* between *brown coal/lignite* and *anthracite*, consisting of a mixture of *banded* and *sapropelic coals* and generally rich in volatile *hydrocarbons*.

Bivalvia/bivalves (lamellibranchs, pelecypods) A class of phylum *Mollusca*; *molluscs* inhabiting a shell made up of two *valves* joined dorsally by a ligament and closed by adductor muscles. Gills are used for both respiration and filter feeding. Range *Cambrian–Recent*.

Bizon A *Silurian* succession in the Mirnyy Creek area of NE Siberia equivalent to the *Ludlow*.

black diamond A variety of massive,

cryptocrystalline diamond valued as an *abrasive.*

black gold 1. *Placer gold* coated with dark material so that its yellow colour is not seen. See also *coated gold.* 2. A colloquial term for oil.

black oil See *asphalt-based crude oil.*

black sand An *alluvial* or *beach placer* with a preponderance of dark *minerals*, usually *magnetite* or *ilmenite.*

black shale A black/dark grey *mudstone* rich in organic *carbon* (>5% by weight), generally formed in *anoxic* marine bottom waters.

black smoker A plume of *hydrothermal fluids* containing finely dispersed black sulphide particles issuing from a vent on an *oceanic ridge.* The sulphides produced include *pyrrhotite, pyrite, marcasite, chalcopyrite* and *sphalerite.* May be the origin of certain *massive sulphide deposits.* cf. *white smoker.*

blackjack A dark, iron-rich variety of *sphalerite.*

Blackwater A *Permian* succession in Queensland, Australia, covering part of the *Wordian*, the *Capitanian*, the *Longtanian* and part of the *Changxingian.*

blairmorite A rare type of *phonolite* containing *phenocrysts* of *analcime.*

blanket bog A type of *bog*, often of *peat*, which drapes upland areas and infills hollows in regions of high precipitation and low evapotranspiration.

blanket sand A horizontally extensive, thin layer of *sandstone.*

-blast Indicative of *in situ* growth during *metamorphism.*

blasto- Descriptive of the complete or partial *replacement* of a *textural* element by crystals growing during *metamorphism.*

Blastoidea/blastoids A subclass of class *Cystoidea*, subphylum *Pelmatozoa*, phylum *Echinodermata*; cystoids consisting of a pentamerally-symmetrical cup (*calyx*) of thirteen plates borne on a stalk. Range M. *Ordovician*–late *Permian.*

blastomylonite A *mylonite* containing crystals, of much larger size than the fine-grained *groundmass*, that grew during *deformation.*

bleaching A weak *wall rock alteration* process in which colour becomes much less intense.

Bleiberg-type lead See *B-type lead.*

blende See *sphalerite.*

Blenheim A *Permian* succession in Queensland, Australia, covering the middle part of the *Wordian.*

blind layer (hidden layer) A layer which cannot be detected using *first arrivals* in *seismic refraction* because its velocity is less than that of the overlying layer and does not generate a *head wave*, or because it is too thin to generate a *first arrival.*

blind lode A *lode* which does not reach the surface.

blind thrust A *thrust fault* which does not reach the surface.

blind valley A steep-sided, river-cut valley which terminates in a steep cliff, common in *limestone* areas with subterranean flow.

blind vein A *vein* which does not reach the surface.

block An angular fragment of rock with a diameter >256 mm. cf. *boulder*.

block caving A large-scale method of underground mining in which the *ore* is undermined and thus collapses into the excavation where it is sufficiently broken up for handling. Surface subsidence is inevitable.

block faulting *Normal faulting* giving rise to *fault blocks*.

blockfield (blockmeer, felsenmeer, stone field) An accumulation of coarse detritus on a level or gently-sloping surface in a mountainous area, comprising local rocks broken up by frost action.

blocking temperature 1. The temperature at which a *thermal magnetic remanence* is acquired and becomes fixed in a grain for at least 20 minutes in the absence of an external *magnetic field*. 2. The temperature at which gases used for *radiometric dating* become locked in a rock.

blocking volume The volume of a grain at which a *chemical remanent magnetization* becomes locked in a rock.

blockmeer See *blockfield*.

bloodstone (heliotrope) A green variety of *chalcedony* containing red flecks of *jasper*.

blowhole A vertical fissure in a *sea cliff* through which *waves* and *tides* cause water to fountain to the surface.

blowout A violent extrusion of gas and oil from a *well*.

blue asbestos (crocidolite) ($Na_2Fe_5Si_8O_{22}(OH)_2$) A blue *asbestiform mineral* with fibres of good flexibility. The most dangerous *asbestos* to health.

blue ground (hardebank) Fresh, resistant *kimberlite* that often outcrops. cf. *yellow ground*.

Blue John A mauve-purple *fluorite* from Derbyshire once used for the manufacture of ornaments.

blue vitriol See *chalcanthite*.

blue-green algae See *cyanobacteria*.

bluehole A circular, steep-sided hole in a *coral-algal reef*, probably formed by *karstic* processes at times of low sea level.

blueschist A low temperature, high pressure, *regionally metamorphosed* rock containing abundant *glaucophane*.

blueschist facies (glaucophane schist facies) A low temperature, high pressure *metamorphic facies* characterized by the presence of the blue *mineral glaucophane*, typically found within a *subduction zone* environment.

bluestone 1. A common name for the imported stones of the Stonehenge monument, England, comprising *dolerite*, *rhyolitic lava*, *tuff* and *sandstone*, which probably came from the Preseli Hills, SW Wales. 2. See *chalcanthite*.

boart See *bort*.

bodden An irregularly-shaped coastal inlet found in the S Baltic, formed by a rise in sea level over uneven topography.

Bodrakian (Belbekian) An *Eocene* succession in the former USSR covering part of the *Lutetian* and the *Bartonian*.

body wave (seismic body wave) A *seismic wave* which travels through the main part of the transmitting medium. cf. *surface wave*.

boehmite (AlO.OH) An *ore mineral* of aluminium found in *bauxite*.

bog (peatland, mire) A wetland comprising accumulations of semi-decomposed plant matter arising from precipitation rather than *groundwater*. cf. *fen*.

bog burst A sudden disruption of a *bog* which releases water and *peat* that flow a considerable distance.

bog iron ore A soft, porous *limonite* found in freshwater environments formed by *precipitation* from iron-rich waters. Currently of no economic relevance.

bogaz An elongate, deep ravine in a *karst* area formed by the widening of large *joints* by *solution*.

Bøggild intergrowth A *microscopic* intergrowth formed by *exsolution* in *plagioclase feldspar*.

boghead coal (torbanite) A *sapropelic coal* composed largely of *algal* matter.

Bohdalec An *Ordovician* succession in Bohemia covering the *Longvillian*, *Marshbrookian*, *Actonian*, *Onnian* and part of the *Pusgillian*.

Bohemian garnet A variety of *pyrope* in which red crystals are developed.

Bohm lamellae *Microscopic fluid inclusions* in planar arrays, possibly formed by the *annealing* of *deformation lamellae*.

bole A *fossil laterite* within a *lava flow*.

bolide The general term for an extraterrestrial body, such as a comet or *asteroid*.

bolide impact hypothesis A mechanism proposed as the cause of *mass extinctions* in the *fossil* record because of the considerable change in environment resulting from the impact of a large body.

Bolindian An *Ordovician* succession in Australia covering part of the *Onnian* and the *Ashgill*.

bolson A trough or *basin* in an arid or semi-arid region with a *playa* at its lowest point and the focus of local *runoff*. Often of *tectonic* origin.

bomb See *volcanic bomb*.

bonanza A mining term for a rich *orebody* containing *precious metals*.

bond clay A highly plastic clay used to bond less plastic materials in ceramics manufacture.

bone turquoise (odontolite) A *fossil* tooth or bone coloured blue by iron phosphate and used as a *gem*.

boninite An *olivine*- and *pyroxene*-bearing, *plagioclase*-poor *pillow lava* with quench-textured *pyroxene* and *accessory magnesiochromite* in a glassy *groundmass*. Found in volcanic *island arcs*.

book structure 1. A *structure* in an *ore* deposit in which *ore* alternates with *gangue*. 2. Structure developed in *mica* crystals in which the *cleavage* is so well developed that the individual foliae have the appearance of pages.

boomer A marine *seismic source* producing *seismic waves* in the band 100–10 000 Hz. A high voltage is discharged through a coil embedded in an epoxy-resin block, producing *eddy currents* in a spring-loaded aluminium plate which rapidly repel it and create compressional waves in the water.

Boomerangian A *Cambrian* succession in Australia covering the upper part of the *Menevian*.

booming dune See *singing sand.*

boracite ($Mg_3ClB_7O_{13}$) A *mineral* which is a source of *borax.*

borax ($Na_2B_4O_5(OH)_4.8H_2O$) An *evaporite mineral* which is an important *industrial mineral.*

bord and pillar mining See *room and pillar mining.*

boreal Of northern hemisphere origin. cf. *austral.*

bore See *tidal bore.*

borehole A hole drilled for exploration or exploitation purposes.

borehole breakout An *in situ stress measurement* by the determination of the *elongation* of a circular borehole in order to determine the orientation and magnitude of near-surface, *in situ stress.*

borehole gravimeter A specialized *gravimeter* used in boreholes to determine the *density* of the *wall rock* over a given vertical interval.

bornhardt A large, domed *inselberg.*

bornite (erubescite, purple copper ore) (Cu_5FeS_4) An *ore mineral* of copper.

borolanite A variety of *nepheline syenite* found in Scotland.

bort (boart) A badly coloured or flawed *diamond* with no value as a *gem.*

Bortonian An *Eocene* succession in New Zealand covering parts of the *Lutetian* and *Bartonian.*

boss A body of *plutonic igneous rock* with a circular plan and steep sides.

botryoidal Occurring as an *aggregate* with rounded surfaces.

bottoming The base of an *orebody.*

bottomset beds The basal units of *cross-stratified* beds formed on the *lee-side* of a structure.

boudin (boudinage) The segmentation of a *competent* layer of rock surrounded by less *competent* material into a series of parallel, elongate *structures* with rectangular to elliptical sections, caused by *extension* parallel to the layer orthogonal to the boudin. May be preceded by the development of a *pinch-and-swell structure* which is subsequently cut into boudins by *fractures.*

boudinage See *boudin.*

Bouguer anomaly A *gravity* measurement corrected for all non-geological sources of variation of *gravity,* i.e. latitude, elevation, topography and, where appropriate, *Earth tide* and *Eötvös* effects.

Bouguer correction The correction to a *gravity* observation for the attraction of the rock mass between the point of measurement and the chosen datum level

boulangerite ($Pb_5Sb_4S_{11}$) A sulphosalt found in low- to medium-temperature *hydrothermal vein* deposits.

boulder A rounded rock fragment with a diameter >256 mm. cf. *block.*

boulder clay (till) *Glacial* debris deposited directly from ice, comprising a wide variety of grain sizes. See *glacial deposition.*

boulder tracing A method of *mineral* exploration, usually in a glaciated terrain, involving the tracing of mineralized *boulders* back to their source.

Bouma sequence A fivefold division

of successive sediments deposited by a waning *turbidity current*, from base to top: A) structureless, most coarse-grained; B) *plane bed* in coarse- to fine-grained *sand*; C) *current ripple laminated* bed in fine *sand-silt*; D) *plane bed* in *silt*; E) structureless to very fine-grained in *mud*. All five units may not be developed.

boundary layer The marginal region of a flow where frictional resistance causes the velocity to decrease near the boundary and *shear stresses* are developed in the fluid.

boundary mapping (contact mapping) A method of *geological mapping* involving the following of a geological contact via a zigzag route, used when *exposure* is good or the contact follows a topographic or vegetational feature.

boundary stratotype A *chronostratigraphic* division comprising a sequence of rocks with standard reference points which are particularly complete at the sequence boundary. See *geological time-scale*.

boundstone A *limestone* in which the grains were bound by an organism or organisms. See also *bafflestone*, *bindstone*.

bourne An intermittent *spring* in *chalk* forming when the *water table* rises sufficiently high for water to flow in a normally *dry valley*.

bournonite ($PbCuSbS_3$) A *sulphide mineral* found in *hydrothermal veins* associated with copper and lead mineralization.

bow-tie effect A feature of an un*migrated seismic reflection* section in which reflection events cross each other. Arises over a *synformal* feature because reflections from different parts of the curved surface are focused onto the same portion of the section. Can be removed by *migration*.

bowenite A yellow/green variety of *serpentine*, sometimes used as a substitute for *jade*.

Bowen's reaction series A series of *minerals* crystallizing from a *magma* of specific chemical composition in which any *mineral* formed early in the series will later react with the melt to form a new *mineral* further down the series.

bowlingite See *saponite*.

box fold A composite *fold* with two *antiformal hinges* lying between two *synformal hinges* or vice versa.

boxstone A hollow *concretion*.

boxwork A honeycomb-like *structure* commonly found in *gossans* which forms when residual *limonite* remains in the cavity resulting when a sulphide grain is *oxidized*.

BP **B**efore **P**resent, by convention taken as 1950.

Brachiopoda/brachiopods A phylum of solitary, bilaterally symmetrical, unsegmented marine invertebrates with a bi*valve*d shell and a complex feeding apparatus (the lophophore). Range *Cambrian–Recent*.

brachyanticline An elongate *periclinal dome* with varying axial *plunge*.

brachydont Descriptive of a tooth with low, short crowns and well-developed roots with narrow canals.

brachysyncline An elongate *periclinal basin* with varying axial *plunge*.

Bradydonti An order of subclass *Elasmobranchii*, class *Chondrichthyes*, superclass *Pisces*; sharks with powerful crushing teeth for eating hard-shelled prey. Range end *Devonian–Permian*.

Bragg Law A law controlling *X-ray diffraction*. $n\lambda = 2d_{hkl}\sin\theta$, where n is an integer, λ the X-ray wavelength, d_{hkl} the spacing of the (hkl) planes of the crystal and 2θ the angle between the incident and diffracted X-ray beams.

braid A multithread channel formed, for example, by the meltwater flow from a *glacier* in a *sandur*.

braid bar An accumulation of sediment causing flow to divide, eventually forming an island at most flow states.

braided river/stream A river/stream that divides and rejoins around *bars* of a width similar to the channel width and with a *sinuosity* of 1–1.3.

braidplain A gently sloping, extensive region covered by *braid bars* and channels.

brammalite A variety of *illite* in which sodium is the inter-layer cation.

branch line The location in an *imbricate fault* system where a *fault* forks and *displacement* is transferred to another *fault*.

Branchiopoda/branchiopods A class of subphylum *Crustacea*, phylum *Arthropoda*; small, bi*valve*d animals enveloped by a carapace. Range L. *Devonian–Recent*.

braunite ($(Mn_2O_3)_3MnSiO_3$) A massive *ore mineral* of manganese.

bravoite ($(Ni,Fe)S_2$) A rare nickel *ore mineral*.

brazilian emerald A green *gem* variety of *tourmaline*.

brazilian peridot *Tourmaline* or *chrysoberyl* with the green colour of *peridot*.

brazilian ruby A red *tourmaline* or pink *topaz*.

brazilian topaz A clear blue variety of *topaz* valued as a *gem*.

brazilianite ($NaAl_3(PO_4)_2(OH)_4$) A rare yellow/green *gem* found in *pegmatites*.

breached anticline An *anticline* whose core has been eroded so that the *fold limbs* form *scarps*.

breaching thrust A *thrust fault* that offsets an existing, structurally higher *fault* or *fold structure*.

bread-crust bomb A *volcanic bomb* with a cracked outer crust and *vesicular* interior.

break-back thrust A *reverse fault* in a *piggyback thrust system* that forms in an existing *thrust structure* rather than nearer to the *foreland*.

break-point bar A permanently submerged *coastal bar* formed near the shore when steep, high energy *waves* break, depositing sediment onshore on the seaward side of the break point and offshore landward of it.

breaker A *wave* that enters shallow water and increases in height until it breaks.

breaker zone The *beach* zone in which *wave* energy is dispersed by breaking.

breast The *face* of a mine working.

breccia A *rudite* with angular *clasts*.

breccio-conglomerate A *rudite* intermediate between *breccia* and *conglomerate*, i.e. with approximately equal numbers of angular and rounded *clasts*.

breithauptite ($NiSb$) A nickel *ore mineral.*

breunnerite A variety of *magnesite* containing about 9% FeO.

breviconic Descriptive of a short *cephalopod* shell which expands rapidly.

brewsterite ($(Sr,Ba,Ca)(AlSi_3O_8)_2 \cdot 5H_2O)$) A rare strontium-barium *zeolite.*

brick clay A clay suitable for making bricks, ranging from unlithified clay to *mudrocks.*

brick earth *Loess* reworked by a river.

Brigantian A *stage* of the *Carboniferous,* 336.0–332.9 Ma.

bright coal The brightest type of *coal* on the scale bright coal, banded bright coal, *banded coal,* banded dull coal, *dull coal.*

bright spot A local increase in the amplitude (i.e. strength) of *reflected seismic waves,* produced by the presence of a strong contrast in *reflection coefficient,* often indicating the presence of *hydrocarbons* in a *reservoir.* cf. *dull spot.*

brine A concentrated aqueous solution of sodium chloride, in nature containing the cations Na^+, Ca^{2+}, K^+ and Mg^{2+} but with Cl^- the dominant anion, which is capable of *leaching* metals from the rocks through which it passes.

Bristol diamond A *quartz*-filled *amygdale* occurring in the *Mesozoic* of the Bristol district, UK.

brittle The property of a material that deforms by *fracture* without appreciable *viscous* or *plastic deformation.*

brittle-ductile shear zone A *shear zone* across which *displacement* is transferred by a combination of *brittle* and *ductile* processes.

brittle-ductile transition The transition from *deformation* by *fracture* (i.e. *cataclasis*) to *deformation* by *crystal plasticity.* Often refers to the depth in the *crust* at which this transition occurs.

brittle mica A *mica* group lacking the alkali content of *biotite* and frequently with calcium replacing magnesium. Includes *chloritoid, clintonite, margarite, ottrelite* and *stilpnomelane.*

brittle mineral A *tenacity* descriptor of a *mineral* that breaks and powders easily.

brittle silver ore A popular name for *stephanite.*

brittle stars See *Ophiuroidea.*

brittle strength The *strength* shown by a material which fails by *fracture.* cf. *ultimate strength.*

brochantite ($Cu_4SO_4(OH)_6$) A *secondary mineral* of copper.

brockram A *Permo-Triassic breccia* of northern England.

bromargyrite A rare *supergene mineral.*

bronzite ($(Mg,Fe)SiO_3$) A brown-green *orthopyroxene.*

bronzitite An *igneous rock* containing *bronzite* and lesser *augite* and calcic *plagioclase,* found in layered intrusions.

brookite (TiO_2) A rare titanium *oxide mineral.*

brousse tigrée A banding of vegetation in which closely spaced trees alternate with sparser bands.

brown asbestos See *amosite.*

brown clay See *red clay.*

brown coal A soft, low *rank coal,* including *lignite* and *sub-bituminous coal,* with a *calorific value* < 19.3 MJ kg^{-1} and a fixed carbon content of 46–60%.

brown iron ore See *limonite.*

brucite ($Mg(OH)_2$) An *industrial mineral* found in *limestones* and where magnesium silicates have been *altered.*

Brunhes A *magnetostratigraphic epoch* of *normal polarity* in the *Pleistocene,* 0.71 Ma–present.

Bryophyta A division of kingdom *Plantae;* spore-producing *plants* including the mosses, liverworts and hornworts. Range U. *Carboniferous–Recent.*

Bryozoa A phylum of largely marine, colonial, moss-like invertebrates in which the animals are commonly housed in calcareous skeletons which make up a colony. Range *Ordovician–Recent.*

bubble pulse The unwanted seismic signal generated by the oscillation of the bubble formed by an underwater explosion.

Bubnoff unit A unit used to quantify the rate of ground loss or slope retreat at right angles to the surface, equal to 1 mm ka^{-1} and equivalent to 1 m^3 km^{-2} a^{-1}.

Buchan zones A series of *metamorphic zones* characterized by the successive appearance of *andalusite, cordierite, staurolite* and *sillimanite,* first described from NE Scotland.

buchite A fine-grained to glassy *hornfels* produced by *thermal* or *contact metamorphism* of a *clay*-rich rock.

buckle fold A *fold* resulting from *buckling.*

buckling (layer parallel shortening) Folding caused by the compression of a layered rock in the direction of the layering. The folded layers maintain their original thickness so the *folds* produced are *parallel* or *concentric.*

buddingtonite ($NH_4AlSi_3O_8$) A rare *feldspar.*

Budnanium A *Silurian* succession in Bohemia comprising the *Kopanina-Schichten* and *Pridoli-Schichten.*

Bugando-Toro-Kibalian orogeny (Kibalian orogeny) An *orogeny* of *Precambrian* age affecting central and E Africa at ~2075–1700 Ma.

buhrstone See *burrstone.*

building stone Any rock suitable for use in construction.

Bulitanian A *Palaeocene* succession on the west coast of the USA covering part of the *Thanetian.*

bulk finite strain The total *finite strain* of a volume of rock calculated assuming that the *strain* is homogeneous.

bulk material A general name for material used in construction, including *clays* for bricks, *limestone* and *shale* for cement, *sand* and *gravel* for concrete, *building stone, road stone, aggregate* and *ballast.*

bulk modulus (incompressibility) An *elastic modulus* defined as the ratio of the *hydrostatic pressure* to the *dilation.*

Bunter A *Permian/Triassic* succession in

Germany equivalent to the *Changxing* and *Scythian*. 2. A traditional British name for the *Scythian epoch* of the L. *Triassic*.

Buntsandstein A *Permian*–L. *Triassic* succession in NW Europe covering part of the *Capitanian*, the *Longtanian* and the *Changxingian*.

buoyancy A term describing the gravitationally-powered ascent of relatively low *density* materials, such as *hydrocarbons* and *evaporites*.

Burdigalian A *stage* of the *Miocene*, 21.5–16.3 Ma.

Burgess Shale An *exceptional fossil deposit* of M. *Cambrian* age in British Columbia where some 120 genera are preserved in aluminosilicates.

burial diagenesis (mesogenesis) The physical, chemical and biological processes acting on a sediment during burial as it is removed from the surface until the onset of *metamorphism* or structural *inversion*.

burial metamorphism *Metamorphism* characteristic of thick sedimentary/volcanic sequences as they are rapidly buried in a sedimentary *basin* or *oceanic trench*.

burmite An *amber*-like *mineral*, a variety of *retinite*.

burnt alum A porous, friable material produced by dehydrating *alum* at dull red heat, used industrially for dyeing, water purification etc.

burrstone (buhrstone) A siliceous rock quarried in N France since at least the 16th century and used for *millstones*.

burst An upward movement of low velocity fluid away from the lower parts of the turbulent *boundary layer*.

Burzyan The oldest *period* of the *Riphean*, 1650–1350 Ma.

Bushveldt Complex A very large *differentiated* igneous complex of *Precambrian* age in South Africa containing vast reserves of chromium, platinum group metals and iron.

Busk Method See *tangent-arc method*.

bustamite ($(Mn,Ca,Fe)SiO_3$) A pink to brown *pyroxenoid*.

butane (C_4H_{10}) A colourless *paraffin series* gas present in *natural gas*.

butte A small isolated hill capped by resistant rock, possibly representing the former land surface.

butte témoin A *butte* situated on the scarp side of a *cuesta* of which it is a remnant.

bysmalith A large igneous intrusion or *pluton* of subcylindrical shape which has forced up and fractured the overlying *country rock*.

byssate Descriptive of a *bivalve* attached to the substrate by tough, horny threads.

bytownite A *plagioclase feldspar*, An_{70-90}.

C

c joint See *cross joint.*

Caen stone A light creamy yellow *limestone* of *Jurassic* age used as a *building stone* in N France and England.

Caerfai The oldest *epoch* of the *Cambrian*, 570–536.0 Ma.

cafemic Descriptive of an *igneous rock* containing calcium, iron and magnesium.

CAI See *colour alteration index.*

Cairngorm stone A semi-precious, *gem* variety of *smoky quartz.*

caking coal (binding coal) Any *coal* that softens, melts and agglomerates on heating and quenching to produce hard *coke.*

calamine See *smithsonite.*

calanque (calas) A coastal inlet with a gorge-like form, possibly a *karstic* and *fault*-controlled valley partially drowned by marine transgression.

calaverite ($AuTe_2$) An *ore mineral* of gold.

calc-alkaline rock A rock with a higher concentration of calcium (CaO) than alkalis ($Na_2O + K_2O$) compared to *alkaline igneous rocks. Plagioclase* is thus the dominant *feldspar.* Such rocks define the *basalt-basaltic andesite-andesite-dacite-rhyolite orogenic andesite* association which characterizes *subduction zone* environments.

calc-flinta (calc-silicate hornfels) A hard, fine-grained, calc-silicate rock formed by the *contact metamorphism* of an impure *limestone.*

calc-schist A metamorphosed, *argillaceous limestone* with *schistosity.*

calc-silicate hornfels See *calc-flinta.*

calc-tufa See *tufa.*

Calcarea A class of phylum *Porifera*; *sponges* with calcareous spicules making up the skeleton. Range *Cambrian–Recent.*

calcarenite A *limestone* with *sand*-grade sized grains.

calcareous aggregate An irregular mass of carbonate grains which have been stuck together, often by microbial processes.

calcareous algae (blue-green algae, cyanobacteria) *Algae* in which calcium carbonate derived from life processes provides a skeleton for the whole or part of the plant. Common in fresh and shallow marine water, where they contribute to *reef* and *bioclastic limestones* and provide both environmental and *biostratigraphic* markers. Range *Cambrian–Recent.*

calcareous ooze (carbonate ooze) A *pelagic*, biogenic sediment above the *carbonate compensation depth* covering very large areas of the ocean floor, comprising the skeletal remains of calcareous microorganisms, e.g. *coccolithoporoids*, *foraminifers* and *pteropods*.

Calcichordata A class of phylum *Echinodermata* which may be ancestral to the *fish*. Range *Cambrian–Devonian*.

calciclastic Descriptive of a *clastic* carbonate rock.

calcification 1. A *pedogenetic* process involving the accumulation of *calcite* or *dolomite* in soils low in moisture. 2. The process whereby organic tissue is converted to *calcite*.

calcilutite A *limestone* with *mud*-grade sized grains.

calcirudite A *limestone* with very coarse grains.

calcisiltite A *limestone* with *silt*-grade sized grains.

calcisphere A small (<500 μm) sphere of *calcite*, commonly found in *Palaeozoic limestones*, comprising a *micrite* coating around a hollow or *sparitic* interior, believed to be of *algal* origin.

calcite ($CaCO_3$) The most common *carbonate mineral*, the principal component of *limestone*.

calcite compensation depth See *carbonate compensation depth*.

calcrete (caliche, duricrust) A powdery, nodular to highly indurated, near-surface terrestrial material mainly composed of calcium carbonate, resulting from *cementation* and the introduction of *calcite* into soil, sediment and rock by *groundwater* in arid to semi-arid regions.

caldera A depression in the Earth's surface with a diameter >1.5 km formed by collapse into a *magma chamber* that has been vacated by the eruption or migration of *magma*.

Caledonian orogeny A mid-*Palaeozoic orogeny* affecting NW Europe and NE North America.

Caledonides The *orogenic belt* formed by the *Caledonian orogeny*.

caliche 1. A nitrate deposit of the Atacama Desert of South America. 2. See *calcrete*.

Californian jade See *californite*.

californite (Californian jade) A compact green variety of *vesuvianite* used as a *gem* substitute for *jade*.

caliper log A borehole logging tool which measures the diameter of the hole; this varies with the *competence* of the horizons penetrated.

calläis An archaeological term for *minerals*, mostly *variscite*, *turquoise* and *malachite*, used to make small blue or green beads in Neolithic and Copper Age times in W Europe.

Callovian A *stage* of the *Jurassic*, 161.3–157.1 Ma.

calomel (Hg_2Cl_2) A white-grey mercury chloride, found in association with *cinnabar*.

calorific value The amount of energy released by the burning of one kilogram of *coal*.

calyx 1. The cup-shaped body of a *pelmatozoan* made up of a number of plates. 2. A bowl-shaped depression on the top of a *coral* skeleton.

cambering The warping and sagging of relatively *competent strata* which overlie less *competent strata*, such as of *clay*, which tend to flow towards adjacent valleys so that a convex top is produced.

Cambrian The lowest *system* of the *Palaeozoic*, 570–510 Ma, during which there was a vast radiation of shelled invertebrates.

camera A chamber within a *mollusc*.

Camerata A subclass of class *Crinoidea*, subphylum *Pelmatozoa*; *crinoids* with a *calyx* of variable form in which all plates are united by rigid sutures. Range *Ordovician–Permian*.

Campanian A *stage* of the *Cretaceous*, 83.0–74.0 Ma.

camptonite An alkali *lamprophyre* containing *phenocrysts* of *barkevikite* and/or *kaersutite amphibole*, *augite*, *olivine* and/or *biotite/phlogopite* in a *groundmass* of calcium-rich *plagioclase*, *amphibole* and *pyroxene* with minor *alkali feldspar*, *feldspathoids*, *apatite*, iron-titanium oxides and *carbonate minerals*.

Canada balsam A mounting medium for *thin sections*, with a *refractive index* of 1.54

Canadian 1. The oldest sub-*period* of the *Ordovician*, 510.0–493.0 Ma. 2. See *Ibexian*.

Canadian asbestos See *chrysotile*.

cancrinite ($Na_6Ca(CO_3)(AlSiO_4)_6 \cdot 2H_2O$) A rare *feldspathoid*.

Candelaria A *Triassic* succession in Nevada, USA equivalent to the *Scythian*.

canfieldite See *argyrodite*.

Canglanpu A *Cambrian* succession in China covering parts of the *Tommotian* and *Atdabanian*.

cannel coal A type of *sapropelic coal* composed of spores or fine organic fragments.

canyon A steep-sided, deep valley cut by a river, often as a result of *rejuvenation*.

cap rock 1. **(roof rock, seal)** An impermeable rock lying above and sealing a gas or oil *reservoir*. 2. A sheath around and over a *salt dome* composed of *gypsum*- or *anhydrite*-bearing *limestone*.

Cape asbestos See *crocidolite*.

Cape ruby See *pyrope*.

capillary fringe (capillary zone) The zone immediately beneath the *water table* into which water can be drawn by capillary action, typically 0.1–3 m thick.

capillary habit A *habit* of *minerals* composed of flexible, thread-like crystals.

capillary pyrite See *millerite*.

capillary zone See *capillary fringe*.

Capitan A *Permian* succession in the Delaware Basin, USA equivalent to the *Capitanian*.

Capitanian A *stage* of the *Permian*, 252.5–250.0 Ma.

caprock model A theory that *waterfalls* occur where soft rock is eroded from under a harder rock, but which does not apply in all cases.

capuliform Cap-shaped.

Caradoc An *epoch* of the *Ordovician*, 463.9–443.1 Ma.

carat (karat) A unit of weight of *diamonds* equal to 0.2 g, different from the measure used to describe the number of parts of *gold* per 24 parts of an alloy.

carbon (C) An element occurring free as *diamond* and *graphite*, and combined in *carbonate minerals*, *hydrocarbons* and gases.

carbon cycle The progression of *carbon* through the surface, interior and atmosphere of the Earth.

carbon-14 dating See *radiocarbon dating*.

carbona A term used in Cornwall, UK for large masses of rich *tin ore*.

carbonaceous chondrite A type of *stony meteorite* containing *chondrules*.

carbonado A variety of *industrial diamond* taking the form of a black, *cryptocrystalline* mixture of *diamond* and *graphite* or amorphous *carbon*.

carbonate buildup A *reef* complex in the geological record or any large *stratigraphic* accumulation of *limestones* and *dolomites*.

carbonate compensation depth (calcite compensation depth, CCD) The level in oceanic sediment at which the rate of *dissolution* of calcium carbonate is the same as the rate of supply. Its level is controlled by the fact that carbonate *dissolution* increases with increasing pressure and decreasing temperature. These parameters increase with depth, but there are many other controlling factors.

carbonate minerals Common *minerals* containing the carbonate anion ($(CO_3)^{2-}$). About 60 carbonate minerals are known, but the most common are *aragonite, calcite, dolomite, magnesite, rhodochrosite* and *siderite*.

carbonate ooze See *calcareous ooze*.

carbonate platform A water-covered, extensive (10^2–10^3 km wide), flat area of *carbonate buildup* developed when a *craton* drowns, especially during phases of high sea level.

carbonate ramp A *carbonate buildup* with the form of a gently sloping ($<1°$) surface passing into progressively deeper water.

carbonate shelf A *carbonate buildup* which is similar to, but less extensive (few to hundreds of km wide) than, a *carbonate platform*.

carbonate-apatite ($Ca_5F(PO_4CO_3, OH)_3$) A variety of *apatite*.

carbonate-hosted base metal deposit A *strata-bound mineral deposit* of lead and zinc, with or without copper, *fluorite* and *barite*, which occurs in thick sequences of *limestone* or *dolomite*. Accounts for the majority of lead and zinc production in the USA and Europe.

carbonation 1. The process of introducing carbon dioxide into water. 2. The process of chemical *weathering* in which *minerals* are replaced by carbonates, commonly by reaction with carbonated water.

carbonatite An *igneous rock* containing $>50\%$ *carbonate minerals*. Occurs as *lava flows*, *dykes* and *sills* and commonly associated with *alkaline igneous rocks* within *rift* systems. Formed by the derivation of carbonate-rich fluids from ascending *magmas* originating from the *partial melting* of *mantle peridotite*.

carbonatization A type of *wall rock alteration* in which *dolomite*, *ankerite* and other *carbonate minerals* form in the *wall rocks* of *epigenetic mineral deposits*.

Carboniferous A *system* of the *Palaeozoic*, 362.5–290.0 Ma.

carbonization The decomposition of organic matter so that only a thin film of *carbon* remains, which may retain features of the original organism.

carborundum An artificial *abrasive* which has largely replaced *corundum*.

Carlin-type deposit A type of *disseminated gold deposit* occurring in carbonate rocks, probably formed by the deposition of *minerals* leached from underlying rocks by *hydrothermal fluids*.

Carlsbad twin A *twin*, commonly found in the *feldspars*, in which the twin axis is the c crystallographic axis.

carnallite ($KMgCl_3.6H_2O$) An *evaporite mineral*.

carnelian (cornelian, carnellian) A red variety of *chalcedony* used as a semi-precious *gem*.

carnellian See *carnelian*.

Carnian A *stage* of the *Triassic*, 235.0–223.4 Ma.

Carnivora An order of infraclass *Eutheria*, subclass *Theria*, class *Mammalia*; meat-eating placental *mammals*. Range *Palaeocene–Recent*.

carnosaur A carnivorous, bipedal, *saurischian dinosaur*.

carnotite ($(K_2UO_2)_2(VO_4)_2.3H_2O$) An *ore mineral* of uranium.

Carpoidea A bizarre group of *calcite*-plated invertebrates similar to *echinoderms* but differing in their lack of radial symmetry. Range L. *Cambrian–Carboniferous*.

carrollite ($CuCo_2S_4$) A *sulphide mineral* of the *thiospinel group*.

carse An *alluvial floodplain* beside a river or *estuary*.

carstone (iron pan) A brown *sandstone* with a *limonite* cement.

cascade fold A *fold* arising from *gravity* collapse. See *gravity collapse structure*.

case-hardening The *induration* of the surface of porous rocks, caused by the infilling of pore spaces by *mineral* cements precipitated from evaporating *meteoric water*, soil moisture or *groundwater* under tropical to sub-tropical conditions.

Cassian A *Triassic* succession in the Alps covering the upper part of the *Ladinian*.

cassiterite (SnO_2) The major *ore mineral* of *tin*.

Castile A *Permian* succession in the Delaware Basin, USA equivalent to the lower part of the *Longtanian*.

Castlemainian An *Ordovician* succession in Australia covering the upper part of the *Early Arenig*.

cataclasis (cataclastic flow) *Deformation* by *fracture*, sliding and rolling of rigid particles without internal *strain* in which the *strength* of the material increases with *confining pressure* and *dilatancy*.

cataclasite A cohesive *fault rock* with a random *fabric* containing 10–50% fragments in a finer-grained *matrix*, probably formed as a result of *cataclasis*. cf. *protocataclasite*, *ultracataclasite*.

cataclastic flow See *cataclasis*.

cataphorite See *katophorite*.

catastrophism The postulate that important changes in the physical environment arise from major events of high magnitude, low frequency and short duration. Essentially the converse of *uniformitarianism*.

catchment The total area of a drainage *basin*.

catena (toposequence) A sequence of soils found successively along a hillslope which are related by similar parental material.

cathodoluminescence (CL) The luminescence generated in *minerals* when bombarded by electrons. The phenomenon is used in *petrographic* studies to elucidate *diagenetic* trends in *sedimentary rocks* and the compositional zoning of crystals in *igneous* or *metamorphic rocks*.

cation exchange capacity (CEC) The amount of exchangeable cations that a *mineral* or soil can absorb at a given *pH*, expressed in mg equivalents per 100 g of material. These cations are mainly held on the surfaces of *colloids*.

cat's eye Any *gemstone* exhibiting *chatoyancy*, e.g. *spinel*, *ruby*.

cattierite (CoS_2) A *sulphide mineral* of the *disulphide group*.

cauldron The *structure* underlying a *caldera*, comprising the subsided block and any *ring dykes*.

cauldron subsidence The subsidence of cylindrical blocks of *country rock* with *magma* filling the space created to form *ring dykes*.

caunter lode See *counter lode*.

caunter vein See *counter vein*.

causse A *limestone plateau* characterized by closed depressions, *caves* and *avens*.

Cautleyian A *stage* of the *Ordovician*, 440.6–440.1 Ma.

cave A hole or fissure in rock, usually large enough for the entry of a person.

cave coral A *speleothem* of *calcite* in the form of a *coral*.

cave drapery A *speleothem* in the form of a sheet, formed by fluid trickling along a dipping ceiling.

cave marble See *cave onyx*.

cave onyx (cave marble) A *speleothem* of *calcite* or *aragonite* in a compact, banded form resembling *onyx*.

cave pearl A *speleothem* of *calcite* or *aragonite* formed by the accretion of layers around a nucleus, such as a *sand* grain.

cavitation erosion *Erosion* caused by the *turbulent flow* of meltwater at high velocity over rough *bedrock* under a *glacier*.

cay An island on a *coral-algal reef* made up of unconsolidated carbonates of *sand* and *gravel* grade.

Cayugan A *Silurian* succession in North America covering part of the *Gorstian*, the *Ludfordian* and the *Pridoli*.

CBED See *convergent beam electron diffraction*.

CCD See *carbonate compensation depth*.

CDP See *common depth point*.

CEC See *cation exchange capacity*.

cedar-tree laccolith A series of *laccoliths* of the same *igneous rock* stacked vertically.

celadonite ($K(Mg,Fe^{2+})(Fe^{3+},Al)Si_4O_{10}(OH)_2$) A green *mica* which is an early *authigenic* mineral in marine *sandstones*.

celerity The velocity with which a wave advances, the product of its frequency and wavelength.

celestine See *celestite*.

celestite ($SrSO_4$) An *industrial mineral* of strontium.

celsian ($BaAl_2Si_2O_8$) A barium *feldspar*.

cement bond log A *geophysical borehole log* which tests the quality of the cement bond between a borehole casing and its surrounding rock by measuring the *attenuation* of a seismic signal.

cementation A *diagenetic* process whereby *authigenic minerals* are precipitated into the pores of sediments, causing them to become consolidated or lithified.

cementstone An *argillaceous limestone* suitable for cement manufacture.

Cenomanian A *stage* of the *Cretaceous*, 97.0–90.4 Ma.

Cenozoic (Kainozoic) The youngest *era* comprising the *Palaeocene* to *Quaternary*, 65.0–0 Ma.

central eruption A *volcanic eruption* from a single *volcanic vent*. cf. *fissure eruption*.

cephalaspids See *Monorhina*.

Cephalopoda/cephalopods A class of phylum *Mollusca*, characterized by a differentiated head with well-developed eyes and a planispirally-coiled, external shell (except for subclass *Coleoidea*), internally partitioned by calcareous septae. Range *Cambrian–Recent*.

cerargyrite (chlorargyrite, horn silver) (AgCl) A *secondary mineral* found in the oxidized zone of silver *veins*.

cerianite ($(Ce,Th)O_2$) A very rare *oxide ore mineral* of thorium.

cerioid Descriptive of a *coral* in which the *corallites* are polygonal and packed together.

cerussite (white lead ore) ($PbCO_3$) An important *ore mineral* of lead.

cesium vapour magnetometer A portable *magnetometer* capable of very precise measurement of the *geomagnetic field*, often used on satellites and aircraft both singly and in pairs as a *magnetic gradiometer*.

Cetacea An order of infraclass *Eutheria*, subclass *Theria*, class *Mammalia*; the whales, dolphins and porpoises. Range *Eocene–Recent*.

ceylonite See *pleonaste*.

cgs The **c**entimetre-**g**ram-**s**econd system of units.

chabazite ($Ca_2Al_2Si_4O_{12}.6H_2O$) A *zeolite* found mainly in cavities in *basaltic* rocks in association with other *zeolites*, *calcite* and *quartz*.

Chadian A *stage* of the *Carboniferous*, 349.5–345.0 Ma.

Chaetetida An order of class *Demospongea*, phylum *Porifera*; *sponges* with a meandroid or *cerioid* basal skeleton with fibrous, tufted, *calcitic* or *aragonitic* walls. Range *Cambrian–Recent*.

Chaetognatha A phylum comprising the arrow worms, first appearing in the *Carboniferous*.

chain silicate See *inosilicate*.

chain structure group A group of *sulphide minerals* characterized by a structure comprising rings, the only member being *realgar*.

chain width error A *planar crystal defect* in which the *displacement* at the fault has a component parallel to the fault plane. cf. *stacking fault*.

chalcanthite (bluestone, blue vitriol) ($CuSO_4.5H_2O$) A *secondary mineral* of copper.

chalcedony (SiO_2) A *microcrystalline* form of *silica*.

chalcocite (redruthite) (Cu_2S) A copper *ore mineral*.

chalcophile Descriptive of an element with a strong affinity for sulphur.

chalcopyrite ($CuFeS_2$) A major *ore mineral* of copper.

chalcosiderite ($CuFe_6(PO_4)_4(OH)_8.4H_2O$) A rare *secondary mineral*.

Chalk An informal name for the U. *Cretaceous* of northern Europe.

chalk A very fine-grained, white, porous *limestone*, common in the *Cretaceous* of western Europe.

Chalmak A *Silurian* succession in the Mirnyy Creek area of NE Siberia covering part of the *Rhuddanian* and the *Aeronian*.

chalybeate Impregnated with iron salts.

chalybite An old name for *siderite*.

chamosite ($(Fe^{2+},Mg,Fe^{3+})_5Al(Si_3Al)O_{10}(OH,O)_8$) An *ore mineral* of iron found in *minette* or *Clinton ironstones*.

Chamovnicheskian A *stage* of the *Carboniferous*, 299.9–298.3 Ma.

Champlainian An *Ordovician* succession in the E USA covering part of the late *Arenig*, the *Llanvirn*, *Llandeilo*, *Costonian*, *Harnagian*, *Soudleyan* and *Longvillian*.

Chandler wobble The oscillation of the location of the Earth's rotational axis, with a period of 435 days, an amplitude of ~100 marc s and a decay time of ~40 years. Possibly caused by changes in the Earth's rotation rate resulting from changes in the magnetic coupling between the *core* and *mantle*.

Changshan A *Cambrian* succession in China covering parts of the *Maentwrogian* and *Dolgellian*.

Changxing A *Permian* succession in China equivalent to the *Changxingian*.

Changxingian The highest *stage* of the *Permian*, 247.5–245.0 Ma.

channel-belt deposit A sediment deposited between successive *avulsions* in a river channel.

channel capacity The maximum *discharge* possible within a river channel.

channel resistance The resistance to flow in a channel arising from particle resistance, the resistance of bedforms, e.g. *ripples*, *dunes*, flow around channel bends and the spill resistance caused by changes in flow pattern at high *Froude numbers*.

channel roughness See *Manning equation*.

channel sample A sample from channels cut across the *face* of an exposed *ore*.

channel storage The capacity of a

channel network to contain a flood *discharge*.

char *Coke* in the form of a granular mass of porous powder which can be briquetted.

chargeability A measure of *induced polarization* in time-domain methods equal to the area under the time-decaying voltage curve over a given time period divided by the steady-state voltage before measurement.

charnockite An *orthopyroxene*-bearing *quartz-feldspar* rock formed at high temperature and pressure, commonly found in *granulite facies metamorphic* terrains.

Charophyta Large, bushy, green *calcareous algae*, mainly found in freshwater ponds and useful *biostratigraphically* from the *Cretaceous* to *Oligocene*, when they underwent great diversification.

chatoyancy A *mineral* property caused by the surface reflection of light to give a silky, banded appearance.

chatter marks Centimetric-scale, curved cracks found on glaciated surfaces, formed by the pressure of irregularly-moving *boulders*.

Chattian The higher *stage* of the *Oligocene*, 29.3–23.3 Ma.

Chautauquan A *Devonian/Carboniferous* succession in E North America equivalent to the *Famennian* and *Hastarian*.

Chebotarev sequence An idealized temporal sequence of changes in *groundwater*, which becomes more charged the longer it is in contact with the *aquifer*. With increasing depth the anion sequence is bicarbonate, sulphate, chloride and the cations change from calcium to sodium.

cheirographic coast A coastline cut by successive deep bays and promontories, caused by *tectonic* activity.

chelate A very strong, multifunctional, claw-like bond formed between an organic molecule and a metal in *chelation*.

chelation The removal by complexing of metal atoms or cations in organic ring compounds by the formation of *chelates* during *weathering*.

Chelicerata/chelicerates A subphylum of phylum *Arthropoda* whose body is divided into a combined head/thorax with usually six appendages, the first used for feeding and the remainder for locomotion and feeding, and an abdomen. No antennae are present. Range *Cambrian–Recent*.

Chelonia An order of subclass *Anapsida*, class *Reptilia*; the turtles and tortoises. Range *Triassic–Recent*.

cheluviation The combination of water containing organic extracts with soil cations by *chelation*. The solution then moves downwards by *eluviation* and transfers metals to a lower level.

chemical and instrumental analysis of minerals Techniques for the identification of *minerals*, quantification of their chemistry and structure and elucidation of their development and *textures*. The methods include *wet chemical analysis*, *X-ray diffraction*, *X-ray fluorescence*, *electron microprobe* and *mass spectrometry*.

chemical remanent magnetization (CRM, crystalline remanent magnetization) A *remanent magnetization* acquired as *ferromagnetic* grains grow through their *blocking volume* in an external *magnetic field*.

chemostratigraphy The use of chemical signatures (e.g. *carbon*, oxygen and strontium isotopes, amino acid residues) in stratal sequences for the purpose of correlation.

chenier plain A coastal plain consisting of progradational, alternate, coast-parallel bands of coarse *clastic* and muddy sediment, occurring in environments undergoing periodic *erosion*.

cheralite ($(Th,Ca,Ce,La,U,Pb)(PO_4, SiO_4)$) A green radioactive *mineral* of the *monazite* group rich in thorium.

Cheremshanskian A *stage* of the *Carboniferous*, 318.3–313.4 Ma.

chert (SiO_2) A granular *microcrystalline* to *cryptocrystalline* variety of *quartz*.

chertification The silicification of a sediment into *chert* by the contemporaneous *dissolution* and *precipitation* of siliceous tests.

cherty iron formation See *banded iron formation*.

Chesterian A *Carboniferous* succession in the USA covering parts of the *Visean* and *Serpukhalian*.

chesterite ($(Mg,Fe)_{17}Si_{20}O_{54}(OH)_6$) A *biopyribole* formed by the *alteration* of *anthophyllite*.

chevron construction An approximate method of constructing the geometry of a non-planar *fault* from the shape of its folded *hangingwall*, which assumes that *heave* is constant and conserved.

chevron fold (accordion fold, zigzag fold) A *fold* with straight *limbs* and a sharp *hinge*.

chevron mark A 'V'-shaped *sole mark*.

Chev'ynskiy A *Permian* succession in the Timan area of the former USSR covering parts of the *Wordian* and *Capitanian*.

Chewtonian An *Ordovician* succession in Australia covering the middle part of the *Early Arenig*.

Chézy equation $v = C/(Rs)$, where v = mean flow velocity in a channel, C = Chézy roughness coefficient, R = *hydraulic radius*, s = channel slope.

chiastolite A variety of *andalusite* containing dark carbonaceous inclusions in the form of a cross.

chicken-wire texture A *texture* resembling a mesh developed in the sulphates which represents the *penecontemporaneous dolomitization* of *sabkhas*. Once taken as evidence of an ancient *sabkha* environment, it is now realized that the *texture* is a common feature of the *diagenesis* of sulphates.

Chientangkiang See *Jiangtangjiang*.

Chile saltpetre ($NaNO_3$) Sodium nitrate, found in Chilean *caliche*.

chilled margin The fine-grained, outer layer of an *igneous body* formed by rapid cooling.

Chilopoda A class of superclass *Myriapoda*, phylum *Uniramia*; the centipedes. Range *Cretaceous–Recent*.

chimney An *orebody* of this shape lying within an *ore*-locating *structure*.

china clay A soft, white, plastic *clay* composed of *kaolinite* with low iron, formed by the *alteration* of rocks rich in *feldspar*.

china stone 1. *Kaolinized granite* with unaltered *feldspar*. 2. Certain very fine-grained *limestones* with a smooth *texture*.

chine A narrow ravine or canyon running to the sea.

chip sample A sample of rock chips taken from a *face*.

Chiroptera An order of infraclass *Eutheria*, subclass *Theria*, class *Mammalia*; the bats. Range M. *Eocene–Recent*.

Chixia A *Permian* succession in China equivalent to the *Sakmarian* and *Artinskian*.

chloanthite ($(Ni,Co)As_{3-x}$) An *ore mineral* of nickel and cobalt.

chlorapatite ($Ca_5(PO_4)_3Cl$) A variety of *apatite*.

chlorargyrite See *cerargyrite*.

chlorastrolite An obsolete name for the fibrous, green variety of *pumpellyite*.

chlorite ($(Mg,Fe)_3(Si,Al)_4O_{10}(OH)_2 \cdot (Mg,Fe)_3(OH)_6$) An important group of *phyllosilicates* found in *metamorphic*, *sedimentary* and *altered igneous rocks*.

chloritization A type of *wall rock alteration* in which *chlorite* ± *quartz* or *tourmaline* forms as the result of the passage of *hydrothermal solutions*.

chloritoid ($(Fe,Mg)_2Al_4O_2(SiO_4)_2(OH)_4$) A *nesosilicate* of the *humite* group.

Choanichthyes A group of *fish* comprising the *Crossopterygii* and *Dipnoi*, which share lungs, nares and paired fins with fleshy lobes. Range L. *Devonian–Recent*.

chocolate block/tablet boudinage *Boudinage* in which a *competent* layer is segmented into equidimensional *structures* by layer-parallel *extension* in two directions.

Chokierian A *stage* of the *Carboniferous*, 328.3–325.6 Ma.

Chokrakian A *Miocene* succession on the Russian Platform covering the upper part of the *Serravallian*.

Chondrichthyes A class of superclass *Pisces*; the sharks, skates and rays. Range *Silurian–Recent*.

chondrite A *stony meteorite* containing *chondrules*.

chondrodite ($Mg_5(SiO_4)_2(F,OH)_2$) A *nesosilicate* of the *humite* group.

Chondrostei An order of subclass *Actinopterygii*, class *Osteichthyes*, superclass *Pisces*; *fish* with a partly cartilaginous skeleton. Range L. *Devonian–Recent*.

chondrule A small, globular mass of *pyroxene*, *olivine* and occasionally *glass* found in certain *stony meteorites*.

chonolith An igneous intrusion which cannot be classified because of its irregular form.

Choquette and Pray classification A scheme for the classification of *porosity* types in carbonate rocks.

Chordata/chordates A phylum comprising animals with a rod of flexible tissue, protected in later forms by a backbone and first appearing in the *Cambrian*.

Chotec A *Devonian* succession in Czechoslovakia covering the upper part of the *Eifelian*.

chott A large, seasonally-flooded desert *basin*.

chrome iron ore See *chromite*.

chrome spinel See *picotite*.

chromite (chrome iron ore) ($FeCr_2O_4$) A *spinel oxide mineral*, the major *ore mineral* of chromium.

chromitite A rock in which *chromite* is the dominant *mineral*.

chron A small unit of geological time.

chronohorizon A thin, characteristic *stratigraphic* interval that can be used for accurate time correlation or as a time reference zone.

chronosequence The development of a soil with time.

chronostratigraphic unit A *stage* or *zone* which, independent of *facies*, represents a layer of specific age.

chronostratigraphy A *geological time scale* represented by a sequence of rocks with standard reference points.

chrysoberyl ($BeAl_2O_4$) A rare *mineral* found in *granitic* rocks, *mica schists* and *sands* used as a *gem*, particularly in the varieties *alexandrite* and *cat's eye*.

chrysocolla ($Cu_4H_4Si_4O_{10}(OH)_8$) A *secondary mineral* of copper found in the oxidized zone of copper deposits.

chrysolite An *olivine* with the composition Fo_{70-90}.

chrysoprase An apple green variety of *chalcedony*.

chrysotile (Canadian asbestos, serpentine asbestos) ($Mg_3Si_2O_5(OH)_4$) A *clay mineral*; a fibrous variety of *serpentine*, possibly with an *asbestiform* habit.

Churchillian orogeny An *orogeny* affecting the central part of the Canadian *shield* in the *Proterozoic* from ~1900–1850 Ma.

chute bar A sedimentary deposit formed at the downstream end of a narrow channel with a swift current.

Cimmeridian See *Kuyalnitskian*.

Cincinnatian An *Ordovician* succession in the E USA equivalent to the *Marshbrookian*, *Actonian*, *Onnian* and *Ashgill*.

cinder cone An accumulation of *scoria* close to a *volcanic vent* resulting from a *strombolian eruption*.

cinders *Vesicular lapilli* composed of dark *glass*.

cinnabar (HgS) The major *ore mineral* of *mercury*.

cinnamon stone See *hessonite*.

CIPW classification An early system for the classification of *igneous rocks* based on *normative compositions*.

CIPW norm A procedure for recomputing the chemical composition of a rock into a group of hypothetical standard *minerals* for simplified comparison between rocks.

circumgranular fracture (grain boundary fracture) A *fracture* in a granular material which goes round the grains. cf. *intergranular fracture*, *intragranular fracture*.

cirque (corrie, cwm) A horseshoe-shaped, steep-walled, valley head caused by *glacial erosion.*

Cirripedia A subclass of class *Maxillopoda*, subphylum *Crustacea*, phylum *Arthropoda*; the barnacles. Range U. *Silurian–Recent.*

citrine (Indian topaz, Madagascar topaz, quartz topaz, yellow quartz) A clear yellow variety of *quartz*, valued as a semi-precious *gem.*

CL See *cathodoluminescence.*

cladistics A taxonomic system applied to the study of evolutionary relationships which proposes that a common origin can be demonstrated by shared characteristics. It assumes that each new *taxon* develops by the splitting of an ancestral lineage into two daughter *taxa.*

cladogenesis The branching into two derivative *taxa* as inferred in *cladistics.*

cladogram A diagram illustrating the branching sequences in a *cladistic* evolutionary sequence.

Cladoselachii An order of subclass *Elasmobranchii*, class *Chondrichthyes*, superclass *Pisces*; *fossil* sharks with an elongate body and two dorsal fins with a spine. Range *Devonian–Carboniferous.*

Claiborne An *Eocene* succession on the Gulf Coast of the USA covering part of the *Lutetian* and the *Bartonian.*

claim An area of land staked out by a person who has certain rights to explore for and exploit *minerals* therein.

Clairault's formula An equation showing how *gravity* varies over the reference *spheroid*, the basis of the *Gravity Formula.*

Clapeyron equation An equation linking the temperature of a *mineral* phase change, the volume change and the pressure.

clarain A type of *banded coal* with bright to semi-bright bands of finely laminated *coal* with a silky *lustre*, often containing thin *vitrain* bands alternating with duller *attrital* material.

Clarence A *Cretaceous* succession in New Zealand comprising the *Urutawan, Motuan* and *Ngaterian.*

clarite A *microlithotype of coal,* consisting of *vitrinite* and *exinite* group *macerals,* which makes up to 20% of *bituminous coals.*

clarodurain A *lithotype of banded coal* with characteristics intermediate between *clarain* and *durain.*

clarofusain A *lithotype of banded coal* with characteristics intermediate between *clarain* and *fusain.*

clast A particle of rock or single crystal which has been derived by *weathering* and *erosion.* The basic building block of a *clastic sediment.*

clast-supported conglomerate (framework-supported conglomerate, grain-supported conglomerate, particle-supported conglomerate) A *rudite* which contains >85% *clasts,* which are mostly in contact. cf. *matrix-supported conglomerate.*

clastation The disintegration of a rock into *clasts* by chemical or physical means.

clastic rock A *sedimentary rock* made up of *clasts*, classified according to *clast* size.

clastic sediment A sediment made up of *clasts.*

clathrates *Hydrocarbon*-bearing ice molecules which make up *gas hydrates*.

clay A sediment with particles $< 4\ \mu m$ in size.

clay ball See *armoured mud ball*.

clay dune An *aeolian dune* composed predominantly of *clay aggregates* rather than *quartz sand* grains.

clay ironstone A sediment composed mainly of *siderite* occurring in thin *nodules* in some *argillites*.

clay minerals *Phyllosilicate minerals* based on composite layers constructed from components with tetrahedrally and octahedrally coordinated cations. Mainly occurring as *microscopic*, platy particles in fine-grained *aggregates* which have varying degrees of *plasticity* when mixed with water. They are hydrous silicates, mainly of aluminium, magnesium, iron and potassium, which lose adsorbed and constitutional water at high temperature to yield refractory materials. The most important groups are *illites*, *kandites*, *smectites* and *vermiculites*.

clay pan (soil crust) A compact *illuvial* layer of clayey material found in the soil zone formed by *leaching* and *eluviation* of the upper soil zone.

claystone A *clastic sedimentary rock* with the composition of *shale* but without its characteristic *lamination* and *fissility*.

clear water erosion The *erosion* of a channel by a river whose sediment load has been removed, e.g. by the construction of a dam.

cleat Closely spaced *jointing* found in *coal seams*, generally in two orthogonal sets normal to the *bedding*.

cleavage 1. A *mineral* property whereby it breaks along regular, crystallographic planes. 2. A *foliation* formed by *deformation* at low *metamorphic grade*, along which a rock splits preferentially.

cleavage fan A radiating pattern of *cleavage* surfaces.

cleavage mullion A *mullion structure* comprising polished or striated cylinders bounded by *cleavage planes*.

cleavage plane A plane along which *fracture* occurs in a rock.

cleavage refraction A change of orientation of *cleavage*, usually across a layer boundary.

cleavage tetrahedron A figure showing the four types of *cleavage*: *crenulation*, *fracture*, *pressure solution* and *slaty*.

cleavage transection The oblique intersection of *cleavage* and a *fold axis*.

cleavage vergence The horizontal direction of rotation through the acute angle from the *cleavage* to an earlier *fabric* within the plane normal to the *cleavage/fabric* intersection.

cleavelandite A white, platy variety of *albite*.

cleveite A variety of *pitchblende* with uranium oxide and *rare earth elements*.

Clifdenian A *Miocene* succession in New Zealand. Equivalent to the *Langhian*.

climatic geomorphology The branch of *geomorphology* covering the formation of landforms by climate.

Climatiformes An order of subclass *Acanthodii*, class *Osteichthyes*, superclass

Pisces; *fish* with bony jaws and a skeleton midway between sharks and bony fish. Range U. *Silurian*–L. *Permian*.

climb The movement of a *dislocation* by the diffusion of atoms normal to the structural displacement at the *dislocation* in the formation of a *crystal defect*.

climbing adhesion ripple structure An *adhesion structure* forming when net deposition occurs by *adhesion* to a surface kept damp by the capillary rise of water.

climbing dune A *dune* which climbs a gentle slope.

climbing ripple A *ripple* which climbs a gentle slope.

clino- Inclined.

clinochesterite ($(Mg,Fe)_{17}Si_{20}O_{54}(OH)_6$) A rare *pyribole*, probably formed by the *alteration* of *anthophyllite*.

clinochlore A variety of *chlorite*.

clinoenstatite ($MgSiO_3$) A *clinopyroxene*.

clinoferrosilite ($FeSiO_3$) A *clinopyroxene*.

clinoform A major sloping depositional surface.

clinohumite ($Mg_9(SiO_4)(F,OH)_2$) A *nesosilicate* of the *humite* group.

clinohypersthene ($(Mg,Fe)SiO_3$) A *clinopyroxene*.

clinojimthompsonite ($(Mg,Fe)_{10}Si_{12}O_{32}(OH)_4$) A *monoclinic pyribole* with triple silicate chains.

clinometer A field instrument used to measure the magnitude of *dip*.

clinoptilolite ($(Na,K,Ca_{0.5})_6(Al_6Si_{30}O_{72}).24H_2O$) A *zeolite*.

clinopyroxene A *monoclinic pyroxene* exhibiting non-parallel *extinction*. cf. *orthopyroxene*.

clinothem A rock unit formed of *strata* which *prograde* gently seawards to deep water.

clinozoisite ($Ca_2Al_3O(SiO_4)Si_2O_7(OH)$) A *sorosilicate* of the *epidote* group.

clint (flachkarren) A *tabular* block of *limestone* in a *limestone pavement*.

Clinton ironstone (Clinton ore) An *oolitic haematite-chamosite-siderite* rock with ~40–50% iron, forming lenticular beds 2–3 m thick. Formed in shallow water conditions.

Clinton ore See *Clinton ironstone*.

clintonite (xanthophyllite) ($Ca(Mg,Al)_{3-2}Al_2Si_2O_{10}(OH)_2$) A *phyllosilicate*; a *brittle mica*.

clitter A type of *blockfield* made up of a scatter of *granite boulders*.

close fold A *fold* with an *interlimb angle* of 30–70°.

cluse A steep-sided valley cutting through a mountain ridge.

Cnidaria (Coelenterata/coelenterates) A large phylum of living and extinct organisms which are the lowest animals with definite tissues, including the *Stromatoporoidea*, *Scyphozoa* and *Anthozoa*.

co-ignimbrite breccia See *lag breccia*.

Coahuila A *Cretaceous* succession on the Gulf Coast of the USA comprising the *Durango* and *Nuevo Leon*.

coal A combustible, organo*clastic*, *sedimentary rock* containing >50% by weight of carbonaceous material and moisture and composed mainly of lithified plant remains.

coal ball A spheroidal mass of, commonly, *calcite*, *dolomite*, *siderite* and *pyrite* in a *coal seam*.

coal basin A sedimentary *basin* containing important *coal seams*.

coal gas (town gas) A gas produced by the distillation of *bituminous coal*, with >50% hydrogen and 10–30% *methane* and a *calorific value* of ~18MJ m^{-3}, used in heating and lighting.

Coal Measures A *stratigraphic* term for the U. *Carboniferous* of western Europe, equivalent to the *Westphalian*.

coal measures A series of *strata* containing economically workable *coal seams*.

coal quality map A contour map showing changes in the nature of a *coal seam* over a mine area or *coalfield*.

coal seam A *bed* of *coal*.

coal tar *Tar* produced by the distillation of *bituminous coal*.

coalfield A region rich in *coal* deposits.

coalification The process by which moist, partially decomposed vegetation such as *peat* is transformed, in response to burial or *tectonic* activity, progressively into *brown coal*, *bituminous coal* and *anthracite* with an increase in *carbon* content and *calorific value* and a decrease in volatiles and moisture.

coarse-tail grading A feature shown by a *graded bed* in which only the larger grain sizes are graded. cf. *distribution grading*.

Coast Range orogeny An *orogeny* affecting the Coast Mountains of British Columbia in the *Jurassic* and early *Cretaceous*, approximately equivalent to the *Nevadan orogeny* of the USA.

coastal aquifer An *aquifer* extending beneath the sea and accessible to seawater.

coastal bar A linear accumulation of mainly *sand* grade sediment lying submerged in the nearshore zone subparallel to the coastline.

coastal notches Horizontal, slot-like recesses formed by *erosion* at the base of a *sea cliff*, normally at the high-water mark, and the main factor in the undermining of the cliff.

coated gold *Native gold* covered by a surface film of *iron oxide* which gives it a rusty or tarnished appearance. See also *black gold*.

coated grain A sedimentary particle made up of successive, concentric layers, e.g. an *oolith*.

coaxial Descriptive of any progressive *deformation* during which the *principal strain* axes do not rotate with respect to reference lines within the rock. 2. Descriptive of *folds* with a common *axial* direction.

cobalt bloom An old name for *erythrite*, a *secondary mineral* of cobalt.

cobalt glance See *cobaltite*.

cobalt pentlandite (Co_9S_8) A *sulphide mineral* of the *metal excess group*.

cobalt vitriol See *bieberite*.

cobaltite (cobalt glance) ((Co,Fe) AsS) A rare *ore mineral* of cobalt.

cobble A rock fragment between 64–256 mm in diameter.

Coble creep *Creep* accomplished by the diffusion of individual atoms along grain boundaries.

Coblencien A *Devonian* succession in France and Belgium covering the *Pragian* and part of the *Emsian*.

Coccolithophoridae/coccolithophoroids Small (<20 μm), unicellular, spheroidal, *planktonic algae* covered in calcareous plates, responsible for extensive marine carbonate production. Range U. *Triassic–Recent*.

coccoliths Minute calcareous plates which cover *Coccolithophoridae*, which contribute greatly to the formation of *chalk* and deep sea *oozes* and which are useful in the fine-scale *stratigraphy* of *pelagic* deposits of *Jurassic* to *Pleistocene* age.

cockpit karst (cone karst, kegelkarst) A *karst* landscape of humid tropical areas comprising star-shaped, closed depressions separated by deep residual hills.

cockscomb pyrites A *twinned* form of *marcasite*.

COCORP **Co**nsortium for **Co**ntinental **R**eflection **P**rofiling. The organization that undertakes deep *seismic reflection* profiling in the USA.

coefficient of friction See *Coulomb criterion*.

coefficient of roughness See *Chézy equation*.

coefficient of work hardening/softening The slope of the *stress-strain* curve as a rock of significant *plasticity* is stressed.

Coelenterata See *Cnidaria*.

Coelolepida An order of subclass *Diplorhina*, class *Agnatha*, superclass *Pisces*; poorly-known, tiny *fish* with a forked tail and small mouth. Range late *Silurian*–L. *Devonian*.

Coelosauria Small (crow-sized), bipedal, *theropod dinosaurs* similar in form to Archaeopteryx. Range U. *Triassic–Cretaceous*.

coenosteum The laminated skeleton of a *stromatoporoid*.

coenozone See *assemblage zone*.

coesite (SiO_2) A high pressure form of *silica*.

coffinite ($U(SiO_4)_{1-x}(OH)_{4x}$) A black, hydrous uranium silicate.

cognate inclusion See *autolith*.

cognate lithic A non-*vesicular clast* found in a *pyroclastic rock* consisting of juvenile material derived from the same, or closely related, *magma* as the *vesicular clasts* in the rock.

cohesion (cohesive strength) The strength of bonding between particles or surfaces; in rock mechanics specifically the inherent *shear strength* of a plane across which there is no *normal stress*.

cohesive strength See *cohesion*.

cokability index A measure of the suitability of a *coal* (only certain *caking coals* of specific *rank*) for making *coke*.

coke A dense, porous product of the *carbonization* of *caking coal* in an oven.

coking coal A *coal* suitable for making *coke*.

col A gap in a *watershed* caused by *glacial erosion*.

colatitude (90°–φ), where φ = latitude.

cold working See *dislocation glide*.

colemanite ($CaB_3O_4(OH)_3.H_2O$) A *mineral* which is a source of *borax*, found in arid areas and saline lakes.

Coleoidea A subclass of class *Cephalopoda*, phylum *Mollusca*; *molluscs* with an internal shell or lacking a shell, including many squids and the octopoids. Range U. *Mississippian–Recent*.

collapse breccia (solution breccia) Chaotic, angular fragments of rocks formed when an underlying *evaporite* layer is removed by *dissolution*.

collinite A component of the *maceral vitrinite* in which no microscopic cell structure is visible.

collision zone A linear belt where two continents or *microcontinents* have collided as the result of their intervening ocean having been consumed by *subduction*.

colloform banding A *texture* found in certain *mineral* deposits in which crystals have grown in a radiating and concentric fashion, possibly resulting from geochemical controls.

colloid A dispersion of extremely fine particles in suspension, possible because the ultra-small size of the particles (1–10 μm) allows the supporting forces to exceed the gravitational forces promoting settling.

collophane (collophanite) Massive, *cryptocrystalline* varieties of *apatite* making up the bulk of *phosphate rock* and *fossil* bone.

collophanite See *collophane*.

colluviation The process by which *colluvium* forms.

colluvium Sediment transported by weakly selective, non-*fluvial* processes such as *mass-wasting* and *slope-wash*. cf. *alluvium*.

colonial coral A *coral* in which the exoskeleton is built by several animals.

colonnade A volcanic layer with *columnar jointing* which is regular and vertical. cf. *entablature*.

Colorado Plateau-type deposit (roll-front-type deposit, sandstone uranium-type deposit, western states-type deposit) A uranium-rich, *sandstone*-uranium-*vanadium base metal deposit*.

Colorado ruby A fiery-red variety of *pyrope*.

Colorado topaz A brown-yellow variety of *topaz*.

colorimetry A method of *wet chemical analysis* in which reagents are added to a solution of the unknown to form coloured compounds. The intensity of the colour is proportional to the concentration of the unknown.

colour alteration index (CAI) A calibration of the colour change of a *conodont* element with temperature used for the assessment of *thermal maturation* of sedimentary *basins* and the heating during thermal and *tectonic* events.

colour index (colour ratio) The total percentage of *mafic* (Mg- and Fe-rich) *minerals* in the *modal analysis* of an *igneous rock*, used as an aid to classification.

colour ratio See *colour index*.

coloured stone mining Mining for *gems* other than *diamonds.*

columbite (niobite) ($(Fe,Mn)Nb_2O_6$) An *ore mineral* of niobium found in *granitic* rocks and *pegmatites.*

columnar jointing The vertical *joint* pattern developed during the cooling of a large body of *volcanic rock* which divides it into regular polygonal columns. Thick bodies may consist of alternating *colonnades* and *entablatures.*

comagmatic Descriptive of *igneous rocks* derived from the same *magma.*

Comanche A *Cretaceous* succession on the Gulf Coast of the USA comprising the *Trinity, Frederiksberg* and *Washita.*

comb See *combe.*

comb structure A *texture* shown by an *ijolite* intruded by *veins,* mineralogically similar to the *host rock,* in which *pyroxene, wollastonite* and *feldspar* exhibit a *prismatic* to *acicular habit* with long *mineral* axes orthogonal to the walls of the *vein.*

combe (coombe, comb, coomb) A small, steep-sided valley which may contain an *ephemeral stream.*

combination trap A combined *structural* and *stratigraphic oil* or *gas trap.*

combined gold (AuTe) *Gold* in combination with tellurium.

comendite A fresh, unaltered, *peralkaline,* silicic, *glassy rock* with $< \sim 12.5\%$ *normative femic minerals.*

Comley Lowest *series* of the *Cambrian.*

comminution The breaking down of material into a fine powder.

commissure A line or plane of junction, such as between the *valves* of a shell.

common depth point (CDP) The location on a reflector which produces *seismic reflections* for a number of different source-receiver combinations. CDP *stacking* generally improves the *signal to noise ratio* of a *seismic record.*

common lead Lead formed by the addition of radiogenic lead to primeval lead, probably with a simple history. cf. *anomalous lead.*

compactibility The property of a sediment that allows *compaction.*

compaction The process of volume reduction and *pore fluid* expulsion within a sediment in response to increasing overburden load, commonly expressed by the change in *porosity.*

companion sand *Sand* which may be present in a *parna.*

comparator electromagnetic method An *electromagnetic induction exploration* method using an artificial, time-varying electromagnetic field.

compensated load force The force on a continental block arising from the *isostatic* compensation of a load, such as at a *continental margin* or *plateau uplift.*

compensator electromagnetic method An *electromagnetic induction method* using a constant transmitter-receiver separation in which the primary electromagnetic field is compensated so that the instrument responds only to the small secondary fields generated by subsurface conducting bodies.

competent Of high relative *strength.*

complex dune See *draa.*

complex twin See *compound twin.*

compliance tensor A complete description, in terms of 36 components of *elastic moduli*, of the relationship between *stress* and *strain* in an *anisotropic* material.

composite intrusion A type of *multiple intrusion* composed of *magmas* of contrasting composition.

composite seam A *coal seam* made up of two or more different *coal* beds juxtaposed when intervening *strata* were wedged out.

composite volcano See *stratovolcano.*

composition plane The plane along which two *twinned* crystals are joined, often the same as the *twin plane.*

compositional layering A set of layers of distinct, different composition, usually applied to *igneous* or *metamorphic rocks* in which the origin of the layering is equivocal.

compositional maturity index The ratio of *quartz* + *chert* grains to *feldspars* + rock fragments in a *sandstone*, providing a measure of maturity as the former pair is relatively resistant to breakdown compared to the latter pair.

compound twin (complex twin) A crystal in which at least two types of *twinning* has occurred.

compressibility The inverse of the *bulk modulus.*

concealed coalfield A *coalfield* hidden beneath other *strata.*

concentration deposit A particularly rich *exceptional fossil deposit*, formed by winnowing, slow deposition rate, *placers* and concentration traps.

concentration factor The degree of enrichment of an element or elements above their normal crustal abundance in the formation of an *orebody*.

concentric fold A *fold* in which the layer thickness perpendicular to its surface is constant and the layer boundaries form concentric arcs of circles.

conchoidal fracture A *mineral fracture* producing smooth, curving surfaces similar to the interior surface of a shell.

conchoidal fringe joint A shorter *fracture* cross-linking *en echelon fractures.*

Conchostracha An order of class *Branchiopoda*, subphylum *Crustacea*, phylum *Arthropoda*; small, bi*valve*d invertebrates enveloped by a carapace. Range L. *Devonian–Recent.*

concordant With margins parallel to the *bedding* or *foliation* of the *country rock.*

concordant coast A coast parallel to the general trend of relief with a consequent straight, regular outline.

concordant intrusion An igneous intrusion which does not cut across the *bedding* or *foliation* of the *country rock.*

concretion A *nodule* without a concentric structure.

concurrent range zone A *zone* based on the co-occurrence of two or more species.

condensate A very light *crude oil* with *API gravity* >50°.

condensed sequence A series of *strata* much thinner than developed elsewhere, probably caused by a diminished sediment supply.

Condoblin A succession in E Australia of middle *Devonian* age.

Condylarthra An order of infraclass *Eutheria*, subclass *Theria*, class *Mammalia*; extinct ungulate-like *mammals*, transitional between the *Ungulata* and *Insectivora*. Range late *Cretaceous*–U. *Miocene*.

cone karst See *cockpit karst*.

cone-in-cone structure A *mineral structure* in the form of a series of nested, concentric cones.

cone of depression The inverted conical form adopted by the *water table* after the extraction of water by pumping.

cone sheet A type of small *ring intrusion* with margins which *dip* inwards, probably towards the centre of the upper part of a *magma chamber*. Possibly emplaced during the *uplift* of a central conical block associated with a pressured *magma chamber*.

confined aquifer See *artesian aquifer*.

confining pressure The radial *stress* applied to a cylindrical unit deformed by axially symmetric loading.

conformability See *conformity*.

conformable Descriptive of a continuous series of *beds*, i.e. with no *unconformities*.

conformity (conformability) The *stratigraphic* continuity of *strata*.

congelifraction (frost shattering, frost splitting, frost weathering, frost wedging) The fragmentation of rock by the expansion and contraction of water freezing and melting in pores, cracks or fissures.

congeliturbation See *cryoturbation*.

conglomerate A *rudite* with rounded *clasts* larger than 2 mm.

conglomerate test A *palaeomagnetic* test for determining the age of a *remanent magnetization*. If the *natural remanent magnetization* of the *clasts* of a *conglomerate* are randomly oriented, the magnetization predates its formation and vice versa.

congruent melting Melting in which the melt has the same composition as the solid. cf. *incongruent melting*.

Coniacian A *stage* of the *Cretaceous*, 88.5–86.6 Ma.

conical fold A *fold* which approximates the shape of a conical surface.

conjugate Referring to a pair, or two sets, of *faults*, *shear zones*, *asymmetric folds*, *kink bands*, *joints* etc. with opposing *dips* (*faults*) or *vergences* (*folds*).

conjunctive use The combined use of *groundwater* and surface water, e.g. in *artificial recharge*.

connate water The water deposited with, and contained within, a sediment, which may be changed in composition during burial e.g. to a *brine*.

Conodontophora/conodonts Marine animals with vertebrate affinities represented in the *fossil* record by scattered elements of their mineralized feeding apparatus. Generally 0.2–2.0 mm in size and composed of *apatite*. Range U. *Cambrian*–uppermost *Triassic*.

Conrad discontinuity A *seismic discontinuity* seen in certain regions of the *continental crust* at a depth of 10–12 km marking the boundary between the *upper* and *lower continental crusts*.

consequent stream A stream flowing in the direction of the original slope of the surface. cf. *anteconsequent, inconsequent, insequent, obsequent, subsequent* and *resequent streams.*

conservation deposit An *exceptional fossil deposit* in which soft tissues or the impressions of soft tissues are preserved.

conservative plate margin A *plate margin* at which *lithosphere* is neither created nor destroyed, i.e. a *transform fault.* cf. *destructive plate margin, constructive plate margin.*

consolidation The processes of *compaction* of a loose material.

constant separation traversing (CST, electric profiling, electric trenching) A *resistivity method* in which the electrodes are maintained at a fixed separation and moved along a profile to map lateral variations in *resistivity* and locate buried three-dimensional bodies of anomalous *resistivity.* cf. *vertical electrical sounding.*

constructive plate margin (accretive plate margin) A *plate margin* at which *lithosphere* is created, i.e. an *ocean ridge.* cf. *conservative plate margin, destructive plate margin.*

contact aureole (metamorphic aureole) A centimetric- to kilometric-scale zone of *contact metamorphism* around an *igneous body.*

contact mapping (boundary mapping) A technique of geological mapping in which a geological contact is followed by zigzagging along it.

contact metamorphism The *thermal metamorphism* developed in rocks adjacent to the contact with an *igneous body,* reflecting the high temperature of the intrusion and possibly the expulsion of *hydrothermal solutions* from the intrusion and the mobilization of *groundwater* in the *country rocks.*

contact resistance The electrical resistance between an electrode and the ground.

contact twin A *twin* whose members are separated by a *composition plane.* cf. *interpenetrant twin.*

contaminated magma A *magma* of *mantle* derivation into which *crustal* material has been added by *assimilation.*

contamination See *assimilation.*

content grading (distribution grading) A gradual change in the entire grain size distribution in a *graded bed.*

continent-continent collision (continental collision) The collision of two *plates* of continental *lithosphere* by the consumption of an intervening ocean by *subduction* beneath one of them to form a *Himalayan-type mountain range.* The buoyancy of continental *lithosphere* allows only very limited *subduction,* but the persisting driving forces may power further *indentation tectonics.*

continental collision See *continent-continent collision.*

continental crust The upper layer of continent-bearing *lithosphere,* bounded at the base by the *Mohorovicic discontinuity* at a depth of ~20–80 km, at which there is a downward increase in density from ~3.0 to ~3.3 Mg m^{-3}. Its average composition is between *granodiorite* and *quartz diorite.* Sometimes divided into upper and lower continental crust by the *Conrad discontinuity.* cf. *oceanic crust.*

continental drift The hypothesis, largely attributed to A. Wegener in the early 20th century, that the continental blocks move relative to each other on geological timescales. Lack of knowledge of the ocean basins prevented the proposal of a cogent mechanism. Superseded by the theories of *sea floor spreading* and *plate tectonics*.

continental platform See *continental shelf*.

continental reconstruction A *palaeogeographic* configuration of continental blocks at some time in the past, determined by jig-saw fitting of their geometric shapes, rotations about *Euler poles*, fitting together *oceanic magnetic anomalies* from either side of an *oceanic ridge* and *palaeomagnetic* and *palaeoclimatic* data.

continental rise The gently-dipping part of the continental margin between the *continental slope* and *abyssal plain*.

continental shelf (continental platform) That part of the sea floor underlain by *continental crust*, taken to lie above a water depth of 200 m.

continental slope The steep slope between the *continental shelf* and the more gentle slope to the *abyssal plain*.

continental splitting The initiation of a new ocean by *rifting* of a continent and subsequent *seafloor spreading*.

continuous reaction series (reaction series) The continuous change in composition of a *solid solution mineral* as it retains equilibrium with a cooling *magma*. cf. *discontinuous reaction series*.

continuous velocity log (sonic log, velocity log) A *geophysical borehole log* in which the *seismic velocities* of the wall-rock are determined by the measurement of the travel time of an ultrasonic pulse from one end of a *sonde* to a *geophone* at the other.

contour current A permanent or semi-permanent *thermohaline* ocean current flowing subparallel to a continental slope, responsible for the deposition of *contourites*.

contourite A muddy or sandy sediment reworked and deposited by a *contour current*.

contracting Earth An early model of Earth behaviour based on its thermal contraction over geological time as the result of cooling. Discounted since the discovery of *radioactivity* and techniques which show the Earth has not contracted significantly over its history. cf. *expanding Earth*.

contraction The decrease in length of a line during *deformation*. cf. *elongation*, *stretch*.

contractional kink band (reversed kink band) A *structure* in which there is *displacement* of the upper side of a *kink band* with respect to the lower in a manner similar to a *reverse fault*. cf. *extensional kink band*.

contributing area The area of a *catchment* contributing to storm *runoff*.

convection The motion within a fluid whereby heat is transferred from one area to another at low *thermal gradient* and approximately constant *viscosity*.

convergent beam electron diffraction (CBED) A type of *transmission electron microscopy* used to determine the symmetry of small crystals.

convergent evolution The tendency of unrelated species to evolve similar structures, physiology or appearance due to the same external factors.

convergent fan (normal fan) A *cleavage fan* radiating from a focus on the concave side of a folded layer. cf. *divergent fan.*

convergent plate margin See *destructive plate margin.*

convolute Descriptive of a coiled shell in which the outer whorls embrace the inner ones so that they are nearly invisible. cf. *evolute.*

convolute lamination An internal sedimentary *structure* in *sand-silt* grade material in which the *lamination* is disturbed or distorted into regular-irregular wavelength and amplitude *folds.* Produced by *liquefaction* with subordinate *fluidization.*

convolution The operation in which a filter acts on a waveform.

Cooley-Tukey method (fast Fourier transform, FFT) An algorithm developed in the late 1960s which performs rapid *Fourier transformation.* The increase in speed over previous techniques made possible the transformation of long signals into the frequency domain, in which many signal processing methods are much simpler.

cooling joint A *joint* in an *igneous body* forming perpendicular to the cooling surface by thermal contraction.

coomb See *combe.*

coombe See *combe.*

coordination number The number of anions surrounding a particular cation and forming a *coordination polyhedron.* Used in the description of *crystal structure.*

coordination polyhedron The shape adopted by the anions around a particular cation. Used with *coordination number* in the description of *crystal structure.*

copaline See *Highgate resin.*

copalite See *Highgate resin.*

Cope-Osborne theory A system of homologizing tooth cusps across the *reptile–mammal* boundary and within the *mammals* which assumed, perhaps incorrectly, that the single pointed reptilian tooth grew extra cusps in order to become a mammalian molar.

Cope's rule The phyletic trend towards increasing body size as an organism evolves.

copper (Cu) A *native element.*

copper glance A mining term for *chalcocite.*

copper pyrites An old name for *chalcopyrite.*

copperas See *melanterite.*

coppice dune See *nebkha.*

coppice mound See *nebkha.*

coprolite A *trace fossil* made up of faecal material.

coquimbite ($Fe_{2-x}Al_x(SO_4)_3.9H_2O$) A hydrated copper sulphate found in some *ore* deposits and *fumaroles.*

coquina A carbonate rock made up of mechanically-sorted debris, particularly shells.

coral-algal reef A marine, largely biogenic, accumulation of calcium carbonate formed from a complex, productive biosystem by *corals* and *algae* plus *bryozoans, gastropods* and serpulid worms. Such reefs develop only in shallow (< 165 m), warm (21°C), saline (30–40 ppt) water free of suspended sediment.

coral sand A *sand*-grade sediment of carbonate grains derived from the *erosion* of a *coral-algal reef.*

coral terrace A terrace of *coral* elevated above the water by *uplift*, such as on a *guyot,* or by a fall in sea level.

Corallinaceae The most important group of the marine *Rhydophyta* (*calcareous* red *algae*). Range *Silurian–Recent.*

corallite The skeleton of a single *coral* polyp.

coralloid Coral-like.

corallum The skeleton of a colonial *coral.*

corals Marine, solitary or colonial, polypoid invertebrates of class *Anthozoa*, phylum *Cnidaria*, represented as *fossils* by their calcareous skeletons which have radial to sub-radial symmetry. Range ?late *Precambrian–Recent.*

cordierite ($(Mg,Fe)_2Al_4Si_5O_{18}.nH_2O$) A hydrated *cyclosilicate* found in *metamorphic rocks.*

core 1. The central part of the Earth extending from the *Gutenberg discontinuity* with the *mantle* at 2900 km to the centre of the Earth at 6370 km. The composition is predominantly iron and nickel, but it must contain some other *mineral* to reduce the *density* to its known value. The outer core, from about 2900–4980 km, is fluid, and the site of generation of the *geomagnetic field*, while the inner core, below 5120 km, is solid. 2. A cylindrical specimen of rock recovered by drilling.

corestone A large *cobble* or *boulder* of relatively pristine rock found within a deep *weathering* profile.

Coriolis force The effect of the Earth's rotation which deflects a body of fluid or gas moving relative to the Earth's surface to the right in the northern hemisphere and the left in the southern hemisphere.

cornelian See *carnelian.*

corner frequency The upper frequency limit of the *seismic waves* generated by an *earthquake*. This can be 0.05 Hz for very large events.

corniche An organic protrusion, often of *calcareous algae*, growing from a steep rock surface near sea level, which forms a narrow walkway at the foot of a *sea cliff.*

Cornish stone A crushed, partially-*weathered granite* with appreciable *feldspar* and *kaolinite*, used in the manufacture of bone china.

cornstone A concretionary *limestone*, usually formed under arid conditions.

corona structure Concentric zone(s) of at least one *mineral*, usually in radial arrangement, surrounding another *mineral*. Formed by reaction with, or as an overgrowth on, the primary *mineral.*

corrasion The mechanical *erosion* by material transported across a surface by water, wind, ice or *mass movement*, the effect on the rock being termed *abrasion.*

correlation The process of comparing two signals, such as *seismic records* or *stratigraphic* logs, to determine the extent of their similarity at various different offsets.

corrie See *cirque.*

corrosion A general term for chemical *weathering*. cf. *corrasion.*

corrugation A large, linear feature on a *fault plane* of unknown origin.

corundum (Al_2O_3) An *oxide mineral* exploited as a *refractory*.

cosmopolitan distribution (pandemic distribution) The worldwide distribution of an organism.

cossyrite See *aenigmatite*.

Costonian The lowest *stage* of the *Ordovician*, 463.9–462.3 Ma.

cotectic Allowing the *crystallization* of two or more solid phases from a single liquid over a finite temperature decrease.

coterminous With shared boundaries.

cottonballs A colloquial term for fine, fibrous crystals of *ulexite*.

Cotylosauria/cotylosaurs An order of subclass *Anapsida*, class *Reptilia*; stem-*reptiles* with skulls pierced only by nose and eye openings. Range L. *Carboniferous*.

coulée See *arroyo*.

coulée flow A very thick (up to 100 m), relatively short, flow of *aa*.

couloir 1. A deep gorge or ravine in the side of a mountain, possibly a result of *avalanche erosion*. 2. A parallel depression between *yardangs* in a desert area.

Coulomb failure criterion (Coulomb-Mohr failure criterion, Coulomb-Navier failure criterion) $\tau = S + \mu(\sigma - p)$ where τ = *shear resistance*, μ = *coefficient of friction*, S = *cohesion*, σ = *normal stress* and p = *pore fluid pressure*. Thus the *shear resistance* to faulting is reduced by the *pore fluid pressure* within the material.

counter lode (caunter lode) A *lode* trending in a different direction from that of the usual direction of the district.

counter vein (caunter vein) A *vein* trending in a different direction from that of the usual direction of the district.

counterpoint bar A river *bar* deposited on the concave side of a channel bend because of a change in flow conditions or meander pattern migration.

country rock (host rock) The rock into which *magma* or *mineralization* is intruded or emplaced.

Couvinien A *Devonian* succession in France and Belgium covering part of the *Emsian* and the *Eifelian*.

covelline See *covellite*.

covellite (covelline, indigo copper) (CuS) An *ore mineral* of *copper*.

cow-dung bomb A *volcanic bomb* of characteristic shape.

crabhole An abrupt depression in the ground surface of centimetric- to metric-scale diameter and 50–600 mm depth, found in sediments susceptible to vertical cracking and horizontal packing.

crack-seal mechanism A mechanism of *vein* filling by repeated cycles of *extensional fracture* followed by *cementation*.

crackle brecciation *Fractures*, usually healed with small *veins* of *quartz* and other *minerals*, that form the *stockwork* mineralization in many *disseminated mineral deposits*, possibly formed by the release of volatiles from the host *magma* on *retrograde boiling*.

crag-and-tail A streamlined ridge comprising a hill of resistant rock and a tail of less *competent* material, formed by *glacier* action.

crater 1. A large, bowl-shaped depression on the Earth's surface at the summit or flank of a *volcano*. 2. A circular depression on a planetary surface caused by *meteorite* impact.

craton The stable interior of a continental *plate*, unaffected by *plate margin* activity since the *Precambrian*.

cratonization (stabilization) The process of the transformation of a *mobile belt* into a *craton* as the *tectonic* and thermal activity of the *mobile belt* ceases.

creep A slow, largely continuous process of *deformation* occurring below the *elastic limit* in response to prolonged *stress* i.e. all *deformation* that is not wholly *elastic*.

creep strain Represented by the empirical expression: $\varepsilon = \varepsilon_e + \varepsilon_1(t) + v_t + \varepsilon_3(t)$, where ε = creep strain, ε_e = instantaneous *elastic strain*, $\varepsilon_1(t)$ = *transient creep*, v_t = *steady state creep* and $\varepsilon_3(t)$ = accelerating *creep*, where t is time.

creep strength The long-term *strength* of a material, i.e. its *strength* at low *strain rate*. cf. *instantaneous strength*.

crenulation Small-scale folding or kinking.

crenulation cleavage (strain-slip cleavage) A *microscopic*-scale *cleavage* formed by the folding of an existing *fabric*, generally resulting from the buckling of a previous *slaty cleavage*.

Creodonta An order of infraclass *Eutheria*, subclass *Theria*, order *Mammalia*; the more ancient order of slow-moving, small-brained, carnivorous, placental *mammals*. Range late *Cretaceous–Pliocene*.

crescent-and-mushroom structure A *fold interference structure* with the form of successive crescents and domes.

crescentic bar A *coastal bar* formed by wave interaction, 100–200 m in length, lying just seawards of low water with horns pointing landwards.

crest 1. The top of a working face of an openpit mine or quarry. 2. The highest part of an *anticline*.

crest line A line joining the topographically highest points on a folded surface.

crest plane A plane joining the *crest lines* of a folded surface.

Cretaceous The youngest *period* of the *Mesozoic*, 145.6–65.0 Ma.

Cretaceous–Tertiary boundary (K-T boundary) The end of the *Mesozoic era* (65.0 Ma) at which time there was a *mass extinction* event, related by some to the environmental effects of *bolide* impact.

crevasse A deep fissure in a *glacier*.

crevasse splay A single or repeated flood event resulting from the breaching of a *levée*, in which coarse sediment is deposited on the landward side.

crianite A *basic igneous rock* comprising intergrown *feldspar*, *titanaugite*, *olivine* and *analcite*.

Crinoidea/crinoids A class of subphylum *Pelmatozoa*, phylum *Echinodermata*; *echinoderms* comprising the stalked sea lilies and unstalked feather stars. Range M. *Cambrian–Recent*.

cristobalite (SiO_2) A high temperature form of *silica.*

critical angle The angle between an incident *seismic wave* and the normal to an interface, across which there is an increase in *seismic velocity* and along which the refracted wave travels as a *head wave.*

critical crack extension force (fracture toughness) A measure of rock *strength* which is independent of experimental procedure.

critical distance The distance from a *seismic source* at which the *head wave* first appears at the surface. cf. *cross-over distance.*

critical flow Flow occurring when the flow velocity in a channel is the same as the wave velocity generated by a disturbance; the *Froude number* is one.

critical reflection A *seismic reflection* at the *critical angle*, at which high amplitude waves are generated.

critical refraction A *seismic refraction* for which the angle of refraction is 90° and a *head wave* is generated.

critical resolved shear stress (Schmid factor) The component of *shear stress* parallel to a slip plane in a *slip system* which controls the activity along that plane.

critical velocity The minimum velocity of a fluid required to entrain a particle.

critically refracted wave See *head wave.*

CRM See *chemical remanent magnetization.*

crocidolite (Cape asbestos) ($NaFe_3^{2+}Fe_2^{3+}.Si_8O_{22}(OH)_2$) A sodic *amphibole* occurring with an *asbestiform* habit as *blue asbestos.*

Crocodilia An order of subclass *Archosauria*, class *Reptilia*; the crocodiles, alligators, caimans and gharials. Range *Triassic–Recent.*

crocoite ($PbCuO_4$) A rare *mineral* found in the oxidized zone of lead deposits.

Croixian A *Cambrian* succession in the E USA equivalent to the *Merioneth.*

crop management factor A term in the *universal soil loss equation.*

cross-bedding See *cross-stratification.*

cross-correlation A measure of the *correlation* between two signals.

cross fold A *fold* whose *axis* trends at an angle to the general *fold* trend in a region.

cross fracture A surface marking of a *joint* surface.

cross joint A *joint* perpendicular to the causative *fold axis.*

cross-lamination A millimetric scale *cross-stratification.*

cross lode A *lode* intersecting a larger one.

cross-over distance The distance along a *seismic refraction* profile at which the *direct seismic wave* from the *shot* is overtaken by the *critically refracted* wave. cf. *critical distance.*

cross-section A diagrammatic representation in the vertical plane of the geology along a profile, usually constructed from a *geological map.*

cross-set The fundamental unit of *cross-stratification.*

cross-slip The movement of a *dislocation* out of its plane. cf. *dislocation glide*.

cross-stratification (cross-bedding, current-bedding, false bedding) The characteristic *bedding structure* produced by the migration of bedforms with inclined depositional surfaces.

cross vein A *vein* intersecting a larger one.

crosscut 1. A mining tunnel cut through *country rock* to intersect an *ore*-bearing *structure*. 2. A horizontal underground tunnel which cuts a *vein* or *orebody* at a high angle. cf. *drift*.

crossite ($Na_2(Mg,Fe^{2+})_3(Fe^{3+},Al)_2Si_8O_{22}(OH)_2$) A sodic *amphibole* intermediate in composition between *glaucophane* and *riebeckite*.

Crossopterygii An order of subclass *Sarcopterygii*, class *Osteichthyes*, superclass *Pisces*; lobe-finned, bony *fish*, possibly the ancestors of the lungfish. Range M. *Devonian–Recent*.

crotovina See *krotovina*.

crucible swelling number A parameter indicating the suitability of *coal* for *coke* manufacture, involving the measurement of the profile of the *coke* residue after heating a small sample of *coal* to 800°C under standard conditions.

crude oil The liquid components of *petroleum*, a mixture of *hydrocarbons* in which the majority belong to the *paraffin* and *naphthene series*.

Crudine A *Devonian* succession in E Australia equivalent to the *Lochkovian*.

crush belt (crush zone) A narrow belt of *crust* in which the rock has been broken and crushed, often by *fault* movement.

crush breccia A cohesive *fault rock* with $<10\%$ *matrix*, containing angular fragments >5 mm in size, formed by the *tectonic* reduction in grain size of the faulted rock. cf. *fine crush breccia*, *crush microbreccia*.

crush microbreccia A cohesive *fault rock* similar to a *crush breccia*, but with fragments <1 mm in size. cf. *crush breccia*, *fine crush breccia*.

crush zone See *crush belt*.

crushing strength A parameter of a *building stone* which describes its mechanical *strength* in sustaining loads and *stresses* in service.

crust The outermost solid layer of the Earth, distinguished chemically from the underlying *mantle* beneath the *Mohorovicic discontinuity*. *Oceanic* and *continental crust* are markedly dissimilar.

Crustacea/crustaceans A subphylum of phylum *Arthropoda*; mainly marine invertebrates whose body is normally divided into head, thorax and abdomen, the head bearing five pairs of limbs, two pairs of antennae and maxillae and one pair of mandibles. Range *Cambrian–Recent*.

crustiform banding Layers of different *mineral* composition formed during the infilling of open spaces, commonly developed in dilatant spaces along *faults* and solution channels in *karst* by the permeation of mineralizing solutions.

cryergic Descriptive of the work of ground ice.

cryoconite A tubular depression in *glacier* ice formed by melting where a dark particle absorbs solar energy.

cryogenic magnetometer (squid magnetometer) A *magnetometer* in which the detecting system is maintained at superconducting temperatures by liquid helium, thus providing greater speed and sensitivity than other types.

cryolite (Greenland spar) (Na_3AlF_6) A fluoride *mineral* found in *pegmatite veins* and exploited as a flux.

cryoplanation A *nivation* process forming a flat *altiplanation terrace*.

cryosphere The part of the Earth comprising ice and frozen ground.

cryoturbation (congeliturbation, geliturbation) The process by which *patterned ground* forms in response to the mixing activities of ice.

Cryptic An *era* of the *Precambrian*, 4560–4150 Ma.

cryptic layering The gradual change in composition of cumulate *minerals* of a *solid solution* through layers in a *cumulate igneous rock*.

cryptocrystalline Formed of crystals visible only under very high magnification.

cryptodome A dome-like area of *uplift* formed by an intrusion of, usually, *andesitic* or *dacitic magma*.

cryptomelane (KMn_8O_{16}) A *mineral* found in manganese *ores*.

cryptoperthite A sub-optical *perthite* formed by the very rapid cooling of *alkali feldspar*.

cryptovolcano A circular structure of highly deformed *strata* lacking direct evidence of formation by volcanic activity.

Cryptozoic See *Archaean*.

crystal class A classification of a crystal according to its symmetry. 32 classes exist, each of which is further subdivided into 7 *crystal systems*.

crystal defect A departure from the usual regular arrangement of atoms in a crystal, comprising *point defects*, *line defects* and *planar defects*.

crystal field See *crystal field theory*.

crystal field theory A theory describing how the energy levels of atoms are perturbed by a regular array of nearby neighbouring negative charges, which produce a *crystal field*.

crystal form See *form*.

crystal group See *crystal system*.

crystal growth The process of *precipitation* following *nucleation* by which a crystal grows spontaneously and is removed from solution.

crystal habit The relative development of faces shown by a crystal; faces large on some crystals may be small or missing on others.

crystal lattice The regular atomic arrangement of a crystalline solid portrayed as a three-dimensional array of regularly spaced points, each of which has the same arrangement of atoms in the same arrangement around it.

crystal-liquid fractionation (fractional crystallization) The most important mechanism of *magmatic differentiation* in which crystals separate from a melt by settling or floating, by flow in a conduit or by the *crystallization* of early *minerals* on the walls of the *magma chamber*.

crystal morphology The shape and appearance of crystals, especially the regularity and symmetry of their faces and varied facial development. Also includes the regular internal arrangement of atoms and the precise geometrical arrangement of crystal features.

crystal plastic deformation A *slip system* specified by crystallographic planes and directions.

crystal settling A mechanism for the formation of a *cumulate igneous rock* by concentration as *minerals* settle out of a *magma*.

crystal structure The regular arrangement of atoms making up a crystalline solid, normally described in terms of *unit cells*, *coordination numbers* and *coordination polyhedra*. Usually determined by *X-ray diffraction*.

crystal symmetry One of seven groups of crystals classified according to common symmetry characteristics.

crystal system (crystal group) The classification of crystals according to their particular types of *crystal symmetry*. The seven crystal systems are *cubic*, *hexagonal*, *monclinic*, *orthorhombic*, *tetragonal*, *triclinic* and *trigonal*.

crystal tuff A *tuff* with crystal fragments more abundant than *lithic* or *glass* fragments.

crystalline massif A portion of *continental crust* made up of *metamorphic* or *igneous rocks* that is stable relative to its surroundings.

crystalline remanent magnetization See *chemical remanent magnetization*.

crystalline rock An imprecise term for an *igneous* or *metamorphic rock*, used in contrast to a *sedimentary rock*.

crystallinity The degree to which a rock exhibits crystal development.

crystallite 1. A very small, often imperfect crystal. 2. A minute inclusion in a *glassy rock*, indicative of incipient *crystallization*.

crystallization The gradual formation of crystals from a liquid.

crystalloblastic texture (hypidioblastic texture) A *texture* formed when several *minerals* crystallize simultaneously during *metamorphism*. Any of the *minerals* may be an inclusion of any other.

crystallographic preferred orientation An alignment or *preferred orientation* of the axes of symmetry of *crystal lattices* in a deformed rock.

crystallographic symmetry elements The symmetry systems which allow the repetition of a basic structural unit in crystals. These are mirror planes, 2-, 3-, 4- and 6-fold rotation axes, centres of symmetry and 3-, 4- and 6-fold inversion axes.

cubanite ($CuFe_2S_3$) A *sulphide mineral* of the *wurtzite group*.

cubic packing One end member of the different ways perfectly sorted spheres can be arranged, in which orthogonal lines join the sphere centres. A sediment with such packing would have 48% intergranular *porosity*. cf. *rhombohedral packing*.

cubic system (isometric system) A *crystal system* whose members have three mutually perpendicular axes of equal length.

cubichnia *Trace fossils* made up of nesting structures.

cuesta An asymmetrical ridge caused by the differential *erosion* of gently inclined *strata*, consisting of a steep scarp face, a well-defined crest and a gentle back slope accordant with the local stratal *dip*.

cuirass A well-cemented *duricrust* covering the land surface, protecting underlying unconsolidated material from *erosion*.

Cullen stone See *Koln stone*.

culm A soft, sooty *coal* found in the *Carboniferous* rocks of SW England.

cummingtonite ($(Mg,Fe)_7Si_8O_{22}(OH)_2$) An *amphibole*.

cumulate igneous rock An *igneous rock* characterized by a framework of touching or interlocking crystals and grains interpreted to have been concentrated by gravitational settling during *magmatic diversification* processes.

Cunningham A *Devonian* succession in E Australia approximately equivalent to the *Emsian* and *Eifelian*.

cupola A small, dome-like protuberance on a larger igneous intrusion.

cupriferous Copper-bearing.

cuprite (Cu_2O) A red, *transparent*, *secondary* copper *oxide mineral*.

cuprouranite See *torbernite*.

Curie temperature (Curie isotherm, Curie point) The temperature above which *minerals* cannot exhibit *ferromagnetic* behaviour and only *paramagnetism* is possible.

Curie–Weiss law 'The *susceptibility* of a *ferromagnetic* material is proportional to the temperature in Kelvin.'

current lineation A *lineation* caused by the movement of sediment or fluid.

current ripple A sinuous, crested *bedform* formed by the transport and deposition of, usually, *sand*-sized particles by unidirectional fluid movement.

current-bedding See *cross-stratification*.

curtain See *drapery*.

curtain-of-fire A line of coalescing *fire fountains* simultaneously erupting along a fissure (i.e. a *fissure eruption*).

cuspate fold profile A *fold* shape with the *fold closure* in one direction having a broad, arcuate shape and in the other direction a cuspoid shape, arising across an interface separating layers of different *competence*.

cut-off An abandoned reach of a river channel, often produced where a meander loop is detached from the active river channel when the neck of the loop is breached. May be occupied by an *oxbow lake*.

cut-off grade The lowest *grade* of *ore* that can be exploited economically from an *orebody*.

cut-off limit See *assay boundary*.

cut-off line (cut-off point) The intersection of a feature with a *fault* surface.

cut-off point See *cut-off line*.

cutinite A *coal maceral* of the *exinite*

group made up of the waxy cuticular coatings of leaves and other plant tissues.

cuvette A non-*tectonic*, depositional *basin*.

cwm See *corrie*.

cyanobacteria (blue-green algae, Cyanophyta) Organisms similar to *calcareous algae* but *prokaryotic*. Range *Cambrian–Recent*.

Cyanophyta See *cyanobacteria*.

cycle of erosion (Davisian cycle of erosion, geographical cycle) The sequence of *denudational* processes existing between the initial *uplift* of an area and its reduction to a plane or *peneplain* close to *base level*.

cyclic evolution The concept that evolution was initially rapid and then followed by a longer phase of more gradual change.

cyclographic trace The trace of a *great circle* on a *stereographic projection*.

cyclosilicate (metasilicate, ring silicate) A *crystal structure* classification in which the *coordination polyhedra* are Si tetrahedra and these form rings when each tetrahedron shares two corners with adjacent tetrahedra.

Cyclostomata An order of subclass *Monorhina*, class *Agnatha*, superclass *Pisces*; the lampreys, hagfishes and slime eels. Range ?*Ordovician–Recent*.

cyclothem A sequence of beds deposited in a single cycle of sedimentation, e.g. in *coal*-bearing *strata* the sequence is *sandstone*, *shale*, *fireclay*, *coal*, *shale*. Cyclothems were believed to reflect a variety of *tectonic*, climatic or sedimentological controls.

cylindrical fold (cylindroidal fold) A *fold* in which the *fold profile plane* has the same orientation everywhere along the *hinge* and the *fold profile* is everywhere the same.

cylindroidal fold See *cylindrical fold*.

cymatogeny The large-scale warping of the *crust* over tens to hundreds of kilometres with little *deformation*, at a smaller scale than *epeirogeny* and less intense than *orogeny*.

cymophane (Oriental cat's eye) A variety of *chrysoberyl* exhibiting *chatoyancy*.

Cyprus-type deposit A type of *volcanic-associated massive sulphide deposit*, often containing *copper*, *gold* and zinc, associated with *basaltic pillow lavas*.

cyrtoconic Descriptive of a coiled *gastropod* shell in the form of a curved, tapering cone.

Cystoidea/cystoids A class or superclass of subphylum *Pelmatozoa*, phylum *Echinodermata*; sessile animals with an ovoid theca comprising irregularly-arranged plates and five food-gathering arms. Range L. *Cambrian*–U. *Permian*.

D

D Letter used to indicate a phase of *deformation*, subscripted to denote each separate phase.

D–layer The lowest 200–300 km of the *mantle*; a heterogeneous layer where there is interaction with metallic *core* material.

dachiardite ($(Na,K)_{1.5}Mg_2(Al_{5.5}Si_{30.5}O_{72}).18H_2O$) A *zeolite* found in *granite pegmatite*.

Dachsteinkalk A *Triassic* succession in the Alps equivalent to the *Norian* and *Rhaetian*.

dacite A *silica-oversaturated*, *intermediate-acid*, *calc-alkaline lava*, intermediate between *andesite* and *rhyolite* in *composition* with *phenocrysts* of *plagioclase*, minor *olivine*, *pyroxene*, *amphibole*, *biotite* and Fe-Ti oxide in a fine-grained siliceous *groundmass*. Occurs both as *lavas* and *pyroclastic rocks*.

Dacryoconarida/dacryoconarids A group of small, shell-bearing, *pelagic molluscs* abundant in the *Devonian*, possibly related to the *Cephalopoda*.

dagala See *kipuka*.

Dala A *Carboniferous* succession in China covering part of the *Bashkirian* and the *Moscovian*.

Daleje A *Devonian* succession in Czechoslovakia covering parts of the *Emsian* and *Eifelian*.

dalmation coast A flooded *concordant coast* characterized by chains of islands and long inlets formed by the inundation of coast-parallel ridges and valleys arising from *tectonic* subsidence and/or sea level rise.

Dalradian The upper division of the *Precambrian* in Scotland and Ireland.

Dalslandian orogeny (Gothian orogeny, Gothic orogeny, Sveconorwegian orogeny) The continuation of the *Proterozoic Grenvillian orogeny* at 1050–1100 Ma into S Sweden and S Norway.

dambo A linear depression without a well-defined stream channel found on old, gently-sloping, land surfaces, especially in the tropics.

danburite ($Ca(B_2Si_2O_8)$) A yellow *tectosilicate*, occurring as an *accessory mineral* in *pegmatites*.

Danian 1. The lowest *stage* of the *Palaeocene*, 65.0–60.5 Ma. 2. A *Palaeocene* succession on the west coast of the USA covering the lower part of the *Danian*.

dannemorite ($(Fe,Mn,Mg)_7Si_8O_{22}(OH)_2$) A rare, manganese-bearing *amphibole*.

daphnite A variety of *chlorite* rich in iron and aluminium.

darcy The unit of *permeability*; defined as the *permeability* which allows a flow of 1 mm s^{-1} of a fluid of *viscosity* 10^{-3} Pa s through an area of 100 mm^2 under a pressure gradient of 0.1 atm mm^{-1}.

Darcy equation $Q = KAdP/dx$, where Q = volume of flow per unit time through a porous medium, K = *hydraulic conductivity*, A = cross-sectional area of medium, dP/dx = *hydraulic gradient*.

dark ruby silver See *pyrargyrite*.

Darriwilian An *Ordovician* succession in Australia equivalent to the *Llanvirn*.

Dasycladales A group of green *calcareous algae*; erect marine plants with a central stem from which branches radiate. Range L. *Cambrian–Recent*.

Datang A *Carboniferous* succession in China equivalent to the *Viséan*.

datolite ($CaB(SiO_4)(OH)$) A *nesosilicate* found as a *secondary mineral* in cavities in *basaltic lavas*.

Datsonian A *Cambrian/Ordovician* succession in Australia covering parts of the *Dolgellian* and *Tremadoc*.

datum A fixed reference point.

daughter element An element formed by the radioactive decay of an existing element.

Davisian Cycle of Erosion See *cycle of erosion*.

Dawson Canyon A *Cretaceous* succession on the Scotian shelf of Canada covering part of the *Cenomanian*, the *Turonian*, *Coniacian* and part of the *Santonian*.

daya A small, *silt*-filled, solutional depression on a *limestone* surface in arid areas of the Middle East and N Africa.

dB See *decibel scale*.

de Broglie relationship $\lambda = h/p$, where λ = electron beam wavelength, h = Planck's constant and p = electron momentum; the basis of *electron microscopy*.

de-asphalting The *degradation* of *crude oil* by the *precipitation* of *asphalt*, probably caused by the introduction of natural gas. Although the quality of the oil is improved, the *reservoir porosity* may be reduced by plugging with *bitumen*.

dead line 1. The depth in an *orefield* below which no economic mineralization is present. 2. The level below which no significant oil is likely to be present.

death assemblage See *thanatocenosis*.

debris avalanche A rapid form of *mass movement* in a narrow channel down a steep slope.

debris fall The near free fall of *weathered* material from a vertical or overhanging face.

debris flow A generally laminar *sediment gravity flow* process in which particles up to *boulder* size are supported principally by their buoyancy in, and the *cohesive strength* of, a sediment-water slurry.

debris streaking A mechanism of *slickenside* formation in which products of the *abrasion* of an asperity accumulate on either side of it in the sense of *displacement*.

debrite A sediment deposited from a *debris flow*.

Debye-Scherrer method An *X-ray powder diffraction* method in which the spatial distribution of cones of diffracted rays is measured on a cylindrical film around the sample in a specialized camera.

Decapoda/decapods An order of subclass *Eumalacostraca*, subphylum *Crustacea*, phylum *Arthropoda*; *arthropods* with ten legs, e.g. lobsters and advanced crabs. Range U. *Devonian–Recent*.

decarboxylation The thermocatalytic decomposition of oxygen-containing *kerogen* to release carbon dioxide during *burial diagenesis,* causing an increase in *carbon* content and a loss of water.

decay constant The constant λ in the equation $-(dN/dt) = \lambda N$, which describes the rate of radioactive decay in terms of the number of radioactive atoms present (N) and the rate of change of that number with time (dN/dt). λ is related to the *half-life* T by $T = 0.693/\lambda$.

decay series The sequence of *daughter elements* produced by the radioactive decay of a parent element.

decaying ripple A type of *wave ripple*.

Deccan traps *Maastrichtian-Palaeocene* age, voluminous *tholeiitic basalts* over 1200 m thick which cover an area of 500 000 km² in west-central India, probably the result of a very large *hotspot*.

decibel scale (dB) A logarithmic scale used to measure the power or amplitude of a signal.

decke See *nappe*.

declination The horizontal angle between *true north* and *magnetic north*.

declivity A *geomorphological* term for the gradient of a slope.

décollement (detachment) A structural discontinuity of *strain*, folding or *fold style* within the Earth, typically the undeformed surface between *strained* or *faulted* areas or the boundary between *allochthonous* and *autochthonous* rocks.

deconvolution The process of undoing a previous *filtering* (*convolution*) operation. An important processing technique applied to *seismic records* which removes from the waveform the deleterious effects of its passage through the Earth during which it increases in length, providing a sharper indication of reflecting interfaces.

decussate texture A *texture* comprising a random arrangement of *tabular* or *prismatic crystals*.

dedolomitization The partial or complete *calcification* of *dolomite*, caused by the reaction of *evaporated-related collapse breccias* with calcium-rich sulphatic waters derived by the *dissolution* of *gypsum*.

deep-focus earthquake An *earthquake* with a depth of *focus* greater than 300 km. cf. *shallow-focus earthquake*, *intermediate-focus earthquake*.

deep lead A buried *alluvial placer deposit*.

deep-sea cone See *submarine fan*.

Deep Sea Drilling Project (DSDP) An international programme, planned by the *Joint Oceanographic Institutes for Deep Earth Sampling*, to drill the Earth in deep water, using the drilling ship D/V Glomar Challenger. Commenced in 1968 and terminated in 1983 when it was superseded by the *Ocean Drilling Program*.

deep-sea fan See *submarine fan*.

deep-sea trench See *ocean trench*.

deep-water fan See *submarine fan*.

deep weathering *Weathering* to a

depth of tens of metres beneath the surface, proceeding as *meteoric water* percolates through pores, *joints* and fissures in the bedrock, dissolving and chemically altering the *minerals* present with only *quartz* unaffected. Moisture eventually penetrates the solid rock along crystal boundaries and *corestones* shrink, leading to the development of *saprolite*. Common on flat landscapes in the humid tropics.

deerite ($(Fe,Mn)_{13}(Fe,Al)_7Si_{13}O_{44}(OH)_{11}$) A black, hydrous iron-manganese silicate found in metamorphosed *shale*, siliceous *ironstone* and impure *limestone*.

deflation The process by which particles are removed from the ground surface by wind action.

deflation lag Large particles remaining after the removal of finer material by *deflation*.

deformation A geological process in which the application of *force* causes a change in geometry, such as the production of a *fold*, *fault* or *fabric*, often associated with *metamorphic* reactions.

deformation band A *tabular* zone of a *crystal lattice* differing in crystallographic orientation from the rest of the *mineral* grain.

deformation history (polyphase deformation) The chronological sequence of *deformation* events in a rock or region.

deformation lamellae Narrow planar zones in *mineral* grains such as *quartz*, *feldspar* and *olivine* which are related to *deformation* of the *crystal lattice*.

deformation mechanism The means by which *deformation* is accomplished at a microscopic scale. The three basic mechanisms are *cataclasis*, *intracrystalline plasticity* and *diffusive mass transfer*.

deformation mechanism map A plot of the progression of *deformation* in coordinates of *stress*, temperature and grain size.

deformation path A line showing successive states in a *deformation history*, plotted in three-dimensional coordinates of, commonly, *strain* and rotation. cf. *strain path*.

deformation twinning A systematic reorientation of part of a *crystal structure* during *deformation* so that the deformed zone is geometrically related to the rest of the crystal. Common in *carbonate minerals*.

degradation The modification of *crude oil* by differential *solution*, *de-asphalting* or *biodegradation*, in which non-*hydrocarbons* and *aromatics* are generally removed preferentially.

delamination The process of *crust detachment* from the *mantle* during *continental collision*, or when an upper crustal layer detaches from a lower layer by *flake tectonics*.

delay time 1. In *seismic refraction*, the difference in *arrival time* of a refracted *seismic head wave* and the travel time if the wave had travelled the distance between *shot* and detector at the velocity of the refractor. Delay times allow the construction of the refractor geometry. 2. In *earthquake seismology*, the difference in *arrival time* of an *earthquake*-generated *seismic wave* and the time predicted by a model of the velocity structure of the Earth, which may allow the model to be refined or a *seismic tomographic* analysis to be made.

delayed recovery The removal of *temporary strain* (*recovery*) a measurable time after removal of the deforming *stress*, characteristic of rocks showing *visco-elastic* behaviour.

delayed runoff 1. *Runoff* that penetrates the subsurface before *discharge*. 2. The water stored temporarily as snow or ice.

delessite An iron-rich, oxidized variety of *chlorite*.

dell A small headwater valley, characteristically choked with sediment and the site of a *swamp*, often located at the head of a gorge on a *plateau*.

Delmontian A *Neogene* succession on the west coast of the USA covering part of the *Messian*, the *Zanclian* and part of the *Piacenzian*.

delta A constructional, triangular-shaped, sediment body, up to thousands of square kilometres in area, where river systems interact with fresh to marine waters and deposit sediments as the flow volume expands and channel flow becomes unconfined.

demagnetization The removal of *natural remanent magnetization* by incremental increases in temperature or the strength of an applied *magnetic field*.

demantoid A green variety of *andradite*.

demoiselle See *earth pillar*.

Demospongea A class of phylum *Porifera* (*sponges*) in which the siliceous spicules have rays at 60° and 120°. Range *Cambrian–Recent*.

demultiplexing The recovery of information from a *multiplexed* signal.

dendritic (arborescent, dendroid) Descriptive of a branching, ramifying or dichotomizing form.

dendrochronology The dating of trees and ancient wood by measuring and counting annual growth rings for, usually, archaeological purposes.

dendroclimatology *Palaeoclimatology* using information obtained from tree growth rings.

dendrogeomorphology A branch of *dendrochronology* used in the understanding of *geomorphological* features, as tree rings are affected by events such as inclination, corrosion of bark, *shear*, burial, exposure, inundation, climate change and nudation which may be the result of such processes as *faulting*, shoreline warping, volcanism, flooding, *mass wasting*, *avalanching*, *glacial* fluctuations, etc.

dendroid See *dendritic*.

density Mass per unit volume (kg m^{-3} or Mg m^{-3}). Density estimation is important in both the *reduction* and interpretation of *gravity survey* data and can be accomplished by direct measurement or in situ. The latter methods include *gamma-gamma* and *gravimeter geophysical borehole logging*, *Nettleton's method* and using the relationship between *seismic velocity* (from *seismic refraction* data) and density.

density contrast The difference in *density* between a body of rock and its surroundings, which controls the magnitude and sign of the *gravity anomaly* over it.

density current A current driven by the density of the fluid, e.g. *turbidity current*.

density log See *gamma-gamma log.*

density of states The number of *molecular orbitals* whose energy lies within a given band.

dentate With tooth-like projections.

denudation Removal of material by *weathering* plus *erosion.*

denudation chronology The reconstruction of the *erosional* history of the Earth's surface using *erosional* remnants where the *stratigraphy* is incomplete.

depletion The loss of water from surface or *groundwater reservoirs* at a greater rate than their *recharge.*

depletion allowance The proportion of income from the exploitation of a deposit not subject to tax, in recognition that it will eventually become exhausted.

depletion drive (dissolved-gas drive, solution gas drive) The mechanism whereby oil in a *reservoir* is driven towards a *well* by the action of dissolved gas in the oil.

depocentre The site of maximum thickness of sediment accumulation in a sedimentary *basin* over a particular period of time.

depositional environment-related diagenesis (eogenesis) *Diagenetic* processes occurring where the composition of the *pore-fluids* is mostly controlled by the overall physical, biological and chemical characteristics of the depositional system.

depositional remanent magnetization (DRM, detrital remanent magnetization) A *natural remanent magnetization* carried by unconsolidated sediment acquired during the deposition of magnetized grains.

depression spring A *spring* forming where the land surface cuts the *water table.*

depth of compensation The depth at which *isostatic* effects cause the pressure due to the overlying rocks to be everywhere equal.

depth of penetration 1. In electrical surveying, the depth at which current flow becomes insignificant, representing the limit to which information is derived. 2. In other geophysical methods, the maximum depth to which reliable information is obtained.

depth zone A zone proposed in the outmoded concept of the existence of a primary correlation between *metamorphic* processes and depth, in which there were considered to be a set of continuous zones parallel to the Earth's surface grading downwards. See also *epizone, mesozone, katazone.*

Derbyshire spar A popular name for *fluorite* or *fluorspar.*

derived fossil See *remanié fossil.*

deroofing The uncovering of a *pluton* by *denudation.*

descloisite ($PbZn(VO_4)OH$) An important *ore mineral* of vanadium, found in the oxidation zone of lead-zinc deposits.

desert armour See *desert pavement.*

desert mosaic See *desert pavement.*

desert pavement (desert armour, desert mosaic, stone pavement) A superficial layer of *pebbles* covering unconsolidated, finer-grained materials in arid regions which protects them from *erosion*.

desert rose (rock rose) A cluster of platy crystals, commonly of *barytes* or *gypsum*, in the crude shape of a flower, created in arid climates by evaporation.

desert varnish A dark, shiny coating, <1 mm thick, found on *pebbles* in arid regions, generally composed of iron and manganese oxides and *clays*. Much thinner than a *weathering rind*.

desiccation crack A *mud crack* formed by the subaerial drying of mud.

desilication The removal of *silica* from a rock by chemical action.

desmine See *stilbite*.

Desmoinsian A *Carboniferous* succession in the USA covering the upper part of the *Moscovian*.

Desmostyla An order of infraclass *Eutheria*, subclass *Theria*, class *Mammalia*; aquatic, hippopotamus-sized *mammals*. Range *Miocene*.

desquammation See *spheroidal weathering*.

destructive plate margin A *plate margin* at which oceanic *lithosphere* is destroyed by *subduction* at an *island arc* or *Andean-type subduction zone*. cf. *constructive plate margin*, *conservative plate margin*.

detachment See *décollement*.

detachment fault A *fault* which marks the *displacement* along a *detachment* horizon or *décollement* plane.

detailed grid mapping *Geological mapping* at scales from 1:100 to 1:10 in which the exposure is marked with a square grid and the detail in each square transferred to graph paper.

detrital Descriptive of a particle, generally of a resistant *mineral*, derived from an existing rock by *weathering* and/or *erosion*.

detrital remanent magnetization See *depositional remanent magnetization*.

deuteric (epimagmatic) Descriptive of *alteration* arising from reaction between primary *magmatic minerals* and *hydrothermal solutions* that separated from the *magma* at a late stage in its solidification.

deuteric reaction The permeation of non-water bearing *minerals* through crystal *fractures* in a newly crystallized *magma* and their reaction with *primary minerals*.

DEVAL See ***deviation from axial linearity***.

development well A *well* used in the development of an oilfield and the production of oil rather than in exploration.

deviation from axial linearity (DEVAL) A small bathymetric offset of an *ocean ridge* crest.

deviatoric stress The *stress* remaining in a *triaxial stress* system when the mean *stress* is subtracted.

devitrification The development of crystals in a *glassy rock*.

devolatilization The loss of volatiles during *coalification*.

Devonian A *period* of the *Palaeozoic*, 408.5–362.5 Ma.

dewatering The removal of *groundwater* in order to lower the *water table* so that work can take place in the dewatered area.

Dewey Lake A *Permian* succession in the Delaware Basin, USA, probably equivalent to the *Changxingian*.

Dewu A *Carboniferous* succession in China equivalent to the *Serpukhalian*.

dextral The sense of movement across a boundary, such as a *fault*, in which the side opposite the observer moves to the right. cf. *sinistral*.

diabanite An iron-rich, aluminium-poor variety of *chlorite*.

diabase An American term for *dolerite*.

diachronism A time-transgressive unit, e.g. a *lithostratigraphic* unit that varies in age.

diadochy The *replacement* of an ion in a *crystal lattice* by another.

diagenesis All physical, chemical and biological processes that occur in a sediment after deposition and before *metamorphism*, during which sedimentary assemblages and their interstitial *pore fluids* react and attempt to reach equilibrium with their evolving geochemical environment.

diallage An obsolete name for a *clinopyroxene*.

dialogite See *rhodocrosite*.

diamagnetism The fundamental, weak magnetism of all substances arising from the orbit of electrons around a nucleus. Superimposed by *paramagnetism* and *ferromagnetism* in more highly magnetic substances.

diamict The general term for *diamicton* and *diamictite*.

diamictite A terrigenous *sedimentary rock* with particle sizes ranging from *clay* to *boulder* size.

diamicton Unconsolidated *diamictite*.

diamond (C) A naturally-occurring, high pressure form of *carbon*, valued as an *industrial mineral* because of its *hardness* and as a *gem*. Diamonds ultimately derive from *kimberlite* and *lamproite* and form *placer deposits*.

diaphthoresis See *retrograde metamorphism*.

diapir A volume of rock rising upwards buoyantly because of its low *density* relative to its surroundings and causing *deformation* of overlying *strata*. Rocks forming diapirs include *evaporites* (commonly *halite*) and *granite*.

diapirism The buoyant ascent of a *diapir*.

Diapsida *Reptiles* with two openings in the cheek region of the skull, including lizards, snakes, *crocodilians, dinosaurs* and *pterosaurs*.

diaspore (AlO.OH) An aluminium *ore mineral* found in *bauxite*.

diastem An *unconformity* in which a pause in sedimentation is marked only by an abrupt change in lithology.

diastrophism A little-used term for large-scale *tectonic deformation* of the *crust*, e.g. *epeirogeny*, *cymatogeny*, *orogeny*, *folding*, *faulting*, *uplift*, subsidence and *plate* movement.

diatom A *microscopic*, unicellular, planktonic *alga* growing in water. Possesses a siliceous cell wall which may contribute to sediment.

diatomaceous earth See *diatomite*.

diatomaceous ooze A *pelagite* made up of the siliceous tests of *diatoms*.

diatomite (diatomaceous earth, kieselguhr) Fine-grained, hydrated *silica* formed by the accumulation of the tests of *diatoms*.

diatreme A vertical *pipe* or funnel-shaped igneous intrusion, 200–2000 m thick and up to 2 km deep, made up of a chaotic *breccia* of blocks of *country rock*, magmatic material and possibly *mantle*-derived *xenoliths* and *xenocrysts* passing down into a *dyke*. A *forceful intrusion* of a mixture of *mafic magma*, *volcanic gases* and *accidental lithic* blocks and *clasts*.

dichotomous Regularly bifurcating.

dichroiscope An instrument for the determination of *pleochroism*.

dichroism See *pleochroism*.

dichroite A *gem*-quality variety of *cordierite*.

dickite ($Al_2Si_2O_3(OH)_4$) A *clay mineral* found in *hydrothermal* deposits.

differential stress The two-dimensional *stress* difference between the maximum and minimum *stresses*.

differential thermal analysis (DTA) A method used in the study of *clay minerals* in which a sample and an inert material are heated. When a temperature difference between the two is observed, the *clay mineral* is undergoing a reaction, and this behaviour can be used in its identification.

differentiation The separation of a *magma* into two or more fractions.

differentiation index The sum of *normative quartz*, *albite*, *orthoclase*, *nepheline*, *kaliophilite* and *leucite*, which quantifies the degree of *differentiation*. The index increases with the quantity of *felsic minerals*.

diffluence A mechanism by which a *glacier* overflows an obstacle at its lowest point.

diffraction The radial scattering of energy at an abrupt discontinuity whose radius of curvature is smaller than the wavelength of the incident wave. Applicable to *seismic waves* and the basis of the *X-ray diffraction* method.

diffraction pattern The interference pattern formed in *X-ray diffraction* by atomic planes in a crystal which is characteristic of the *mineral*.

diffractometer An instrument allowing the distribution and intensity of *diffracted* radiation to be measured, e.g. in *X-ray diffraction* methods.

diffuse reflectance spectroscopy A technique for the quantitative measurement of *streak*.

diffusion The transport of matter by the mixing of molecules and ions by thermal agitation.

diffusion coefficient See *Fick's law of diffusion*.

diffusion in sediments The tendency of random molecular and ionic movements in *pore-fluids* to reduce chemical gradients.

diffusive mass transfer A *deformation mechanism* involving the movement of material by the solid state *diffusion* of vacancies, atoms, molecules or ions, controlled by temperature and *strain rate*.

diffusivity A measure of the rate of heat *diffusion* through a material, defined as the ratio of *thermal conductivity* to the product of specific heat and *density*.

digenite (Cu_9S_5) A copper *ore mineral.*

dikaka A *sand dune* covered by scrub or grass.

dike See *dyke.*

dilatancy The volume increase of a granular material when it is subjected to *stress* and approaching *failure.* Caused by changes in crack and pore distribution in the rock and important in *earthquake prediction* as it changes the ratio of *P wave* to *S wave* velocity, the *electrical conductivity*, the ground level, the *groundwater* level and affects the acoustic emissions from *microearthquakes.*

dilatancy-diffusion theory An *earthquake prediction* model for the build-up of *strain* prior to an *earthquake* in which *microcracks* increase in number before locking up; an influx of water then leads to *failure.* cf. *dilatancy-instability theory.*

dilatancy-instability theory An *earthquake prediction* model for the build-up of *strain* prior to an *earthquake* in which *microcracks* increase in number and then avalanche to relax slightly the build up of *stress.* cf. *dilatancy-diffusion theory.*

dilatant zone The open space formed in the more steeply-dipping part of a *pinch-and-swell structure* arising from *normal faulting.*

dilatation See *dilation.*

dilatancy-hardening The strengthening of a rock caused by a reduction in *pore-fluid pressure* due to *dilatancy.*

dilation (dilatation) The ratio of change in volume to initial volume. Negative dilation indicates a volume decrease and vice versa.

dilation vein A *vein* formed by the infilling of an existing void caused by a *fault* or fissure.

dilution gauging A technique for the estimation of *stream flow* by introducing a tracer and timing its passage over a known distance.

diluvialism An early form of *catastrophism* which related the shaping of the landscape to Noah's Flood.

diluvium An obsolete term for unconsolidated sediment which could not be explained by *fluvial* or marine action.

dim spot A reduction in amplitude of *seismic waves* on a *seismic reflection* profile indicative of a reduction in *reflection coefficient.* cf. *bright spot.*

dimension stone A *building stone* dressed into regularly-shaped blocks.

dimorphism The state of an element or compound existing in two forms, e.g. *diamond* and *graphite.*

Dinantian 1. The older sub-period of the *Carboniferous*, also called the *Mississippian*, 362.5–322.8 Ma. 2. A *Carboniferous* succession in NW Europe equivalent to the *Tournaisian* and *Viséan.*

dinocyst An organic-walled vesicle of a *dinoflagellate.*

dinoflagellates *Microscopic*, unicellular *protists* with *dinocysts* 30–60 μm across, of importance in the *biostratigraphy* of the *Jurassic* to *Pleistocene.*

dinosaur A general term for the orders *Saurischia* and *Ornithischia*, subclass *Archosauria*, class *Reptilia* which dominated the terrestrial ecology of the *Mesozoic*, many of which achieved great size.

diopside ($CaMgSi_2O_6$) A white to light green *clinopyroxene.*

dioptase (emerald copper) ($Cu_6(Si_6O_{18}).6H_2O$) A rare, green *cyclosilicate*.

diorite A medium- to coarse-grained intrusive *igneous rock* containing *plagioclase* more *Ab*-rich than Ab_{50} and <20% *quartz*, *amphibole* and/or *pyroxene*. It is equivalent to *andesite* in composition and grades into *tonalite* with increased *quartz* and into *monzonite* with >10% *alkali feldspar*. Found in *island arc* settings.

Diorite model A type of *porphyry copper deposit* characterized by the presence of zones of *potassic alteration*, *sericitization*, *intermediate argillic alteration* and *propylitic alteration* with increasing distance from the mineralization.

dip The inclination of a planar surface, measured in the vertical plane perpendicular to its *strike*. cf. *apparent dip*.

dip-angle system An *electromagnetic induction method* based on the measurement of the orientation of the resultant of the primary (due to the source) and secondary (due to an electrically anomalous body) electromagnetic fields. cf. *phase component system*.

dip fault A steep *fault* whose *strike* is parallel to the *dip* of the *bedding*.

dip isogon A line joining points of equal *dip* in a folded sequence, which can be used to define the shape of a *fold profile* and to distinguish different *fold styles*.

dip moveout The *moveout* generated by a dipping reflector. cf. *normal moveout*.

dip separation An offset of a planar feature in the vertical plane normal to the *fault*. This can be resolved into *heave* and *throw*. cf. *strike separation*.

dip-slip The movement parallel to the *dip* of a planar surface such as a *fault*.

dip slope The topographic slope parallel to the *dip* of the *bedding* and generally at an angle lower than it.

diphotic zone The level in a water body where sunlight is faint and little photosynthesis can take place. cf. *aphotic zone, photic zone*.

Diplopoda A class of superclass *Myriapoda*, phylum *Uniramia*; the millipedes. Range *Silurian–Recent*.

Diplorhina (Pteraspidomorpha) A subclass of class *Agnatha*, superclass *Pisces*; jawless *fish*, perhaps similar to the modern hagfish. Range U. *Silurian*–U. *Devonian*.

dipmeter log A *geophysical borehole log* in which formation *dip* and *strike* are measured by taking four micro*resistivity* readings around the borehole.

Dipnoi An order of subclass *Sarcopterygii*, class *Osteichthyes*, superclass *Pisces*; the lungfish. Range L. *Devonian–Recent*.

dipole field The *magnetic field* due to two *magnetic poles* of identical strength and opposite polarity. The *geomagnetic field* approximates such a field.

dipyre ($m$$Na_4(Al_3Si_9O_{24})Cl$ + $n$$Ca_4(Al_6Si_6O_{24})CO_3$) A member of the *scapolite* series containing 20–50% *meionite*, found in *regionally metamorphosed* rocks.

direct runoff The *runoff* of precipitation falling directly on saturated soil and unable to infiltrate.

direct seismic wave A *seismic body*

wave taking the most direct route from *seismic source* to detector.

direct-shipping ore (lump ore) *Ore* requiring no *benefication* before transportation.

directional fabric The *fabric* in an *igneous, sedimentary* or *metamorphic rock* comprising the alignment of *minerals* caused by motion during the formation of the rock.

dirt band A thin bed of inorganic rock in a *coal seam.*

discharge See *stream flow.*

discharge hydrograph See *hydrograph.*

discoaster A *coccolith* with the form of a *stellate* shield, usually built of six radiating rays.

disconformity (lacuna) An *unconformity* marked by evidence of *erosion*, across which there is no change in *dip.* cf. *hiatus.*

discontinuous reaction series A sequence of *mineral* reactions which occur at specific temperatures as a *magma* cools. The higher temperature species dissolve at that temperature and the new *mineral* remains until the next reaction temperature is reached. cf. *continuous reaction series.*

discordant 1. Descriptive of an *igneous rock* which cross-cuts *bedding* or *foliation.* 2. Descriptive of *unconformable strata.*

discordant coast A coast where the structural grain runs transverse to the coastline. cf. *concordant coast.*

discovery well A *well* in which oil or gas was revealed in a new location.

dish structure A slightly concave-up *structure* in *sandstone* marked by a 0.2–2 mm thick, *clay*-rich coating produced by the upward escape of *pore fluid.*

disharmonic folds *Folds* whose *fold style* changes from layer to layer, probably reflecting the different *rheologies* of the folded layers.

dislocation A surface across which there is a loss of continuity, e.g. a *fault*, a termination of a half-plane in a *crystal lattice.*

dislocation climb A *deformation* process by the formation and movement of *dislocations* in a *crystal lattice* such that they move out of their original lattice planes.

dislocation creep A type of *creep* in which parts of crystals glide past each other along crystalline *dislocations.*

dislocation glide (cold working) A *deformation* process by the formation and movement of *dislocations* within a *crystal lattice* such that each *dislocation* remains in its own lattice plane.

dislocation line A line normal to the *displacement* of a structure at a *dislocation* in a crystal.

dismicrite A *limestone* mainly composed of *micrite* with patches or lenses of *sparite*, caused by disturbance of the original lime *mud* by *algae* or escaping gas.

dispersion The dependence of propagation velocity on wave frequency, characteristic of *surface seismic waves.* The general increase in velocity with depth means that lower frequencies travel faster than higher frequencies.

dispersive pressure The fluid condition in which grains are supported above a bed in a dispersed state due to grain collisions and interactions, which give rise to a viscous force with a strong vertical component.

displaced terrane (exotic terrain) A provably *allochthonous terrane*.

displacement The relative distance moved across a line or plane.

displacement plane The plane at right angles to the walls of a *shear zone*, containing the *shear direction*.

displacement pressure The smallest capillary pressure required to force *hydrocarbons* into the largest interconnecting pores of a water-wet rock.

displacement vector See *slip vector*.

displacive transformation The rapid transformation of one *polymorph* into another by the expansion, distortion or rotation of *coordination polyhedra* without bonds being broken. cf. *reconstructive transformation*.

disseminated deposit A generally low *grade* deposit in which *ore minerals* are dispersed throughout a *host rock*, e.g. *diamonds* in *kimberlite*, *porphyry copper deposit*.

dissipative beach A low gradient *beach* protected by a *bar* which absorbs and dissipates much wave energy. cf. *reflective beach*.

dissolution A *diagenetic* process by which a solid is dissolved in an aqueous *pore fluid* leaving behind a pore space in the *host rock*.

dissolved gas drive See *depletion drive*.

distal Descriptive of a feature far from its source. cf. *proximal*.

disthene See *kyanite*.

distribution coefficient See *partition coefficient*.

distribution grading (content grading) A feature shown by a *graded bed* in which there is a gradual change in the entire grain size distribution. cf. *coarse-tail grading*.

disulphide group A group of *sulphide minerals* characterized by the presence of anion pairs such as S_2^{2-}, AsS^{2-}.

diurnal variation The daily variation in the *geomagnetic field* affecting all the *geomagnetic elements*.

divergent erosion The difference between *erosion* in low latitudes, where chemical *weathering* affects *planation surfaces* to a greater extent than steeper slopes, and mid-latitudes, where the opposite effects attain.

divergent evolution The evolutionary radiation of organisms.

divergent fan (reverse fan) A *cleavage fan* which radiates from a focus on the convex side of the folded layer. cf. *convergent fan*.

diversity A measure of taxonomic variety in terms of species, genera etc.

divide See *watershed*.

diving seismic wave A *seismic wave* that follows a curved path between the *seismic source* and detector, caused by a progressive increase in *seismic velocity* with depth.

Dix formula A formula by which the *interval velocity* of a *seismic wave* can be calculated for a given depth interval between reflectors.

djurleite ($Cu_{1.96}S$) A *sulphide mineral* derived from a *chalcocite* structure.

Dobrotiva An *Ordovician* succession in Bohemia equivalent to the early *Llandeilo.*

dog-tooth spar A form of *calcite* in the shape of a canine tooth.

Dogger The middle *epoch* of the *Jurassic*, 178.0–157.1 Ma.

dogger A metric-scale flattened, ovoid, calcareous or ferruginous *concretion* in *sand* or *clay.*

dolerite (diabase) A fine- to medium-grained *mafic igneous rock*, mineralogically and chemically equivalent to *basalt*, commonly forming minor intrusions.

Dolggellian A *stage* of the *Cambrian*, 514.1–510.0 Ma.

doline (shakehole) A circular to oval, simple closed depression found in *karst* terrain, formed by solution, *cave* collapse, *piping* or subsidence.

dolocrete A *duricrust* cemented by *dolomite.*

dololithite A *dolostone* comprising detrital fragments of *dolomite* derived by *weathering* from an existing rock.

dolomite ($CaMg(CO_3)_2$) A *carbonate mineral* found in *magnesian limestone*, formed by *dolomitization.*

dolomitization The formation of *dolomite* or a *dolostone* by *replacement* of the calcium of a calcium carbonate precursor by magnesium.

dolostone A rock made up of *dolomite.*

domain A subdivision of a larger area or volume that is more homogeneous than the whole.

dome 1. A volcanic feature formed by the accumulation of *magma* above a *volcanic vent.* 2. An *antiform* with a circular to subcircular *outcrop* pattern.

dome-and-basin structure A *fold interference structure* formed by early upright *folds* whose *axes* and *axial planes* make a large angle with later *folds.*

dome dune A circular to subcircular *dune*, 0.1–1 km in diameter, with a poorly developed or absent *slip-face.*

domichnia *Trace fossils* formed from dwelling structures.

dominance diversity A term used in *palaeoecology* to describe the relative abundance of taxa within a sample, calculated in different ways, such as by use of the *Shannon-Weaver dominance diversity equation.* cf. *equitability.*

donga A *gully* or *arroyo*, particularly in southern Africa, formed by the *erosion* of surficial deposits by *runoff* and *piping.*

doorstopper technique See *overcoring.*

dormant volcano A volcano which is currently inactive but which has erupted within historical time. cf. *extinct volcano.*

Dorogomilovskian A *stage* of the *Carboniferous*, 298.3–295.1 Ma.

Dott classification A classification scheme for *sandstones.*

double zigzag structure A *fold interference structure* formed by overfolded early *folds* whose *axial planes* make a large angle with later *folds,* but whose *axes* make a small angle with these, and where the later flow direction is oblique to the earlier *axial planes.*

down-plunge projection (down-plunge view) The reconstructed profile of a *fold structure* constructed at right angles to the *plunge* of the *fold axis.*

down-plunge view See *down-plunge projection.*

downhole geophysical survey See *geophysical borehole logging.*

downthrow The *displacement* of one side of a *dip-slip fault* relative to the other.

downthrown Descriptive of the side of a *fault* with relative downwards movement. cf. *upthrown.*

Downtonian A *series* of the early *Devonian* in British *stratigraphy*, approximately contemporaneous with the *Gedinnian.*

downward continuation The computation, from the *potential field* measured at a certain level, of what the field would be at a lower level. Based on the solution of *Laplace's equation.* cf. *upward continuation.*

downwash A process involved in the formation of *grèze litée* in which a half-fluid mixture of fine sediment is spread over a stony layer.

downwearing The *erosion* of the slopes at the top of a hill or *escarpment* more rapidly than the lower slopes. cf. *backwearing.*

draa (complex dune, 'uruq) A large *dune* formed by a regional wind pattern.

drag fold 1. A *minor* or *parasitic fold*, probably formed by *shear* in an *incompetent* layer between two *competent* layers folded by *flexural slip.* 2. A *fold* produced by *fault drag.*

drag force The *force* exerted by a fluid on a surface in a direction parallel to the flow which controls settling behaviour and, with *lift force*, the initiation of sediment movement.

drainage coefficient The amount of *runoff* per unit area in 24 hours.

drainage density The average length of stream channel per unit area of a drainage *basin*, giving a measure of the degree of *fluvial* dissection.

drainage network A hierarchical system of channel links within a drainage *basin.*

drainage ratio The ratio of *runoff* to precipitation over a given period of time.

drainage well A *well* used to drain excess superficial water into an *aquifer.*

drapery (curtain) A *tabular* or folded *speleothem* that hangs from *cave* ceilings or wall projections with a curtain-like appearance.

dravite ($NaMg_3Al_6B_3Si_6O_{27}(OH,F)_4$) A brown, manganese-bearing *tourmaline*, sometimes used as a *gem.*

drawdown The loss of head of pressure around a *well* which is being pumped.

dreikanter A *ventifact* of *pebble* size moulded into a three-faceted form by wind *abrasion.* cf. *einkanter, zweikanter.*

Dresbachian A *Cambrian* succession in the E USA covering the lower part of the *Maentwrogian.*

Dreuss A *Jurassic* succession in Utah/Idaho, USA, covering part of the *Callovian.*

driblet See *spatter.*

drift 1. A horizontal underground tunnel following a *vein* or parallel to the *strike* of an *orebody*. cf. *crosscut*. 2. A gradual change in the reading of a stationary geophysical instrument, such as a *gravimeter*, with time. 3. Unconsolidated superficial sediment. 4. An accumulation of sediment on the ocean floor transported by ocean currents.

drill hole A metalliferous mining term for borehole.

drill string All equipment within a *drill hole* during drilling.

drilling fluid See *drilling mud*.

drilling mud (drilling fluid) An oil- or water-based fluid containing clay, lime or *barite* forced down a *drill hole* during drilling to cool and lubricate the bit, seal the sides of the borehole and prevent *blowouts*.

dripstone See *stalagmite*.

drive A tunnel driven along or near an *ore* deposit.

DRM See *depositional remanent magnetization*.

dropstone A *clast* dropped through the water column into soft sediment, typically released from ice.

dross Inferior or worthless *coal*.

drowned placer See *beach placer*.

drumlin A rounded hummock of *glacial till*.

druse A cavity into which *euhedral* crystals in the *host rock* project.

drusy Containing cavities, often lined with crystals.

dry gas Natural gas composed almost entirely of *methane*. cf. *wet gas*.

dry lake See *playa lake*.

dry placer A *placer* that cannot be exploited for lack of water.

dry valley A valley seldom, if ever, occupied by a stream.

dry wash See *arroyo*.

DSDP See *Deep Sea Drilling Project*.

DTA See *differential thermal analysis*.

ductile A *tenacity* descriptor of a *mineral* that can be drawn into a wire.

ductile deformation *Deformation* resulting in *macroscopically* continuous *strain*; *deformation* in which there is a large, non-elastic, *permanent strain* before *failure*.

ductile flow *Flow* at a *macroscopic* scale.

ductile stringer A *fault gouge* feature in a *fault rock* made up of hard inclusions drawn out in the direction of *shear*.

ductility The phenomenon of deforming by *ductile deformation*.

duff Fine *coal* of too low a *calorific value* for direct sale.

Dulankara An *Ordovician* succession in Kazakhstan covering parts of the *Marshbrookian*, *Actonian* and *Onnian*.

dull coal The dullest type of *coal*. See *bright coal*.

dumortierite ($Al_7O_3(BO_3)(SiO_4)_3$) A *nesosilicate* found in aluminium-rich *metamorphic rocks*.

dune 1. An accumulation of unconsolidated material (*sand*, *clay*, *gypsum* or carbonate) shaped by the wind into a distinguishable landform. 2. A bedform resulting from transport and deposition in a current under a particular range of flow conditions.

dune-bedding The large-scale *cross-stratification* developed in *dunes*.

dune grass A grass used in *dune stabilization* in a temperate environment with adequate rainfall.

dune ridge A *dune* inundated by the sea and separated from the land to enclose a *lagoon*.

dune stabilization The artificial prevention of *erosion* or immobilization of a *dune* for engineering purposes.

dungannonite A *corundum*-bearing *diorite* containing *nepheline*.

Dunham classification A classification scheme for *limestones*.

dunite (peridotite) A medium- to coarse-grained *ultramafic rock* comprising >90% *olivine*, often of *mantle* origin.

Duntroonian An *Oligocene* succession in New Zealand covering the middle part of the *Chattian*.

duplex An *imbricate structure* in which *faults* branch from an underlying *floor thrust* and join a common, higher-level *roof thrust*.

durability Resistance to *weathering*.

durain A *lithotype* of *banded coal* made up of hard, grey-black bands with a dull to *greasy lustre*.

Durango A *Cretaceous* succession on the Gulf Coast of the USA equivalent to the *Berriasian*, *Valanginian* and *Hauterivian*.

duricrust A hard, *mineral*-cemented crust occurring in *weathered* material or the soil zone, commonly composed of *alcrete*, *calcrete*, *dolocrete*, *ferricrete*, *gypcrete*, *salcrete* or *silcrete*. Formed by the mobilization and deposition of chemicals during *deep weathering*.

duripan A *silica*-cemented soil layer.

durite A *microlithotype of coal* made up of *inertinite* and *exinite macerals*.

Durlston Beds A *Cretaceous* succession in England covering the lower part of the *Berriasian*.

duroclarain A *lithotype of banded coal* intermediate between *durain* and *clarain*.

dust Solid particles <0.08 mm in diameter suspended in the air, originating from many possible sources and sometimes deposited as *loess*.

dust storm A phenomenon in which *dust* reduces visibility to <1000 m.

Dvur An *Ordovician* succession in Bohemia covering parts of the *Pusgillian* and *Cautleyan*.

Dyfed The middle sub-*period* of the *Ordovician*, 476.1–472.7 Ma.

dyke (dike) A *tabular*, near-vertical, minor igneous intrusion that cuts across horizontal to gently dipping planar structures in the *host rock*.

dyke swarm A set of *dykes*, generally subparallel, with a common origin.

dynamic correction The correction for *moveout* time on a *seismic reflection* section, applied before *stacking* a *common depth point* gather.

dynamic equilibrium (dynamic homeostasis) A self-regulating system in which any change in the energy status results in a change in the system variables to regain equilibrium.

dynamic homeostasis See *dynamic equilibrium.*

dynamic metamorphism *Metamorphism* cause by intense localized *stress.*

dynamic recrystallization The formation of new crystal species as a result of *deformation* or *tectonism.*

dynamothermal metamorphism *Regional metamorphism* at high temperature and pressure.

dyscrasite (Ag_3Sb) A silver *ore mineral.*

dyscrystalline Descriptive of a poorly-crystalline *igneous rock.*

dystrophic lake A lake poor in nutrients and oxygen and rich in undecomposed plant matter. cf. *eutrophic lake, oligotrophic lake.*

E

e plagioclase An intermediate *plagioclase* with a fine-scale, slab-like micro*texture* detectable by *X-ray* or *electron diffraction.*

Eagle Ford A *Cretaceous* succession on the Gulf Coast of the USA covering part of the *Cenomanian* and the *Turonian.*

Early Imbrian A *period* of the *Hadean,* 3850–3800 Ma.

Early Llandeilo A *stage* of the *Ordovician,* 468.6–467.0 Ma.

Early Llanvirn A *stage* of the *Ordovician,* 476.1–472.7 Ma.

earth flow A rapid type of *mass movement* of unconsolidated material down a slope. Usually occurs due to an increase in *pore fluid* pressure, which reduces the friction between particles.

earth hummock A type of *patterned ground* in which rounded hummocks form an irregular net-like pattern as the result of *frost heaving.*

Earth movement A general term for *deformation* of the *crust.*

Earth Observation System A multi-satellite project between NASA, the European Space Agency and other countries planned to give improved information on the global *lithosphere,* hydrosphere and atmosphere.

earth pillar (demoiselle) A column of earthy material capped by a *boulder* which protects it from *erosion,* typical of *badland* and *morainic* areas.

Earth Resources Technology Satellite (ERTS satellite) The original name of a *Landsat satellite* used in *remote sensing.*

Earth tide *Deformation,* on a centimetric scale, of the solid Earth by the gravitational attractions of the Moon and Sun. Provides information on the internal *rigidity* of the Earth and requires correction in *gravity reduction.*

earthquake A sudden release of accumulated *stress* along a subsurface planar *discontinuity* according to the *elastic rebound theory.*

earthquake engineering (engineering seismology) The study of hazards arising from *earthquakes* to man-made structures.

earthquake focus (focus, hypocentre) The location of origin of an *earthquake,* usually assumed to be a point but in reality usually a *fault plane* of finite lateral extent.

earthquake intensity (intensity) A subjective measure of the strength of the effects of an *earthquake* at and around the *epicentre,* expressed on a scale of I to X or XII. cf. *earthquake magnitude.*

earthquake magnitude (magnitude) A measure of the amount of energy released by an *earthquake* estimated from the amplitude of the *seismic waves* it produces and expressed on a logarithmic scale, e.g. *local magnitude, body wave magnitude, surface wave magnitude.* cf. *earthquake intensity.*

earthquake mechanism See *elastic rebound theory.*

earthquake prediction An attempt to make a short-term estimate of the time, place and *magnitude* of an *earthquake*; also long-term forecasting of the probability of *earthquakes* of a given *magnitude* in a given time for a given region. Many prediction methods have been devised but, as yet, only some six successful predictions have been made.

earthquake swarm A prolonged series of *earthquakes* of small to moderate *magnitude* without a single 'main' event.

Earth's magnetic field See *geomagnetic field.*

earthy cobalt See *asbolane.*

earthy lustre The non-metallic *lustre* of porous *mineral aggregates* such as *clays.*

Eastonian An *Ordovician* succession in Australia covering the *Harnagian, Soudleyan, Longvillian, Marshbrookian, Actonian* and part of the *Onnian.*

ecdysis The moulting process by which some animals shed the exoskeleton or outer skin.

Echinodermata/echinoderms A phylum of marine invertebrates with a spiny *calcite* endoskeleton, a water vascular system and pentameral symmetry. Range L. *Cambrian–Recent.*

Echinoidea/echinoids A class of subphylum *Eleutherozoa,* phylum *Echinodermata* in which the body is enclosed in a globular to discoid *test* of interlocking calcareous plates which carry movable appendages. Range *Ordovician–Recent.*

echo dune A depositional *dune* forming on a steeper slope than a *climbing dune.*

echo sounder (fathometer) An instrument for determining water depth by measuring, near the sea surface, the travel time of an acoustic pulse reflected from the sea bed.

echogram (fathogram) A graph of seafloor topography made by an *echo sounder.*

eckermannite ($Na_3(Mg,Li)_4(Al,Fe)Si_8O_{22}(OH,F)_2$) A rare blue-green, alkali *amphibole* found in some *plutonic alkaline igneous rocks.*

eclogite A coarse-grained *metamorphic rock* comprising pink, *pyrope*-rich *garnet,* green *omphacite* ± *kyanite,* of deep-seated origin.

eclogite facies A *metamorphic facies* of high pressure and medium temperature, characterized in *basaltic* rocks by the presence of *omphacite* and *pyrope.*

economic basement The level below which there is minimal probability of finding an economic *mineral resource.*

economic geology Geological studies for the exploration and exploitation of materials which can be profitably utilized by man.

economic yield The maximum rate at which water can be extracted from an *aquifer* without damaging water quality or creating a deficiency.

ecostratigraphy The study of the occurrence and evolution of *fossil* communities through time, especially with application to *stratigraphic* correlation.

ecosystem The interdependence of species with themselves and their environment.

ecotone The narrow transition zone between different communities.

ecotope The habitat of an organism. cf. *biotope*.

ectinite A *metamorphic rock* developing without introduction or loss of its component *minerals*.

eddy current Loops of electric current induced to flow in a conducting body by a time-varying *magnetic field*. The measurement of the *magnetic fields* generated by eddy currents in subsurface conductors forms the basis of the *electromagnetic induction methods* of *geophysical exploration*.

eddy viscosity The component of resistance to *deformation* in a fluid arising from the generation of eddies in a *turbulent flow*.

edelopal A variety of *opal* with a very brilliant play of colours.

edenite ($NaCa_2Mg_5AlSi_7O_{22}(OH)_2$) A variety of *hornblende*.

Edentata An order of infraclass *Eutheria*, subclass *Theria*, class *Mammalia*; toothless *mammals* including aardvarks, armadillos and sloths. Range U. *Palaeocene–Recent*.

edge coal A *coal seam* of high inclination.

edge dislocation A *line defect* in a crystal formed when an additional half-plane of atoms is inserted and the surrounding atoms adjust position to accommodate it.

edge water The water in a saturated *reservoir* rock surrounding an oil pool.

Ediacara The younger *epoch* of the *Vendian*, 590–570 Ma.

Ediacara fauna Complex animals of *Proterozoic* age ($\sim$ 670 Ma) with unusual soft-body preservation found in New South Wales, Australia.

Edrioasteroidea/edrioasteroids A class of subphylum *Eleutherozoa*, phylum *Echinodermata*; *echinoderms* with a discoidal to cylindrical exoskeleton of irregular, flexible, polygonal plates. Range L. *Cambrian*–L. *Carboniferous*.

EDX See *energy* ***d****ispersive* ***X****-ray analysis*.

effective elastic thickness An expression of *flexural rigidity* (usually of the *lithosphere*) in terms of the thickness of an ideal elastic *plate* behaving in the required manner.

effective permeability The ability of a rock to allow the passage of a fluid in the presence of other fluids, e.g. oil and water.

effective porosity The percentage of a given mass of rock or soil consisting of interconnecting interstices.

effective stress The difference between applied *normal stress* and *pore fluid pressure*.

efficiency of a water well The ratio of *aquifer loss* to total *drawdown* of a *well*.

effusive eruption A *volcanic eruption* characterized by a lack of explosive activity, caused by a *magma* low in volatiles.

effusive igneous body See *extrusive igneous body*.

Egyptian blue A fused mixture of *quartz*, *lime* and copper *ore* ground to fine powder and used in antiquity as a pigment.

Eh Oxidation potential, a measure of the electron concentration of a system in internal equilibrium.

Eifelian A *stage* of the *Devonian*, 386.0–380.8 Ma.

einkanter A *ventifact* with one face. cf. *dreikanter*, *zweikanter*.

EIS See *environmental impact statement*.

ejecta The solid material thrown from a *volcano* or impact *crater*.

Ekman layer The thickness over which an *Ekman spiral* is representative.

Ekman spiral A graphical representation of current velocity distribution with depth caused by applied wind *stress* on the water surface. The spiral shape arises from the *Coriolis effect* and frictional resistance of the underlying water.

Ekman transport The net *displacement* from an *Ekman spiral*.

elaeolite (eleolite) A massive variety of *nepheline*.

Elasmobranchii A subclass of class *Chondrichthyes*, superclass *Pisces*; sharks and related *fish*. Range M. *Devonian*–*Recent*.

elasmosaurs Extinct, long-necked, aquatic *reptiles*.

elastic bitumen See *elaterite*.

elastic constants (elastic moduli) Constants that define the elastic properties of an isotropic medium and which control the *seismic wave velocity* of the medium.

elastic deformation (elasticity) *Deformation* that is instantaneously and totally recoverable.

elastic limit (yield point, yield stress) The *stress* above which *elastic behaviour* is no longer followed.

elastic mineral A *tenacity* descriptor indicating a *mineral* that bends and returns to its initial shape on release of pressure.

elastic moduli See *elastic constants*.

elastic-plastic Descriptive of a material which undergoes *elastic deformation* below the *yield stress*, at which it behaves with *plasticity*. cf. *rigid-plastic*.

elastic rebound theory A model for the mechanism of an *earthquake* whereby forces progressively distort a body of rock until its *strength* is exceeded and energy is suddenly released in a catastrophic event.

elastic strain The change in shape resulting from *elastic deformation*.

elastic wave A wave which vibrates its host medium without causing permanent *deformation*, e.g. a *seismic wave*.

elastica A type of *fold profile* in which the *fold angle* is negative, e.g. an *anticline* with limbs converging downwards.

elasticity 1. See *elastic deformation*. 2. Of an *artesian aquifer*, referring to the presence of compressed water, so more is present than would be the case under atmospheric pressure.

elastoviscous deformation (Maxwell model) *Deformation* comprising *elastic* and *viscous* behaviour in series, i.e. the application of *stress* causes instantaneous *elastic strain* followed by flow at constant *strain rate*.

elaterite (elastic bitumen) Solid *bitumen* resembling dark brown rubber, elastic when fresh.

elbaite A lithium-rich variety of *tourmaline*, which may be green, pink or blue.

electric calamine See *hemimorphite*.

electric drilling See *vertical electrical sounding*.

electric log A *geophysical borehole log* in which *electrical resistivity* is measured down a borehole using a variety of different electrode configurations to study the nature of the zone invaded by *drilling mud filtrate* and the pristine formations present.

electric profiling See *constant separation traversing*.

electric trenching See *constant separation traversing*.

electrical conductivity The reciprocal of *electrical resistivity*, unit S m^{-1}.

electrical resistivity The resistance in ohms between the opposite faces of a unit cube of material, unit ohm m.

electrode configuration The arrangement of electrodes in electrical surveying. See *Schlumberger configuration*, *Wenner configuration*.

electrode polarization See *overvoltage*.

electrolytic polarization 1. The build-up of charge on metal electrodes connected to a DC source, which can be avoided by the use of *non-polarizing electrodes*. 2. See *membrane polarization*.

electromagnetic induction methods (electromagnetic methods, EM methods) *Geophysical exploration* methods based on energizing the subsurface with time-varying electromagnetic fields. These induce *eddy currents* to flow in subsurface conductors which generate their own fields and which can be detected at the surface, e.g. *VLF*, *INPUT®*, *AFMAG*.

electromagnetic methods See *electromagnetic induction methods*.

electron microprobe (microprobe) An instrumental analytical technique in which the chemistry of small phases and intergrowths are examined by use of a focused electron beam in a vacuum acting on a polished section.

electron microscopy The use of a beam of high-energy electrons to form images of the surface or internal structure of a material, based on the *de Broglie relationship*.

electron spin resonance (ESR) A dating method based on the detection of unpaired electrons resulting from ionizing radiation and/or heating.

electrum A natural *gold-silver* alloy.

eleolite A massive variety of *nepheline*.

Eleutherozoa A subphylum of phylum *Echinodermata*, distinguished by the absence of a stalk. Range *Ordovician–Recent*.

elevation correction The correction applied to *gravity* and *magnetic survey* data to compensate for the varying elevations of observations.

elevation energy The potential energy of a mass of water with respect to its elevation above a datum.

Ellesmerian orogeny An *orogeny* in *Devonian* times affecting the Canadian Arctic.

elongation The relative change in length of a line with respect to its original length during *deformation*.

elongation lineation A *lineation* formed by a set of parallel, elongate objects in a deformed rock.

Elsonian orogeny An *orogeny* during the *Proterozoic* from 1500–1400 Ma affecting the eastern Canadian *shield*.

Elster glaciation The first of the four *glacial* periods of northern Europe in the *Quaternary*.

elutriation A natural process by which *clastic* particles are separated by grain size, either in water or *pyroclastic flows*.

eluvial placer A *placer* formed by the *creep* of material down a slope.

eluviation The movement of soil material through the soil zone, resulting from *throughflow* or *leaching*.

eluvium *In situ weathered bedrock*.

elvan A Cornish mining term for a *dyke* cutting *granite*.

EM methods See *electromagnetic induction methods*.

Embrithopoda An order of infraclass *Eutheria*, subclass *Theria*, class *Mammalia*; an order whose only representative is Arsinotherium, a huge, horned form. Range L. *Oligocene*.

Embry and Clovan classification An expanded form of the *Dunham classification* of *limestones*.

emerald A deep green *gem* variety of *beryl*.

emerald copper See *dioptase*.

emery A natural *abrasive* comprising *corundum* and *magnetite*, formed by the *thermal metamorphism* of ferruginous *bauxite*.

emplacement The intrusion of an *igneous rock* body into an envelope of *country rock*.

Emsian A *stage* of the *Devonian*, 390.4–386.0 Ma.

emu **E**lectro**m**agnetic **u**nit, the *cgs* unit of magnetism.

en echelon An arrangement of parallel lines or planes in which each is of finite length and displaced laterally from its neighbours in a consistent sense.

enantiomorphism The existence of two chemically identical crystals which are mirror images.

enantiotrophy The conversion of one *polymorph* to another at a critical temperature and pressure. cf. *monotropy*.

enargite (Cu_3AsS_4) An *ore mineral* of *copper*.

endellite An American term for *halloysite*.

enderbite A *charnockitic* rock comprising *quartz*, *antiperthite*, *hypersthene* and *magnetite*.

endichnia *Trace fossils* occurring within the preserving bed.

endlichite A variety of *vanadinite* containing arsenic.

endogenetic (endogenic, endogenous) Originating within the Earth. cf. *exogenetic*.

endogenic See *endogenetic*.

endogenous See *endogenetic.*

endometamorphism See *endomorphism.*

endomorphism (endometamorphism) The *alteration* of the composition of a *magma* by the *assimilation* of *country rock.*

endoskarn A *skarn deposit* formed by the *replacement* of an intrusion. cf. *exoskarn.*

endrumpf A *peneplain* reduced to a flat or gently undulating landscape by *erosion.*

energy dispersive X-ray analysis (EDX) An *X-ray diffraction* technique in which the whole X-ray spectrum is examined in a single measurement, which has a higher detection limit than *wavelength dispersive analysis.*

engineering geological map A map showing units defined by their engineering properties.

engineering geology The application of geological information, techniques and principles to the design, construction and maintenance of engineering works.

engineering seismology See *earthquake engineering.*

englacial (intraglacial) Within ice.

enhanced oil recovery (tertiary oil recovery) Techniques used for the recovery of oil after normal pumping and reinjection are no longer effective, by such methods as steam injection to mobilize viscous oil.

ensialic Within or on *continental crust.*

enstatite ($MgSiO_3$) The magnesian end-member of the *orthopyroxene solid solution* series.

enstatitite A *pyroxenite* composed almost entirely of *enstatite.*

entablature A volcanic layer with fan-like *joints,* commonly alternating with *colonnades.*

enterolithic Descriptive of a *sedimentary structure* in the form of ropy folds, formed by the crumpling of an *evaporite* resulting from its swelling on *hydration.*

entrainment equivalence Descriptive of grains in a sediment bed when they begin to move at identical fluid *shear stress.*

enveloping surface The surface which would join the *crest* or *trough lines* of a set of *folds.*

environmental engineering geology The branch of *applied geomorphology* which covers the study of features and processes in relation to environmental management and engineering.

environmental impact statement (EIS) A summary of information about the impact an action, such as an engineering project, will have or has had on the environment.

Eocambrian The late *Precambrian,* approximately equivalent to the *Riphean.*

Eocene An *epoch* of the *Palaeogene,* 56.5–35.4 Ma.

eogenesis See *depositional environment-related diagenesis.*

eolian See *aeolian.*

eon A large division of geological time comprising a number of *eras.*

eonothem The largest *chronostratigraphic* unit.

Eosuchia An order of subclass *Lepidosauria*, class *Reptilia*; a brigade of poorly known fossil *reptiles* which may or may not be related. Range U. *Permian*–U. *Triassic*.

Eötvös balance An early form of *gravimeter*.

Eötvös correction The correction applied to *gravity survey* data collected from a moving platform (ship or aeroplane) to take account of the component of E–W motion which reinforces or decreases the centripetal force arising from the Earth's rotation, which, in turn, decreases the gravitational attraction of the Earth.

epeiric sea (epicontinental sea) A shallow inland sea.

epeirogenesis See *epeirogeny*.

epeirogeny (epeirogenesis) Very large-scale *tectonic* movements which cause *uplift*/subsidence of the *continental* and *oceanic crust* without significant *deformation*, *regional metamorphism* or intrusion.

ephemeral stream A stream or river which does not flow at all times of the year. cf. *perennial stream*.

epi- Upon, above.

epibole See *acme zone*.

epicentral angle The angle subtended at the centre of the Earth by the *epicentre* of an *earthquake* and the location of its detection.

epicentral distance Multiply *epicentral angle* by 111.1 to give the distance along a *great circle* route in kilometres from *epicentre* to recorder.

epicentre The location on the surface vertically above the *focus* of an *earthquake*.

epichnia A *trace fossil* occurring on the top of the preserving bed.

epiclastic Descriptive of sedimentary material redeposited from an existing sediment.

epicontinental On a continent.

epicontinental sea See *epeiric sea*.

epidiorite A *metamorphic rock* with a *granular texture* derived from a *basic igneous rock* but containing the same *minerals* as *diorite*.

epidosite A *metamorphic rock* comprising *epidote* and *quartz*.

epidote (pistacite) ($Ca_2(Al,Fe)Al_2(SiO_4)(SiO_7)(O,OH)_2$) A group of apple-green, hydrated calcium aluminosilicate *sorosilicates*.

epidote-amphibolite facies See *amphibolite facies*.

epidotization An *alteration* process whereby the *feldspar* of, commonly, a *basic igneous rock* is *albitized* with the separation of *epidote* and *zoisite*.

epifaunal Descriptive of an organism which lives on the floor of an ocean, lake or river, either attached to a larger organism or free-moving. cf. *infaunal*.

epigene At or near the Earth's surface.

epigenesis Changes affecting *sedimentary rocks* after compaction, excluding *weathering* and *metamorphism*.

epigenetic Descriptive of a deposit forming after its host rock. cf. *hypogene*.

epigenetic drainage See *superimposed drainage*.

epilimnion The upper, oxygenated, circulating layer of a stratified lake, from <10 m to >50 m thick. cf. *hypolimnion.*

epimagmatic See *deuteric.*

epimorph The natural cast of a *mineral.*

epistilbite ($CaAl_2Si_6O_{16}.5H_2O$) A white to colourless *zeolite.*

epitaxy A cement overgrowth different in *mineralogy* from the main grain.

epithermal deposit An *epigenetic deposit* formed at low temperatures (50–200°C) near the Earth's surface (<1500 m).

epizone A *depth zone* of moderate temperature and low *hydrostatic pressure.* See also *mesozone, katazone.*

EPMA **E**lectron **p**robe **m**icro**a**nalysis.

epoch A third order geological time unit.

epsilon cross-stratification A lateral, *accretionary,* sedimentary *structure* with gently-inclined *bedding* surfaces dipping approximately perpendicular to the *palaeocurrent* direction.

Epsom salts See *epsomite.*

epsomite (Epsom salts) ($MgSO_4.7H_2O$) An *evaporite mineral,* also found as an efflorescence on the walls of *caves* or workings.

equal-angle net See *Wulff net.*

equal-area net See *Schmidt net.*

equal-area projection See *Schmidt net.*

equant See *equidimensional.*

equidimensional (equant) With the same, or nearly the same, dimensions in all directions.

equigranular Descriptive of a *texture* in which grains are all of about the same size.

equipotential surface A surface on which the *potential* is constant.

era A first order geological time unit.

erathem A first order *chronostratigraphic* unit.

erg 1. **(koum, sand sea)** A region covered by *dunes* or *sand sheets.* 2. The *cgs* unit of energy or work.

Erian A *Devonian* succession in E North America covering the lower part of the *Givetian.*

erionite ($K_2NaCa_{1.5}Mg(Al_9Si_{27}O_{72}).28H_2O$) A *zeolite* found in *fractures* in *rhyolitic* and *basaltic* rocks.

erodibility The resistance of a soil to the entrainment and transport of its particles by an agent of *erosion,* controlled by its mechanical and chemical properties.

erosion The process whereby particles are detached from rock or soil and transported away, the principal agents being ice, wind and water.

erosion control practice factor A term in the *universal soil loss equation.*

erosion surface A flat plain resulting from *erosion* and representing the final phase of a *cycle of erosion.*

erosional sheltering A mechanism of *slickenside* formation whereby debris is deposited in the direction of slip behind an asperity on a *fault* surface.

erosivity A measure of the potential ability of an eroding agent, such as rainfall or wind, to cause *erosion*, based on its kinetic energy.

Erqiao A *Triassic* succession in China equivalent to the *Rhaetian*.

erratic A stone transported by a *glacier* and deposited far from its point of origin.

ERTS satellites See *Earth Resources Technology Satellite*.

erubescite See *bornite*.

erythrite (cobalt bloom) ($Co_3(AsO_4)_2.8H_2O$) A pink *secondary mineral* of cobalt.

escape tectonics See *indentation tectonics*.

escarpment (scarpslope) The steeper slope of a *cuesta*. cf. *dip slope*.

esker (osar) An elongated ridge of stratified *gravel*, probably formed by streams flowing beneath or on a *glacier*.

essential mineral A *mineral* whose presence or absence determines the name of a rock. cf. *accessory mineral*.

essexite An *alkaline gabbro* composed of *plagioclase*, *hornblende*, *biotite* and *titanaugite* with minor *alkali feldspar* and *nepheline*.

estuary A partly enclosed body of water open to the sea where fresh and sea water intermix.

Etalian A *Triassic* succession in New Zealand equivalent to the *Anisian*.

etchplain A broad, erosional land surface in the tropics formed by the *deep weathering* of *crystalline rocks* where *erosion* has not caused deep incision.

eu- Well, good, rich in.

euclase ($BeAl(SiO_4)(OH)$) A rare beryllium *cyclosilicate*.

eucrite 1. An obsolete term for a coarse-grained, commonly *ophitic*, *basic gabbro* of deep-seated origin containing *plagioclase* (near *bytownite*), *ortho-* and *clinopyroxene* and *olivine*. 2. A stony *meteorite* of *basaltic* composition.

eucryptite ($LiAlSiO_4$) An *inosilicate* formed as an *alteration* product of *spodumene*.

eucrystalline Descriptive of an *igneous rock* which is well crystallized.

eudialite See *eudialyte*.

eudialyte (eudialite) ($Na_4(Ca,Fe,Ce,Mn)_2ZrSi_6O_{17}(OH,Cl)_2$) A pink-red or yellow-brown, complex, hydrated sodium-calcium-iron zirconosilicate found in some *alkaline igneous rocks*.

eugeosyncline A *geosyncline* with abundant *magmatic* activity.

euhedral (automorphic, idiomorphic) Descriptive of a grain with a fully-developed *crystal form*.

eukaryote An organism with cells possessing a distinct nucleus. cf. *prokaryote*.

Euler pole (pole of rotation) See *Euler's theorem*.

Euler's theorem 'Any motion on the surface of a sphere can be defined in terms of an angular rotation about a specific axis, the points at which the axis intersect the surface being known as *Euler poles*.' Used in describing *plate* movements across the globe.

eulysite A *metamorphic rock* containing

iron and manganese silicates, e.g. *hedenbergite*, *fayalite*, *almandine*, *spessartine*.

Eumalacostraca A subclass of class *Malacostraca*, subphylum *Crustacea*, phylum *Arthropoda*, characterized by a non-b*ivalve* shell, six abdominal somites and a tail formed by the last somite and telson, including shrimps, crabs and lobsters. Range *Devonian–Recent*.

euphotic zone See *photic zone*.

Euryapsida A subclass of class *Reptilia*, characterized by only one temporal opening; possibly not a natural grouping. Range *Permian*–U. *Cretaceous*.

Eurypterida/eurypterids A class of subphylum *Chelicerata*, phylum *Arthropoda*; large (up to 2 m), predatory invertebrates with a small prosoma whose last appendage may be modified as a swimming paddle, a long episthosoma of 12 somites and a telson in the form of a spine or paddle. Range *Ordovician–Permian*.

eustasy Global change in sea level.

eutaxitic texture A *texture* of flattened glassy discs (*fiamme*) in an *ashy matrix* seen in *welded tuffs*.

eutectic Descriptive of a mixture of at least two substances which have crystallized simultaneously.

Eutheria (placentals) An infraclass of subclass *Theria*, class *Mammalia*; the placental *mammals*. Range *Cretaceous–Recent*.

eutrophic lake A lake containing much dissolved plant nutrient and a seasonal lack of oxygen in the lowest layer due to significant amounts of decaying organic matter. cf. *dystrophic lake*, *oligotrophic lake*.

eutrophism The process whereby a lake is rejuvenated by an increase in plant nutrients so that *algae* bloom on the surface, preventing light penetration and oxygen absorption.

euxenite ($(Y,Er,Ce,La,U)(Nb,Ti,Ta)_2(O,OH)_6$) A massive, brown-black, niobate, tantalate and titanate of yttrium, erbium, cerium, thorium and uranium found in *granite pegmatites*.

euxinic Descriptive of an environment of restricted water circulation where *anaerobic* conditions attain.

evaporative drawdown The loss in *brine* volume and its lowering in level as evaporation takes place in an *evaporite basin*.

evaporative pumping The upward movement of *groundwater* towards a *deflation* surface caused by severe evaporation at the sediment-air interface.

evaporite A rock made up of *mineral*(s) formed by *precipitation* from concentrated *brines*.

evaporite basin A low-lying area where evaporation exceeds fluid input so that *brines* are sufficiently concentrated for the *precipitation* of *evaporite minerals* to take place.

evaporite deposit A deposit from which various salts can be recovered, principally *anhydrite*, *borax*, *carnallite*, *celestine*, *colemanite*, *gypsum*, *halite*, *kernite*, nitrates, *sylvite* and *trona*.

evaporite-related collapse breccia Chaotic, angular rock fragments forming when an underlying *evaporite* layer is removed by *dissolution*.

event stratinomy The *event stratigraphy* of individual *beds.*

event stratigraphy The recognition, study and correlation of the effects of significant physical events, e.g. marine *transgressions, volcanic eruptions,* in the expectation that truly synchronous horizons could be defined.

evolute Descriptive of a coiled shell in which all whorls are exposed. cf. *convolute.*

evolution path The trend in *organic matter diagenesis* as *kerogen* decreases in oxygen and hydrogen content.

ex- Out of, not having.

Excelsior A *Triassic* succession in Nevada, USA equivalent to the *Anisian* and *Ladinian.*

exceptional fossil deposit (fossil lagerstätte) A deposit in which *fossils* are exceptionally well preserved, often including soft tissues, or exceptionally rich.

excess mass The difference in mass between an anomalous body of rock (e.g. an *orebody*) and the *country rock* which would otherwise occupy its volume. Can be estimated from *gravity anomalies* using *Gauss' theorem.*

exchangeable sodium percentage A property of certain *clays* referring to the percentage of sodium that is readily lost by *dissolution.* If this is high the *clay* is very susceptible to *erosion.*

exfoliation (desquamation, onion-skin weathering, spheroidal weathering) The *weathering* of *boulders* by the spalling of surface layers, millimetres to a few metres in thickness, probably arising from the release of *lithostatic pressure* on exhumation, by *weathering* or the growth of salt crystals just below the surface of the rock.

exhalite A chemical deposit, principally on the seafloor, formed mainly from *hydrothermal* exhalations such as *black smokers.*

exichnia A *trace fossil* occurring as infillings of the preserving *bed* with another substrate.

exinite A *coal maceral* group made up of small organic particles such as algae, spores, cuticles etc. and relatively high in hydrogen and volatiles.

Exmoor A *Permian* succession in Queensland, Australia, covering part of the *Artinskian,* the *Kungurian* and part of the *Ufimian.*

exogenetic (exogenic, exogenous) Originating at or near the Earth's surface. cf. *endogenetic.*

exogenic See *exogenetic.*

exogenous See *exogenetic.*

exoskarn A *skarn deposit* developed in metasediments. cf. *endoskarn.*

exotic terrane See *displaced terrane.*

expanding Earth A model suggesting the Earth has expanded significantly over geological time, explaining the relative movements of the continents by the dismemberment of a once complete shell of *continental crust.* Modern measurements have shown that the Earth has not expanded at the required rate. cf. *contracting Earth.*

expansive soil A soil which shrinks and swells with changing moisture content, such as one containing *montmorillonite.*

exploitation well A *well* sunk in a proven deposit.

exploratory well A *well* sunk in the hope of finding new oil or gas accumulations. cf. *development well.*

explosion breccia An *igneous breccia* formed by explosive volcanic activity.

explosive index The percentage of *pyroclastic* material in the products of a *volcanic eruption.*

exposed coalfield A *coalfield* in which *coal* crops out at the surface. cf. *concealed coalfield.*

exposure A surface where *in situ* rock is seen free from soil or vegetation cover.

exposure mapping (outcrop mapping) A type of *geological mapping* in which every *exposure* is visited and its limits plotted, along with relevant topographic features.

exsolution The development of two or more compositionally different phases from a *solid solution*, usually as cooling takes place.

extended Griffiths failure criterion (Griffiths-Murrell failure criterion) An extension of the two-dimensional *Griffiths failure criterion* into three dimensions.

extension An increase in the length of a line during *deformation.*

extension joint (tension joint) A *joint* forming by *tensile failure* perpendicular to the least *principal stress.*

extensional cleavage A set of planes oblique to an existing planar structure so that *displacements* on the *cleavage* cause net *extension* parallel to the existing planes.

extensional crenulation cleavage A *cleavage* formed when one set of *shear bands* is developed more strongly than the other.

extensional fault A *fault* across which *extension* has occurred.

extensional fault system A set of related *faults* on which individual *displacements* give a net *extension* in the system as a whole.

extinct volcano A *volcano* that has not erupted within historical time.

extinction 1. The disappearance of a group of organisms. 2. The phenomenon of least illumination at a particular orientation of a crystal when viewed in *thin section* by a microscope with polarized light and crossed polars.

extraformational conglomerate A *conglomerate* whose *clasts* originate mostly outside the *basin* of deposition.

extraversion The mechanism whereby the *supercontinent* proposed in the *SWEAT hypothesis* rifted and the resulting continents moved apart.

extrusion tectonics See *indentation tectonics.*

extrusive igneous body (effusive igneous body) An *igneous body* emplaced at the surface.

exudatinite A *coal maceral* of the *exinite* group of secondary origin, found in cavities in other *macerals*, which was soft and mobile at some stage during *coalification.*

eyot A small island on the bend of a river.

F

F Letter used to indicate a phase of *foliation* formation, subscripted to denote each separate phase.

f joint See *en echelon fracture*.

Fa A method of indicating *olivine* composition as a percentage of *fayalite*, e.g. Fa_{10} indicates a composition of 10% *fayalite*, 90% *forsterite*.

Fa Lang A *Triassic* succession in China equivalent to the *Ladinian*.

fabric The pervasive features of a rock.

fabric cross Orthogonal *fabric* axes of *monoclinic* symmetry.

fabric diagram (petrofabric diagram) A *stereogram* showing the components of a *petrofabric*.

fabric domain A region of a rock that is homogeneous with respect to the orientation of a *fabric element*.

fabric element A feature of a rock that contributes to its *fabric*, e.g. *cleavage*, *fracture*, *lineation*, grain shape, grain boundaries and *crystallographic* orientations.

fabric skeleton Lines linking the highly populated parts of a *stereogram* showing *fabric elements*, which define *preferred orientations*.

fabric symmetry The class of symmetry shown by a *fabric element*.

face A mining and quarrying term for the exposed rock surface, excluding the *back* or floor.

face method A mapping method used in quarries and *opencast mines* in which geological data are plotted in plan view or at a reference datum measured accurately near the *toe*, and data from higher levels are estimated.

facet An element of the surface of a crystal or cut *gem*.

facies All lithological and palaeontological features of a particular *sedimentary rock*, from which depositional environment may be inferred.

facies map A map illustrating lateral changes in the lithology of a *formation*, *group* or *system* within a sedimentary *basin*, which allows complex *stratigraphic* data to be presented and which may be used to construct *palaeogeographic* or palaeoenvironmental conditions.

facing The direction in which *beds* become younger. See also *younging*.

fahl ore See *tetrahedrite*.

fahlband A band of *metamorphic rock* carrying disseminated sulphides that are more abundant than *accessory minerals* but too few to form an *orebody*.

fahlerz See *tetrahedrite*.

failed rift See *aulacogen*.

failure criterion The relationship between *principal stresses* which gives the condition for *failure*.

failure Loss of *strength.*

failure strength (failure stress) The *stress* at which *failure* occurs.

failure stress See *failure strength.*

fairfieldite ($Ca_2(Mn^{2+},Fe^{2+})(PO_4)_2.2H_2O$) A white, hydrated phosphate of calcium and manganese found in *granite pegmatites.*

fall velocity See *settling velocity.*

falling dune A steep-sided *stationary dune* forming on the *lee slope* of an obstacle such as a hill where *sand* collects.

false bedding See *cross-stratification.*

false-colour composite A colour image of *remote sensing* data made by combining images of several *spectral bands*, some of which are beyond the visible spectrum.

false ruby See *pyrope.*

famatinite (Cu_3SbS_4) A copper sulphosalt found in low- to medium-*grade* copper deposits.

Famennian The highest *stage* of the *Devonian*, 367.0–362.5 Ma.

Famennien A *Devonian* succession in France and Belgium covering parts of the *Frasnian* and *Famennian.*

FAMOUS **F**ranco-**A**merican **M**id-**O**cean **U**ndersea **S**tudy. A detailed study of the Mid-Atlantic ridge between 36.5–37°N.

fan A slope of detritus increasing in width down the slope.

fan shooting A method of *seismic refraction* surveying in which the detectors are sited at similar distances from the *seismic source* from which they radiate out in an arc. Provides information on the three-dimensional form of a refractor.

fanglomerate A *rudite* deposited in an *alluvial fan.*

faro A small, elongate *reef* enclosing a *lagoon* up to 30 m deep, forming on the rim of a *barrier reef* or *atoll.*

fasciculate Descriptive of a compound *coral* whose *corallites*, although associated, are sufficiently spaced to avoid mutual interference.

fathogram See *echogram.*

fathometer See *echo sounder.*

faujasite ($(Na_2,Ca,Mg)_{32}(Al_{64}Si_{128}O_{384}).256H_2O$) A rare colourless or white *zeolite.*

fault A discontinuity surface across which there has been *shear displacement.*

fault-bend fold A *fold* produced in the *hangingwall* by the movement of a *fault* over a non-planar *fault* surface.

fault block A body of rock partly or completely defined by *faults* and differing in elevation from its surroundings.

fault breccia A non-foliated, incohesive *fault rock* with $>30\%$ visible fragments surrounded by a *matrix.*

fault drag The bending of a marker across a *fault.*

fault gouge An incohesive *fault rock* with $<30\%$ visible fragments surrounded by a *matrix.*

fault inlier An *inlier* created by a *fault* crossing a valley.

fault plane The plane along which a *fault* acts.

fault plane solution See *focal mechanism solution.*

fault propagation The process by which a *fault* extends along its length.

fault reactivation The re-use of a *fault* in a later phase of *deformation*.

fault rock A rock produced by the action of a *fault*.

fault set See *multiple faults*.

fault zone A *tabular* volume containing many *faults* and *fault rocks*.

faunule A *fossil* fauna from a small area or *stratigraphic* range.

fayalite (Fe_2SiO_4) The iron-bearing end member of the *olivine solid solution* series.

FDSN See ***F**ederation of **D**igital **S**eismic **N**etworks*

feather edge The intersection line on a map of a *stratigraphic* boundary with a higher boundary such as an *unconformity*, which marks the zero *isopachyte* of the *strata* between the boundaries.

feather fracture See *en echelon fracture*.

feather joint (pinnate fracture) A minor *joint* adjacent to a larger *fracture* and intersecting it at an acute angle.

feather ore A *plumose* or *acicular* form of *jamesonite*.

feather structure See *plume structure*.

Federation of Digital Seismic Networks (FDSN) A global network of *seismometers* based on digital recording which has superseded the *World-Wide Standardized Seismograph Network*.

Feixianguan A *Triassic* succession in China covering the *Griesbachian* and part of the *Nammalian*.

feldspars Framework aluminosilicates of sodium, potassium and calcium, the most abundant *mineral* group in the *crust*. Common feldspars are *solid solutions* of the three end-member components *anorthite*, *albite* and *orthoclase*. Combinations predominantly of *albite/anorthite* are termed *plagioclase* and combinations of *albite/orthoclase* termed *alkali feldspar*.

feldspathic greywacke See *arkosic wacke*.

feldspathic wacke See *arkosic wacke*.

feldspathization See *potassic fenitization*.

feldspathoids (foids) A group of aluminosilicate *minerals* with a variety of framework structures, similar to the *feldspars* but containing less *silica*. Characteristic of *undersaturated alkaline igneous rocks*.

felsenmeer See *blockfield*.

felsic Containing at least one of the light-coloured *minerals* ***fe**ldspar*, ***l**enad* or ***si**lica* as the major component of the *mode*. cf. *mafic*.

felsitic Descriptive of *granular*, *cryptocrystalline aggregates* formed by the *devitrification* of *glass*.

femic Ferromagnesian.

fen A mire resulting from *groundwater* rather than precipitation.

fenester (fenster) A *tectonic* 'window' through which rocks below a *thrust sheet* have been exposed by *erosion*.

fenestrae (birdseyes) Millimetric-sized, rounded to planar to irregular voids in a *sedimentary rock* which may be partly or completely infilled by sediment or a cement. Formed by desiccation or air entrapment.

Fengshan A *Cambrian* succession in China covering the upper part of the *Dolgellian*.

fenite *Country rock* which has been subject to *metasomatism* by the *emplacement* of *alkaline* or *carbonatite intrusive rocks*.

fenitization The process of forming a *fenite*. See *potassic fenitization, sodic fenitization*.

fenster See *fenester*.

feral relief A landscape in which valley sides are deeply dissected by *insequent streams*, related to rapid *runoff*.

ferberite ($FeWO_4$) The iron end-member of the *wolframite mineral* series.

fergusite An *alkaline syenite* comprising large *pseudoleucite* crystals in a *matrix* of *aegirine-augite, olivine, apatite, sanidine* and *iron oxides*.

fergusonite ($(Y,Ce,Nb)NbO_4$) An *ore mineral* of the *rare earth elements* found in *granite pegmatites*.

fermentation The *anaerobic* process by which bacteria metabolize oxygen-containing organic matter, liberating hydrogen and carbon dioxide, which takes place during shallow organic matter *diagenesis*.

ferrallitization A *pedogenetic* process involving the accumulation of sesquioxides of iron and aluminium under humid tropical conditions.

ferricrete (laterite) An iron-rich *duricrust* often formed in *deep weathering* profiles in humid tropical and subtropical conditions.

ferrierite ($(Na,K)Mg_2(Al_{5.5}Si_{30.5}O_{72}.18H_2O)$ A *zeolite*.

ferriferous (ferruginous) Containing iron.

ferrimagnetism A form of *ferromagnetism*, typical of most natural crustal magnetic *minerals*, in which *magnetic domains* are oppositely magnetized but of different strength, giving rise to strong *spontaneous magnetism*.

ferrimolybdite ($Fe_2(MoO_4)_3.8H_2O$) A bright yellow, soft molybdate produced by the *alteration* of *molybdenite*.

ferro- Iron-bearing.

ferroactinolite ($Ca_2Fe_5Si_8O_{22}(OH)_2$) A dark green *amphibole*.

ferrocarbonatite A *carbonatite* containing *siderite*.

ferrogabbro A *gabbro* containing iron-rich *olivine*.

ferrohastingsite ($NaCa_2(Fe^{2+})_4AlAl_2Si_6O_{22}(OH)_2$) A variety of sodic, iron-rich *hornblende*.

ferrohedenbergite ($(Ca,Fe)Si_2O_6$) A *clinopyroxene* found in *quartz syenites, granophyres* and *ferrogabbros*.

ferrohortonolite An *olivine* with the composition Fo_{30-10}.

ferromagnesian Containing iron and magnesium.

ferromagnetism Strong magnetic behaviour resulting from internal quantum-mechanical exchange or super-exchange forces which cause electron spins to become coupled but quench the coupling between the magnetization associated with the electron orbits. This results in a strong *spontaneous magnetization*.

ferropseudobrookite ($FeTi_2O_5$) A rare iron-titanium *oxide mineral*.

ferrosilite ($Fe_2Si_2O_6$) A *pyroxene* found cementing *sandstones* and *pelites*.

ferruginous See *ferriferous*.

fersmannite ($Na_4Ca_4Ti_4(SiO_4)_3(O,OH,F)_3$) A rare, titanium-bearing *nesosilicate*.

fetch The extent of open water across which a *wave*-generating wind blows, determining the height and energy, and hence the *erosional* and depositional potential, of the *waves*.

fiamme The irregular, flattened, *glassy* discs in a *welded tuff* which determine *eutaxitic texture*.

fibre growth A *slickenside* formation mechanism in which elongate crystals grow on a *fault* surface in the movement direction.

fibroblastic A type of *metamorphic fabric* in which the grains are of equal size and fibrous *habit* due to solid-state *crystallization* during *metamorphism*.

fibrolite See *sillimanite*.

fibrous fracture The *fracture* seen in *minerals* giving a fibrous appearance.

fibrous texture A *texture* with the appearance of a mass of fibres, as shown by *asbestos*.

Fick's law of diffusion A law controlling *diffusion in sediments*: $J_i = -D_i dC/dx$, where J_i = mass of component i transported per unit area per unit time, D_i = *diffusion coefficient*, dC/dx = concentration gradient.

field capacity A descriptor of a soil in which gravitationally-driven draining has ceased.

Filicopsida A class of division *Trachaeophyta*, kingdom *Plantae*; the ferns and their relatives. Range U. *Silurian–Recent*.

filiform Thread-like.

Fillipovskiy See *Iren'skiy*.

filter pressing A possible mechanism of *magma differentiation* in which melt separates from a crystal-rich *magma* by draining or by being pressed out, perhaps by the overlying weight of the *magma* body.

filtering The process of modifying a waveform, usually to suppress or enhance certain information, such as the improvement of a *seismic record* by the removal of *noise* frequencies.

fine crush breccia A rock similar to a *crush breccia*, but with fragments between 1 and 5 mm in size. cf. *crush breccia, crush microbreccia*.

fineness An expression of the quality of *native gold* in ppt, i.e. 1000 is pure *gold*.

finger lake A long, narrow lake in a deep trough, probably the result of *glacial erosion*.

finite rotation The actual rotation about a *Euler pole* required to bring two linear features (e.g. *ocean ridges*) together. cf. *instantaneous rotation*.

finite strain (total strain) The total change in shape of a deformed body relative to its shape before *deformation*.

fiord See *fjord*.

fire damp A combustible gas contained in *coal* comprising a mixture of *methane* and other *hydrocarbons*.

fire fountaining A type of *volcanic*

eruption in which low *viscosity magma* erupts continuously and rises a few hundred metres as an incandescent jet, typical of *Hawaiian eruptions*.

fire opal A variety of *opal* with a brilliant orange colour.

fireclay (refractory clay) An *underclay* rich in *kaolinite* with commercial application as a refractory because of its low *mica* and iron content.

firn See *névé*.

first arrival (first break) The earliest seismic signal to be recorded from a particular *seismic source*.

first break See *first arrival*.

first order fold A *fold* larger than a *second order fold* and which folds the *enveloping surface* of *second order folds*.

fish 1. See *Pisces*. 2. An instrument package towed behind a ship.

fishscale dune pattern See *aklé*.

fissility The property of a fine-grained rock which has surfaces of weakness along which it splits easily. A more general term than *cleavage* in that it implies no causative mechanism.

fission track dating A dating method based on measuring the concentration of fission tracks generated during the decay of ^{238}U. Also used to study thermal history and rates of *uplift*.

fissure eruption The eruption of *magma* at several points along an elongate volcanic conduit. cf. *central eruption*.

fissure vein An old term for *vein*.

fixed carbon The *carbon* left after all volatiles have been expelled from *coal*.

fjord (fiord) A deep, narrow, *glacial* trough inundated by the sea.

flachkarren See *clint*.

flagstone A *sandstone* containing *mica*, which enhances its *fissility*.

flake graphite A commercial term for flat, platy grains of *graphite* disseminated through a *metamorphic rock*.

flake mica (ground mica, scrap mica) Fine-grained *mica* that is a by-product of certain processing operations and which is used in various industrial processes.

flake tectonics The process at a *subduction zone* of the detachment or *delamination* of an *upper crustal* layer from a subducting *plate* and its emplacement on the overriding *plate*.

flame structure A *structure* at the base of a *bed* comprising upward-pointing fingers or wedges of sandy or silty sediment penetrating a finer-grained substrate because of load pressure. Often associated with *load casts*.

Flandrian transgression The rise in sea level during *Holocene* and late *Pleistocene* times caused by the melting of *ice sheets* of the last *glacial*.

flap A type of *recumbent syncline* formed by *gravity collapse*.

flaser A *mud* lens preserved in the trough of a ripple. There is gradation from flasers to *linsen*, both of which form in aqueous environments in which slack water and *wave* activity alternate.

flaser bedding A type of *heterolithic bedding* comprising discontinuous, curved lenses of *mud* or *silt* that were deposited in troughs or draped over the *ripples* in *cross-laminated sands*.

flaser gabbro A *cataclastic gabbro* in which *chlorite* or *mica* flakes curl around *augen* of *quartz* and *feldspar*.

flash A water-filled depression caused by surface subsidence.

flat 1. A horizontal to subhorizontal *replacement orebody*. 2. The part of a *fault* that does not cut across datum surfaces, such as *bedding*. cf. *ramp*.

flatening (flatening strain) A shape change in which there is *contraction* along only one *principal strain* direction.

flatening strain See *flatening*.

flatiron A small, steep-sided, triangular *mesa*.

flatjack A thin metallic membrane inserted in rock to measure *in situ stress*.

flexible mineral A *tenacity* descriptor of a *mineral* that will bend and remains bent on release of pressure.

flexural flow A *fold mechanism* in which there is layer-parallel *shear* in the *fold limbs* without distortion in the *hinge zone*.

flexural rigidity A term describing the resistance of an elastic beam, such as the *lithosphere*, to *flexure*.

flexural slip A *fold mechanism* in which there is *flexural flow* by discontinuous layer-parallel *shear* distributed between the rock layers and *interlayer slip* on their bounding surfaces.

flexure The bending of an elastic beam in order to support a load. The *lithosphere* can act in this way to support a mountain range, *ice sheet* or sediment load.

Flinn diagram A graph used to illustrate and analyse the shapes of *strain ellipsoids*. cf. *Hsu diagram*.

flint (SiO_2) A term used for *microcrystalline silica* found in the *Chalk*, equivalent to *chert* in other rocks.

flint clay A *microcrystalline* rock composed mainly of *kaolin* which forms a very hard, non-plastic *fireclay*.

flinty crush-rock An old term for *pseudotachylite* or *ultramylonite*; a fine-grained to *cryptocrystalline cataclastic* rock appearing similar to *flint* and often showing intrusive relationship to the host rock.

float *Eluvial* material.

floatstone A coarse-grained *limestone* with *matrix*-supported *clasts*, of which $>10\%$ exceed 2 mm in size.

flocculation The aggregation of *clay* particles into randomly oriented lumps.

flood basalt An extrusion of low *viscosity basaltic magma* of very large volume.

flood lava A *lava flow* contributing to a *plateau lava*.

flood routing The estimation of the shape of a *hydrograph* anywhere along a river during flooding.

floodplain An area of land periodically inundated by floodwater.

floor thrust (sole thrust) The basal *thrust* from which *faults* branch in an *imbricate fan*.

Floran A *Cambrian* succession in Australia covering parts of the *Solvan* and *Menevian*.

flos ferri A variety of *aragonite* resembling *coral*.

flotation (froth flotation) A method of concentrating *minerals* by selective flotation in which the *mineral* attaches to bubbles blown through a mixture of ground *ore*, water and a frothing agent, and rises to form a surface froth.

flour gold The very finest-grained *placer gold*.

flow The permanent *deformation* that has a continuous *strain* distribution.

flow banding (flow layering) Layering produced by *flow* in an *igneous* or *metamorphic rock*.

flow cleavage A type of *slaty cleavage* in which the rock's *fabric* is believed to reflect *recrystallization* accompanied by solid-state *flow*, so that original sedimentary *bedding* may be lost.

flow competence The maximum particle size capable of being transported in the *bedload*.

flow folding (shear folding) The formation of a *fold* by *shear* or bulk *flow* of rock in a direction oblique or normal to the layering.

flow foliation (flow lineation) A more general term than *flow banding* for describing a line or surface produced by *flow*.

flow layering See *flow banding*.

flow lineation See *flow foliation*.

flow regime A classification of flows based on the frictional resistance experienced. See also *lower flow regime*, *upper flow regime*.

flow separation The detachment of the *boundary layer* from a surface through generation of adverse pressure gradients close to that surface, commonly arising from an abrupt change in bed geometry. Important in the generation of *turbulent flow*.

flow visualization Techniques used to make flow structure visible to direct observation or photography, e.g. smoke in an airflow, dye in a stream.

flower structure 1. A *structure* produced by local changes in direction in a *strike-slip fault system* where *faults* have opposed *dips*, leading to alternate zones of elevated and depressed blocks arising from local zones of *transpression* and *transtension*. 2. A coherent flow pattern within, for example, a channel; an area of large-scale, turbulent recirculation.

flowstone A type of *speleothem* deposited by water flowing over the walls or floor of a *cave*.

floxoturbidite A poorly-graded sediment deposited from a *turbulent gravity-induced flow*.

fluid inclusion An inclusion of fluid inside a crystal, which can be used to determine the pressure of formation of the crystal and hence the depth of formation.

fluidity The inverse of *viscosity*.

fluidization A process whereby the vertical escape of fluid from a *granular aggregate* exerts sufficient drag to support the grains against *gravity*.

flume 1. A deep, narrow gorge containing a turbulent stream. 2. An apparatus used to reconstruct and study fluid flow and sediment transport.

fluorapatite ($Ca_5(PO_4)_3F$) The commonest variety of *apatite*.

fluorescence The property of emitting light when light of a certain wavelength is absorbed.

fluorine test dating A dating method for bone by determining its fluorine content, which increases with time as it replaces calcium.

fluorinite A *coal maceral* of the *exinite* group believed to originate from plant oils and fats.

fluorite (CaF_2) A *mineral* used as a flux in smelting.

fluorspar Material with sufficient *fluorite* to be exploited commercially.

flushed zone The annulus around a borehole from which *pore fluids* have been flushed out and replaced by *drilling mud filtrate*.

flute A type of *sole mark*; a feature of turbulent *erosion* with a bulbous depression upstream.

fluting A type of differential *erosion* in which the surface of a coarse-grained rock is made ridged or corrugated.

fluvial (fluviatile) Pertaining to a river or stream.

fluviatile See *fluvial*.

fluvioglacial See *glaciofluvial*.

fluviokarst A *limestone* landscape produced by the combined action of *fluvial erosion* and *limestone dissolution*.

fluxgate magnetometer A continuous-reading *magnetometer* for measuring the strength of the *geomagnetic field*, now largely replaced by more modern instruments.

fluxion structure A finely lensoid or banded *structure* produced by the *elongation* of grains of contrasting composition at high *shear strains*, characteristic of *mylonite* and *ultramylonite*.

flysch A thick sedimentary deposit deposited by a *turbidity current* and originating from the *erosion* of rapidly rising fold mountains in the early stages of *orogeny*. cf. *molasse*.

Fo A method of indicating *olivine* composition as a percentage of *forsterite* e.g. Fo_{10} indicates a composition of 10% *forsterite*, 90% *fayalite*.

focal depth The depth to the *focus* of an *earthquake*.

focal mechanism solution (fault plane solution) The identification of the nature of the faulting responsible for an *earthquake* and the orientation of the *nodal planes* from recordings at *seismographs* distributed around the globe.

focal sphere A hypothetical sphere centred on an *earthquake focus* which facilitates representation of *focal mechanism solution* data on a *stereogram*.

focus See *earthquake focus*.

fodinichnia *Trace fossils* formed from feeding structures.

foidite A general term for a *volcanic rock* with >60% *feldspathoids* by volume of the *felsic* constituents.

foidolite A general term for a *plutonic* rock with >60% *feldspathoids* by volume of the *felsic* constituents.

foids See *feldspathoids*.

fold A curved or angular shape of an originally planar geological surface.

fold amplitude Half the distance between the *enveloping surfaces* of the *crests* and *troughs* of a *fold train*.

fold angle See *fold interlimb angle*.

fold axial plane A planar surface defined by the successive positions of *fold hinges* through a layered sequence.

fold axial surface A curved plane defined by the successive positions of *fold hinges* through a layered sequence, the more general form of *fold axial plane*.

fold axial trace The line of intersection of a *fold axial surface* with the topographic or some other defined surface.

fold axis The orientation of a *fold hinge*.

fold belt A large-scale group of related *folds*, probably forming part of an *orogenic belt*.

fold class A classification of *folds* based on *dip isogons*: in Class 1 they converge downwards, in Class 2 they are parallel and in Class 3 they diverge downwards.

fold closure The direction in which a *fold hinge* lies with respect to its *limbs* or *axial plane*; e.g. a *fold* may close eastwards or downwards.

fold core The part of a folded layer closest to the *fold hinge* zone.

fold crest A *fold hinge* zone which is concave downwards. cf. *fold trough*.

fold culmination An elevated zone on a *fold hinge* or *crest* which is of variable height, or a high point on a folded surface with *dome-and-basin structure*.

fold depression The location on a *fold* where the *fold hinge line plunge* of a *non-cylindrical fold* causes the *hinge* zone to be depressed.

fold hinge The location of greatest curvature of a folded surface.

fold hinge line The line in a folded surface linking the points of maximum curvature.

fold inlier An *inlier* created by a *fold* crossing a valley.

fold interference structure A *structure* formed when a folded surface is deformed by a later set of *folds*. See *dome-and-basin structure*, *double zigzag structure*, *crescent-and-mushroom structure*.

fold interlimb angle The angle between *fold limbs*.

fold limb The part of a *fold* between the *hinges*.

fold mechanism A method whereby a *fold* forms, i.e. the *deformation* distribution and history during folding.

fold mullion A *mullion structure* formed by *fold hinges*.

fold nappe A large *asymmetric fold structure* with a subhorizontal *axial surface* and *hinge line*, i.e. a *recumbent fold*.

fold nose The *hinge* zone of a *fold*.

fold orientation The direction of a *fold* in three-dimensions, defined by the trend of the *axial surface* and the *plunge*.

fold plunge The *plunge* of a *fold axis*.

fold profile The trace of a *fold* in a plane perpendicular to the *fold hinge* at any point along its length.

fold style Various geometric features of a *fold profile* considered useful in classification and indicating *fold mechanism*.

fold symmetry The similarity in size and shape of *fold limbs*.

fold system A group of related *folds*, possibly of different size and geometry, that formed together.

fold test A *palaeomagnetic* test for determining the age of a *remanent magnetization* in a *fold*. If there is less scatter of directions after correcting for the folding than before, the magnetization is pre-folding.

fold tightness The definition of a *fold* in terms of the *interlimb angle*. See *gentle fold, open fold, isoclinal fold*.

fold train An isolated surface curved to form alternating concave upward and downward regions.

fold trend The orientation of a *fold axis* or *fold hinge* in the horizontal plane.

fold trough A *fold hinge* zone which is concave upwards. cf. *fold crest*.

fold wavelength The separation of two adjacent *fold crests* or *troughs* measured along the length of the *fold train*.

folding frequency See *Nyquist frequency*.

foliation A repeated or penetrative planar feature in a rock which may be defined by *fabric*, compositional layering or pervasive *fracture*. Most commonly used for *metamorphic fabrics*, e.g. *cleavage, schistosity, gneissosity*.

Folk classification A scheme for the classification of *limestones*.

fool's gold A colloquial term occasionally applied to *pyrite* or *chalcopyrite* or a mixture which could be mistaken for *gold*.

footwall The wall lying beneath a horizontal or inclined *fault* or *orebody*.

footwall ramp A *ramp* in which truncation of *bedding* or other datum surface is seen in the *footwall*.

Foraminiferida/forams An order of single-cell *protozoans* characterized by an ectoplasm of fine, granular pseudopodia and an endoplasm enclosed in a *test* of varying composition, with a single chamber or many chambers connecting through an opening. Of major importance to *biostratigraphy* and environmental analysis of *Devonian* to *Recent* sediments.

force That which produces motion in a body; mass times acceleration. *SI* unit: *Newton*, the force required to give a mass of 1 kg an acceleration of 1 m s^{-2}

forced oscillation The vibration in a solid or quasi-solid not arising from a natural resonance.

forceful emplacement (forceful intrusion) An *emplacement* in which *country rock* is actively deformed.

forceful intrusion See *forceful emplacement*.

forearc basin An elongate *basin* between the *trench* and volcanic region of a *subduction zone*.

foredeep basin See *foreland basin*.

foredune A linear, coastal *dune* found behind and parallel to the *backshore* zone of a *beach*.

foreland The undeformed marginal region bordering an *orogenic belt*.

foreland basin (foredeep basin) A *tectonic basin* on the *foreland* of an *orogenic belt* and genetically related to it, ascribed to *flexure* caused by the load of the belt.

foreland thrust belt A *thrust belt* of discrete *thrust faults* in the *foreland*.

forelimb The limb of an *asymmetric fold* with the greater *dip*. cf. *backlimb*.

forensic seismology See *nuclear explosion seismology*.

forereef facies A central sedimentary *facies* of a *coral-algal reef* after growth and *cementation*.

foreset bed The inclined surface within a *cross-laminated* bed. cf. *toeset bed, topset bed*.

foreshock A small to moderate *magnitude earthquake* which precedes a large, shallow *earthquake*. cf. *aftershock*.

form (crystal form) A set of crystal faces which are equivalent to each other by the *point group* symmetry or which can be generated by operating the symmetry elements of the *point group* on one starting face. See also *general form, special form*.

form line A line on a map indicating the general direction of the *strike* of a *fold*.

form surface A planar surface that intersects the ground surface as a *form line*, used in structural mapping.

formation A grouping of *beds* used in *lithostratigraphy*; the smallest unit mappable on a reasonable scale. See also *bed, member, group, supergroup*.

formation evaluation The integration of physical data on rocks during oil exploration, commonly achieved by *geophysical borehole logging*.

fornacite (furnacite) ($(Pb,Cu)_3((Cr,As)O_4)_2(OH)$) A basic, copper-lead chromarsenate found associated with *dioptase*.

forsterite (Mg_2SiO_4) The magnesium-bearing end member of the *olivine solid solution* series.

forsterite marble (ophicalcite) A *marble*-like rock produced by the *contact metamorphism* of *silica*-bearing, *dolomitic limestone*.

foskorite A *magnetite-olivine-apatite* rock found in some carbonate-*alkaline* igneous complexes, sometimes mined for its iron and phosphorus content.

fossil 1. The trace of an organism buried naturally and subsequently preserved permanently. 2. Of great age.

fossil assemblage An association of *fossils* without ecological relations. cf. *fossil community*.

fossil community An association of ecologically interrelated *fossils*. cf. *fossil assemblage*.

fossil fuel A fuel originally of organic origin.

fossil placer A lithified *placer*.

foundry sand A refractory, cohesive, porous *sand* suitable for forming moulds for metal castings.

fourchite An *intrusive igneous rock* comprising *essential* titanium-bearing *augite* and *kaersutite* in a *matrix* of *analcite* or *glass*.

fourier analysis A means of expressing a waveform in terms of a combination of sine waves.

fourier transform A method of changing between a waveform expressed as a time-variable amplitude to one expressed as a frequency-variable amplitude. Many *filtering* operations are more conveniently and efficiently accomplished in the frequency domain.

fowlerite A zinc-rich variety of *rhodonite*.

foyaite A variety of *nepheline syenite* with equal amounts of *nepheline* and *potash feldspar* and a subordinate *mafic mineral*, e.g. *aegirine*.

fractional crystallization See *crystal-liquid fractionation*.

fractionation factor (α) A term used in determining temperature and environment from stable isotopes of carbon and oxygen in sediments, e.g. for *calcite*: $10^3\ln\alpha = 2.78(10^6T\text{-}2)\text{–}2.89$, where T is temperature in Kelvin and $10^3\ln\alpha$ is approximately the difference between the oxygen isotope composition of the *mineral* and that of the fluid from which it formed.

fractography The study of *fracture* surfaces, which provides information on *fracture* propagation.

fracture 1. In *structural geology*, a discontinuity across which there has been separation, e.g. *joint*, *fault*. 2. In *mineralogy*, the breaking of *minerals* which does not occur along particular crystallographic directions, e.g. *conchoidal fracture*, *fibrous fracture*, *hackly fracture*, *splintery fracture*, *uneven (irregular) fracture*.

fracture cleavage A *cleavage* defined by closely spaced *fractures*.

fracture porosity A *porosity* developed as a result of the presence of *fractures* in a rock.

fracture toughness A measure of the resistance of a material to *brittle failure* by the spreading of cracks.

fracture zone A zone of past or present *transform fault* movement within the *oceanic crust*, generally represented by a topographic depression up to several thousands of kilometres in length and sometimes accompanied by flanking ridges.

fragipan A brittle, acidic, cemented horizon of *clay*, *silica*, iron, aluminium or organic matter between the soil and the underlying *bedrock*, often the result of *periglacial* activity.

fragmental Descriptive of a *texture* of a *sedimentary rock* in which the *clasts* are broken particles.

fragmentation level The level in a *volcanic vent* at which *vesicles* can explosively tear apart the *magma* to form *pyroclasts* suspended in up-rising gas.

framestone An *authochthonous*, organically-bound *limestone* in which the organisms formed a framework during deposition.

francolite A variety of *apatite*.

Franconian A *Cambrian* succession in the E USA covering parts of the *Maentwrogian* and *Dolgellian*.

franklinite ($(Zn,Fe,Mn)(Fe,Mn)_2O_4$) A rare, zinc-rich *spinel*.

Frasch process A method of extracting subsurface *native sulphur* by melting it with hot water and blowing the melt to the surface.

Frasnian A *stage* of the *Devonian*, 377.4–367.0 Ma.

Frasnien A *Devonian* succession in France and Belgium covering parts of the *Givetian* and *Frasnian*.

Frederiksberg A *Cretaceous* succession on the Gulf Coast of the USA covering part of the *Albian*.

free air anomaly A *gravity* measurement to which a *latitude* and *free air correction* (± an *Earth tide* and an *Eötvös correction*) have been applied. Provides an indication of the degree of *isostatic* compensation of a broad feature. cf. *Bouguer anomaly*.

free air correction The correction applied to a *gravity* measurement to account for the decrease in *gravity* with height in free air. cf. *Bouguer correction*.

free face The wall of an *outcrop* which is too steep for debris to rest on it.

free-milling gold *Gold* in an *orebody* in *native* form that is easily amalgamated or cyanided for recovery.

free oscillation The resonation of a body, such as the Earth, at particular harmonics when excited by an event, such as an *earthquake* of *magnitude* ≥ 7.5. The oscillations provide information on the physical properties of the interior.

free water An American term for *held water*.

freestone A fine-grained stone that can be cut and worked without fracturing.

freeze-thaw A type of *weathering* in response to alternate freezing and thawing.

freibergite A *silver*-bearing *ore mineral*; a variety of *tetrahedrite*.

French chalk A variety of *talc* with a compact form used to mark cloth or remove grease stains.

Frenkel crystal defect A *crystal defect* formed when an atom is transferred from its normal position to an immediately adjacent interstitial site not normally occupied.

friable Descriptive of a material that can be disintegrated into grains by finger pressure.

fringe joint See *en echelon fracture*.

fringing reef A type of *coral-algal reef* bordering an island or continent with a flat surface exposed at low tide.

frondescent mark A type of *sole mark* that was modified by flowage.

frontal ramp A *ramp* trending normal to the tectonic transport direction in a *thrust system*.

frost heave A type of *mass movement* in which soil is moved upwards by the migration of water which expands on freezing.

frost shattering See *congelifraction*.

frost splitting See *congelifraction*.

frost weathering See *congelifraction*.

frost wedging See *congelifraction*.

froth flotation See *flotation*.

Froude number (Fr) A dimensionless number which quantifies the influence of *gravity* on a flow: $Fr = U/\sqrt{(gL)}$, where U = mean flow velocity, g = *gravity*, L = a length term, usually mean flow depth. When $Fr < 1$, the influence of *gravity* is pronounced and the flow is *subcritical*, when $Fr > 1$ inertial forces predominate and the flow is *supercritical*, and $Fr = 1$ indicates a *standing wave* or *hydraulic jump*.

fuchsite A chrome-rich variety of *muscovite*.

fucoid An old term for a *trace fossil* formed from a burrow.

fugichnia *Trace fossils* formed from escape structures.

fulgurite A hollow tube of *glass* (predominantly *lechatelierite*) formed by the action of lightning on *quartz sand.*

fulje A deep parabolic depression between closely interlocking *dunes.*

Fuller's earth An absorbent *clay* composed of calcium *montmorillonite* (in the USA *attapulgite* and *sepiolite*) used in decolourizing oils etc.

fumarole A *volcanic vent,* distinct from those erupting *magma*, from which *volcanic gases* escape.

functional morphology An attempted interpretation of the function of organs or structures in *fossils.*

fundamental strength The maximum *stress* a material can sustain indefinitely at a given temperature and *confining pressure.*

Fungi One of the three kingdoms of multicellular organisms, along with the *Metaphyta* and *Metazoa*; organisms which feed by ingesting organic matter. Range *Precambrian–Recent.*

furnacite See *fornacite.*

fusain A *lithotype of banded coal* composed of soft, friable material similar to charcoal.

fusibility The property of being capable of conversion from solid to liquid by heating.

fusiform Shaped like a spindle.

fusinite A *coal maceral* of the *inertinite* group with a high *reflectance.*

fusite A *microlithotype of coal* composed of *fusinite.*

fusoclarain A *lithotype of banded coal* intermediate between *fusain* and *clarain.*

G

G See *gravitational constant.*

g See *gravity.*

gu See *gravity unit.*

gabbro A coarse-grained, *basic igneous rock* composed of *plagioclase* with $An_{>50}$, *pyroxene* with *clinopyroxene* > *orthopyroxene* and *accessory olivine, quartz* and *nepheline.*

Gabbs A *Triassic* succession in Nevada, USA equivalent to the *Norian* and *Rhaetian.*

gadolinite ($(YFeBe_2SiO_4)_2O_2$) A *rare earth element ore mineral* found in *pegmatites.*

gahnite (zinc spinel) ($ZnAl_2O_4$) A zinc-aluminium *spinel* found in crystalline *schists, contact metamorphosed limestones*, high temperature *replacement ore* deposits, *granite pegmatites* and *placer deposits.*

gaining stream A watercourse that receives water from an *aquifer* because its valley cuts the *water table.* cf. *losing stream.*

Gal The *cgs* unit of *gravity*, 1 Gal = 1 cm s^{-2} = 10 *gravity units.*

galaxite ($MnAl_2O_4$) A rare, black manganese-aluminium *spinel.*

galena (lead-glance) (PbS) The major *ore mineral* of lead.

galena group A group of *sulphide minerals* characterized by a *halite* structure, with *cubic* close *packing* of anions in planes parallel to (111) and both cations and anions in regular octahedral six-fold coordination.

gallery A horizontal tunnel or passage in a mine.

Gallic A division of the *Cretaceous*, 131.8–88.5 Ma.

gamma(γ) The *cgs* subunit of *magnetic field* strength, $1\gamma = 10^{-5}$ Gauss = 10^{-9} *Tesla.*

gamma-gamma log (density log) A *geophysical borehole log* in which the scatter of artificial gamma rays provides a measure of the *density* of the adjacent wallrock.

gamma ray log A *geophysical borehole log* which measures natural gamma radiation, which is most prevalent in the presence of *clay minerals.*

gamma ray spectrometer. An instrument which measures gamma rays in a similar way to a *scintillation counter*, but which can discriminate between rays produced by uranium, thorium and potassium-40.

gangue The unwanted material with which *ore minerals* are associated.

ganister A highly leached *sandstone* or *siltstone* beneath a *coal seam.*

Gard notation A method of describing the unit cell relationship in *polytypism.*

garnets Cubic *minerals* with the general formula $A_3^{2+}B_2^{3+}Si_3O_{12}$, where A = magnesium, iron, manganese or calcium and B = aluminium, iron or, rarely, chromium. Characteristic of *metamorphic rocks*, but also found in some *igneous rocks* and as detrital grains in sediments. Common garnets are *almandine*, *andradite*, *grossular*, *pyrope*, *spessartine* and *uvarovite*.

garnierite ($(Ni,Mg)_3Si_2O_5(OH)_4$) A green, nickel-rich, *serpentine group mineral*.

gas cap The gas above an oil accumulation in an oil *reservoir*.

gas cap drive The pressure exerted by a *gas cap* as oil is removed from a *reservoir* which drives it towards the *well*.

gas hydrate An accumulation of *clathrates* formed under low temperature and high pressure.

gas–oil ratio (GOR) The volume of gas as it exists in a *reservoir* relative to the volume of oil.

gas pool An accumulation of *natural gas* in a single *reservoir* and *gas trap*.

gas trap A stratal arrangement that can trap gas in the same manner as an *oil trap*.

Gasteropoda See *Gastropoda*.

gastrolith A stone found in the stomach of an animal to aid food processing or to counteract its buoyancy when in water.

Gastropoda/gastropods (Gasteropoda) A class of phylum *Mollusca* in which the anterior part of the foot is developed into a head and a helically-coiled shell protects the organism. Range *Cambrian–Recent*.

Gault A *Cretaceous* succession in England covering the upper part of the *Albian*.

Gauss 1. A *magnetostratigraphic epoch* of *normal polarity* in the *Neogene*, 3.34–2.42 Ma. 2. The *cgs* unit of *magnetic field* strength. cf. *Tesla*.

Gauss' theorem 'The outward flux of the force of attraction over any closed surface in a gravitational field is equal to 4π times the mass enclosed by the surface.' Used to determine *excess mass* from *gravity* measurements.

gaylussite ($Na_2Ca(CO_3)_2.5H_2O$) An *evaporite mineral* precipitated from saline lakes.

geanticline A broad area of *anticlinal uplift*.

Gebbie A *Permian* succession in Queensland, Australia, covering the upper part of the *Artinskian*.

Gedinnine A *Devonian* succession in France and Belgium equivalent to the *Lochkovian*.

gedrite ($Na_{0.5}(Mg,Fe)_2(Mg,Fe)_{3.5}(Al,Fe^{3+})_{1.5}Si_6Al_2O_{22}(OH)_2$) An *amphibole* found in *metamorphic* and *metasomatically*-altered *rocks*.

gehlenite ($Ca_2Al_2SiO_7$) A *feldspathoid* of the *melilite* group.

Geiger counter (Geiger-Müller counter) An instrument for detecting beta rays by their ionization of a gas.

Geiger-Müller counter See *Geiger counter*.

geikielite ($MgTiO_3$) A rare, titanium-bearing *oxide mineral*.

gelifluction A type of *solifluction* taking place in *periglacial* environments underlain by *permafrost*.

geliturbation See *cryoturbation*.

gem (gemstone) A *mineral* or organic material with an intrinsic value because of its beauty, durability or rarity.

gemstone See *gem*.

general diagenetic equation An equation whose solution can quantify the rates of bacterial decomposition, *compaction*, *diffusion*, *bioturbation*, *adsorption*, *ion-exchange*, *dissolution* and *precipitation*.

general form A crystal *form* which has no particular relationship with the symmetry operators present in the *special form*.

generalized reciprocal method A method of interpreting *reversed seismic refraction profiles* in terms of an undulating refractor, with more general application than the *plus-minus method*.

geniculate twin (knee twin) A *twin* in which the *twin plane* changes the crystal shape so that it has the appearance of a knee joint.

gentle fold A *fold* with an *interlimb angle* of $> 120°$.

geo A narrow, linear cleft running inland from a *sea cliff*.

geo- Pertaining to the Earth.

geobarometry The use of *fluid inclusions* or the chemistry of *mineral* systems to determine the pressure of deposition or of a subsequent event.

geobotanical survey *Geochemical* prospecting utilizing metallophilic plants, plant poisoning and the identification of anomalous metallic concentrations in plant tissue.

geochemical anomaly An abnormal concentration of elements with respect to the background level.

geochemistry The study of the chemistry of the Earth's constituents.

geochronology The measurement of geological time.

geocline A succession of *strata* with a low, uniform *dip*.

geocronite ($Pb_5(Sb,As)_2S_8$) A white sulphosalt *mineral* of lead.

geode A cavity lined with crystals which project towards its centre.

geodesy The science of the precise measurement and mapping of the Earth's surface.

geodynamics The study of the dynamic processes which affect, or have affected, the solid Earth, generally those related to *tectonics*.

geoelectric section A one-dimensional section obtained by *vertical electrical sounding* or a similar *electromagnetic induction method* showing how *electrical resistivity* varies with depth; it can often be interpreted geologically.

geographical cycle See *cycle of erosion*.

geoid The *equipotential surface* of the gravitational field of the Earth represented by the sea level surface, usually taken as the *datum* plane in *gravity reduction*.

geologic time-scale An absolute time-scale made up of standard *stratigraphic* divisions based on rock sequences.

geology The study of the solid Earth.

geomagnetic correction The correction to magnetic survey data for the variation with latitude and longitude of the *geomagnetic field*, often applied using the *International Geomagnetic Reference Field*.

geomagnetic dipole field The 80% of the *geomagnetic field* that can be ascribed to the field of a single, fictitious, magnetic dipole. cf. *geomagnetic non-dipole field*.

geomagnetic elements (magnetic elements) Descriptors of the strength and orientation of the *geomagnetic field* in terms of its total field, vertical field, horizontal field, *inclination* and *declination*.

geomagnetic event (geomagnetic excursion) A short period of constant geomagnetic polarity, usually < 10 000 years.

geomagnetic excursion See *geomagnetic event*.

geomagnetic field The *magnetic field* of the Earth.

geomagnetic non-dipole field The part of the *geomagnetic field* (~20%) not accounted for by the *geomagnetic dipole field*.

geomagnetic polarity time-scale A time-scale constructed by making use of changes in polarity of the *geomagnetic field* that occur at intervals from several per Ma to ~50 Ma. The time-scale is complete from 160 Ma.

geomagnetism The study of the *geomagnetic field* and its use in *geophysical exploration* for magnetic *minerals*.

geomorphic Concerning the form of the Earth or its surface features.

geomorphological threshold A change in a landform initiated by changes in the morphology of the landform itself with time.

geomorphology The study of the form of the ground surface and the processes which shape it.

geopetal fabric A *fabric* formed by the partial infilling of the bottom part of a cavity with sediment under the influence of *gravity*.

geophone An instrument used on land to detect the arrival of seismic energy, commonly by using a moving coil technique to convert ground movement into a varying voltage. cf. *hydrophone*.

geophysical anomaly A perturbation from the norm in a measured field, usually resulting from a change in the physical properties of the subsurface, e.g. *gravity anomaly, magnetic anomaly*. In order to be able to recognize such anomalies, all non-geological sources of variation in the field are removed by a *reduction* process.

geophysical borehole logging (downhole geophysical survey, geophysical well logging, well logging, wire-line logging) The recording of the properties or characteristics of the rock formations traversed by measuring apparatus in a borehole, which largely obviates the necessity of the expense of coring. The principal techniques utilized are *electric logs, induction logs, self potential logs, radioactivity logs, sonic logs, temperature logs* and *dipmeter logs*.

geophysical exploration See *applied geophysics*.

geophysical well logging See *geophysical borehole logging*.

geophysics The application of the methods and techniques of physics to the study of the Earth and the processes affecting it.

geopressure See *abnormal pressure*.

geopressuring See *overpressuring*.

Georgiev A *Jurassic* succession in W Siberia equivalent to the *Kimmeridgian*.

geostrophic current A current in which the pressure gradient and *Coriolis forces* are in balance.

geosyncline An elongate trough with a great thickness of sediment which subsequently becomes an *orogenic belt*. Now an obsolete term since such features can be put into their *plate tectonic* setting.

geotectonic Relating to major Earth *structures*, such as *orogenic belts*, *cratons*, *basins*, and the processes forming them.

geotherm A curve showing the variation of temperature with depth.

geothermal Concerning the flow of heat from the interior of the Earth to the surface.

geothermal brine A saline solution within *geothermal systems* and *hot springs*, the major constituents being Na, K, Ca, Mg and Cl, possibly with other metals which may become of economic importance.

geothermal energy Energy obtained from a *geothermal system* by pumping out hot water or pumping water through hot rocks and back to the surface (*hot dry rock concept*).

geothermal field An area of high *heat flow* that can produce *geothermal energy*.

geothermal gradient The rate of change of temperature with depth in the Earth.

geothermal system A circulating *groundwater* system activated by a high *geothermal gradient*.

geothermometry The determination of the temperature of formation of a *mineral* deposit. This can be accomplished by the use of *fluid inclusions* in *minerals* such as *fluorite*, *baryte* or *quartz*, inversion points in *polymorphs*, the resolution of *exsolution textures*, equilibrium *mineral assemblages* and stable isotopes.

gersdorffite (nickel arsenic glance) ($NiAsS$) A rare *ore mineral* of nickel.

geyser A vent from which hot water and steam are violently and periodically ejected at the surface in a volcanic area, caused by the heating of *groundwater* by subsurface *magma*.

geyserite (sinter) A variety of *opal* found in *geysers* and *hot springs*.

ghost reflection A *seismic multiple* generated by a *seismic reflection* at the base of the *weathered layer*.

Ghyben-Herzberg relationship A relationship allowing the determination of the thickness of a lens of fresh *groundwater* overlying salt water from their densities and the height of the top of the lens above sea level.

giant gas field A field with *natural gas* reserves exceeding a volume of $\sim 140 \times 10^9$ m^3.

gibber A desert plain covered with a layer of *pebbles* or *boulders*.

gibbsite (hydrargillite) ($Al(OH)_3$) An *ore mineral* of aluminium found in *bauxite* deposits.

Gilbert A *magnetostratigraphic epoch* of *reversed polarity* in the *Pleistocene*, 5.3–3.34 Ma.

Gilbert delta A steep-fronted *delta* formed where there are steep nearshore slopes, inflow velocities are large and outflows are dominated by inertia. Usually forms in an area of little reworking by waves or *tide*.

gilbertite A fluorine-rich variety of *muscovite*.

gilgai Micro-relief characterized by small hummocks separated by shallow troughs.

gilsonite See *uintaite*.

Gilyakian A *Cretaceous* succession in the far east of the former USSR equivalent to the *Turonian*.

gipfelflur A summit plane shown by mountain peaks of similar elevation.

girdle distribution A pattern of points around a *great circle* on a *stereogram*.

Gisbornian An *Ordovician* succession in Australia covering part of the late *Llanvirn*, the *Llandeilo* and part of the *Costonian*.

gismondine (gismondite) ($CaAl_2Si_2O_8.4H_2O$) A rare *zeolite* found in *amygdales* in *basaltic lava*.

gismondite See *gismondine*.

Givetian A *stage* of the *Devonian*, 380.8–377.4 Ma.

Givetien A *Devonian* succession in France and Belgium covering the lower part of the *Givetian*.

glaci- (glacio-) Pertaining to *glacier* ice.

glacial 1. Adjective referring to a *glacier*. 2. A period of glaciation.

glacial control theory A theory that suggests that marine platforms abraded during low sea levels associated with glaciation were subsequently sites of *coral-algal reef* growth.

glacial deposit A sedimentary feature deposited by a *glacier*, composed mainly of *till*. Includes *braids*, *drumlins*, *eskers*, *kames*, *moraine* and *sandurs*.

glacial erosion *Erosion* by *glacier* ice by the processes of *frost shattering*, meltwater flow and *plucking* to form features such as *arêtes*, *cirques*, *fjords*, *hanging valleys*, *horns*, *knock-and-lochan* topography and *roches moutonnées*.

glacial plucking See *plucking*.

glacial proximal trough A trough eroded by *glacier* ice, running water or wind which increases in velocity around a rock body.

glacier A mass of ice and snow which deforms and *flows* under its own weight if sufficiently thick.

glacimarine Referring to sediments produced by the interaction of a *glacier* with the sea.

glacio- See *glaci-*.

glaciofluvial (fluvioglacial) Referring to *glacial* meltwater activity.

glaciolacustrine Concerning *glacial* lakes or their deposits.

glaciotectonism The production of landforms and *structures* by the *deformation* of soft rock and *drift* as a consequence of the movement of *glacier* ice.

glacis An *erosional pediment*, sometimes in several generations, in arid regions.

glance A mining term for an *opaque mineral* of high reflectivity.

glass Amorphous matter formed by the rapid cooling of *magma*.

glass sand A *sand* with even-sized grains which, because of a high *silica* content (95–99.8%) and low alumina content (<4%), is suitable for glass making.

glassy rock A rock consisting partly or wholly of *glass*, usually formed when *magma* is quenched too rapidly for *crystallization* to occur, but also forming in a *fault rock* by *dynamic metamorphism* (*pseudotachylite*) or by *thermal metamorphism* (*buchite*).

Glauber salt See *mirabilite*.

glauberite ($Na_2Ca(SO_4)_2$) Sodium-calcium sulphate, an *evaporite mineral*.

glauchroite ($CaMnSiO_4$) A rare manganese-bearing *olivine*.

glaucodot ((Co,Fe)AsS) A rare *ore mineral* of cobalt.

glauconite ($(K,Na,Ca)_{0.5-1}(Fe^{3+},Al,Fe^{2+},Mg)_2(Si,Al)_4O_{10}(OH)_2.nH_2O$) A *clay mineral* found as an *authigenic mineral* in *sedimentary rocks*.

glaucony A green, marine, sedimentary *facies* characterized by the presence of *glauconite* which develops on *continental margins* and bathymetric highs.

glaucophane ($Na_2Mg_3Al_2Si_8O_{22}(OH)_2$) A blue-black *amphibole* occurring in *metamorphic rocks*.

glaucophane schist facies See *blueschist facies*.

Gleedonian A *stage* of the *Silurian*, 425.4–424.0 Ma.

glei See *gley*.

gleittbretter See *microlithon*.

glendonite A *stellar pseudomorph* of *calcite* after *ikaite* formed under *glacial* temperatures.

gley (glei) A waterlogged soil horizon formed by *gleying*.

gleying A *pedogenetic* process involving waterlogging and the development of *anaerobic* conditions so that organic decomposition is slow and ferrous iron forms.

glide (translation gliding) The movement of a *dislocation* through a *crystal lattice* along a *slip plane*.

glide twinning *Twinning* that occurs when a *stress* forces part of a crystal to move into a *twin*-related orientation.

glimmerite An *ultrabasic rock* composed mainly of *mica*.

Global Positioning System (GPS) A constellation of artificial satellites which allows accurate three-dimensional positioning by radio interferometry. Used in measuring the rate of *plate* movements and in studies of continental *deformation*.

Globigerina ooze A *pelagite* composed of the tests of the *foraminifera Globigerina*.

globulite A tiny, spheroidal *crystallite* found in *glassy rocks*.

glomerocryst A cluster of *phenocrysts*.

glomeroporphyritic texture The *texture* of a rock containing *glomerocrysts*.

GLORIA **G**eological **Lo**ng **R**ange **I**nclined **A**sdic, a long range *side-scan sonar* system.

glory hole A large open pit from which *ore* is, or has been, extracted.

gloss coal The highest *rank lignite*; black and compact with a *conchoidal fracture* and glossy *lustre*.

Glossopteris flora A cold-climate flora widespread in *Gondwanaland* before its breakup at ~180 Ma.

gmelinite ($(Na_2,Ca)(Al_2Si_4O_{12}).6H_2O$) A *zeolite* found chiefly in the *amygdales* of *basaltic lavas* associated with other *zeolites*.

gnamma A *basin* on a rock surface, especially *igneous rock* and *sandstone*, produced by *weathering*.

gneiss A *metamorphic rock* characterized by *gneissosity*.

gneissic layering The compositional layering found in a *gneiss*.

gneissoid With a *gneiss*-like *texture* or *structure* unrelated to *metamorphic* processes.

gneissosity A *foliation* of compositional layering or lensoid structure found in high-*grade metamorphic rocks* and deformed *igneous rocks*, caused by the *deformation* of an existing *texture* or *structure*, *metamorphic segregation* or both.

Gnetopsida A class of division *Tracheophyta*, kingdom *Plantae*; an artificial grouping of land *plants*. Range L. *Jurassic–Recent*.

goethite ($\alpha FeO.OH$) A 'rust-like' hydrated oxide of iron produced by the *weathering* of iron *minerals*.

gold (Au) A *native metal*, the commonest *ore mineral* of gold.

gold dust Fine specks of *gold* found in *placer deposits*.

golden beryl A clear yellow, *gem* variety of *beryl*.

Goldschmidt's rule 'The number of naturally-occurring *minerals* in a rock is equal to the number of components where a given *mineral assemblage* is stable over a range of temperature and pressure.'

Gondwana See *Gondwanaland*.

Gondwanaland (Gondwana) The southern *supercontinent*, comprising Africa, Malagasy, India, Sri Lanka, Australasia, Antarctica and South America, which probably formed over 2000 Ma ago and began to split some 180 Ma ago.

Goniatitida/goniatites A subclass of class *Cephalopoda*, phylum *Mollusca*; *ammonoids* with a trilobed suture. Range M. *Devonian*–end *Permian*.

goniometer An instrument for the accurate measurement of the *interfacial angles* of crystals.

gonnardite ($Na_2Ca((Al,Si)_5O_{10})_2.6H_2O$) A rare *zeolite* found in *vesicles* in *basaltic* rocks.

GOR See *gas–oil ratio*.

Gorstian A *stage* of the *Silurian*, 424.0–415.1 Ma.

goshenite A colourless *gem* variety of *beryl*.

goslarite (white copperas) ($ZnSO_4.7H_2O$) A rare hydrated zinc sulphate formed by the decomposition of *sphalerite* and found as a wall coating in lead mines.

gossan (iron hat) A mass of *limonite* and *gangue* resulting from the *oxidation* by percolating surface waters of *outcrops* of sulphide deposits, generally characterized by *boxwork*.

Gothian orogeny See *Daslandian orogeny*.

Gothic orogeny See *Daslandian orogeny*.

Gotlandian An old term for the *Silurian* once used in continental Europe.

gouge 1. The *clay* filling of a *vein*. 2. See *fault gouge*.

GPR See *ground-penetrating radar*.

GPS See *Global Positioning System*.

grab sample A random, possibly hurried, sample of mineralized ground with no statistical validity, taken simply to check the type of mineralization.

graben A *fault block* downthrown between *normal faults*.

grade 1. Of *coal*, a little used classification based on the amount of *ash* (i.e. impurity) present. 2. In *geomorphology*, descriptive of the condition of *dynamic equilibrium*, e.g. the exact balance between sediment load in a channel and the amount that can be moved by the river. 3. **(tenor)** Of *ore*, the concentration of a metal in an *orebody*. 4. See *metamorphic grade*.

graded bed A layer in which there is a gradual, abrupt or step-wise vertical and/or lateral change in the grain size distribution. Develops in *aeolian* or fluid systems in response to changes in flow velocity or sediment supply which allow the deposition of different grain size populations.

grading See *sorting*.

gradiometer An instrument measuring the horizontal or vertical gradient of a *potential field*, most commonly applied to magnetic surveys.

grahamite A member of the *asphaltite* group with a high *density* and high *fixed carbon*.

grain boundary fracture See *circumgranular fracture*.

grain boundary gliding A form of *creep* by movement along grain margins.

grain boundary sliding A *deformation* mechanism involving *displacement* of grains relative to each other along grain margins.

grain fall lamination A *lamination* produced in an *aeolian* system by the fall of *sand* from suspension in zones of *flow separation*, i.e. as *foreset* and *toeset* stratification.

grain flow The *gravity*-driven movement of a sediment supported by grain–grain contact in which the grains move separately and non-cohesively.

grain size coarsening The *recrystallization* of aggregates of grains into a larger grain size, commonly occurring when temperature increases.

grain-supported conglomerate See *clast-supported conglomerate*.

grain surface texture A *textural* property reflecting the *abrasive* and *corrosive* processes which affected a *clast* during *erosion*, transport, deposition and *diagenesis*.

grainstone A little used term for a *clastic rock*.

grammatite See *tremolite*.

Grampian event A major episode of *deformation* in Scotland in the *Cambrian*.

granite A coarse-grained *igneous rock* composed of > 20% *quartz* and *feldspar* of which *plagioclase* and *alkali feldspar* are present in approximately equal amounts.

granite-greenstone terrain A type of *Archaean crust* composed of associated *granite batholiths* and *greenstone belts*, typically in the *greenschist facies*.

granite porphyry (porphyritic microgranite) A *granitic* rock containing *phenocrysts*, usually of *feldspar* and/or *quartz*, in a medium- to fine-grained *groundmass*.

granitic With the *mineral* composition of *granite*.

granitic layer An outmoded term for the upper *continental crust*. cf. *basaltic layer*.

granitization A postulated *metasomatic* process whereby a rock is converted to *granite*.

granitoid A term for any *granitic* rock.

granoblastic texture The *texture* of a *metamorphic rock* with equal-sized grains.

granodiorite A coarse-grained *igneous rock* composed of > 20% *quartz* and *feldspar* of which *plagioclase* makes up > 67% of the total *feldspar*.

granophyre A fine- to medium-grained, commonly *porphyritic*, acidic, *felsic rock* characterized by a *groundmass* containing intergrown *quartz* and *alkali feldspar*.

granophyric texture A small-scale *graphic texture*.

Granton Shrimp Bed An *exceptional fossil deposit* of L. *Carboniferous* age near Edinburgh, Scotland, where the soft parts of *conodonts*, shrimps, worms and hydroids are preserved in phosphate.

Grantsville A *Triassic* succession in Nevada, USA equivalent to the *Anisian* and *Ladinian*.

granular texture A *texture* comprising *mineral* grains of approximately equal size.

granulation The grinding, during *deformation*, of crystals into smaller, equal-sized grains, which can rotate relative to each other.

granule A particle of 2–4 mm diameter.

granulestone A *clastic rock* made up of *granules*.

granulite A *metamorphic rock* formed in the high-temperature, high-pressure *granulite facies*, characterized by a *mineral assemblage* of *plagioclase* and *pyroxene* ± *garnet*, *quartz*, anhydrous aluminosilicates, *alkali feldspar*, *calcite* and *forsterite*-rich *olivine*, commonly with a *crystalloblastic fabric*. Common in *Archaean shield* areas and probably formed at a high *metamorphic grade* in the lower *continental crust*.

granulite facies A *metamorphic facies* found at great depths and temperatures > 650°C in which the characteristic *mineral assemblage* of a *basic igneous rock* is *clinopyroxene* + *hypersthene* + *plagioclase*.

granulite-gneiss belt A very high-*grade metamorphic rock* association of the *Archaean* made up of *quartzite*, *limestone*, *pelite*, *banded iron formation*, *amphibolite* and layered *anorthosite* complexes.

granulitic texture (intergranular texture) A *texture* comprising crystals of *augite* and/or *olivine* between laths of *plagioclase*.

grapestone A rock formed of *ooids* which have been agglutinated, often by microbial processes, into irregular masses.

graphic texture An intergrown *texture* of *quartz* and *alkali feldspar* visible in hand specimen, probably formed by the simultaneous *crystallization* of both *minerals*.

graphite (C) A soft, grey-black, low pressure form of *carbon*.

graphitization The conversion of amorphous *carbon* to *graphite* by heat.

Graptolithina/graptolites A class of extinct colonial animals, probably of phylum *Hemichordata*, composed of a chitinous exoskeleton with individuals inhabiting tubes (thecae) which overlap in single or double rows (stipes) along branches originating from a small, conical cup (sicula). Range M. *Cambrian*–M. *Devonian*.

gravel A sediment of variable composition with a grain size larger than *sand*, i.e. 2 mm (4.75 mm in engineering terminology)–20 mm.

gravimeter (gravity meter) A small, portable instrument for measuring changes in *gravity*.

gravitational constant (G, universal gravitational constant) The constant of proportionality in *Newton's law of gravitation*, $6.67 \times 10^{-11}\ m^3kg^{-1}s^{-2}$.

gravitational load (gravitational pressure) The pressure exerted by *gravity* on a subsurface area.

gravitational potential (U) The scalar defined: $U = GM/r$, where G = the *gravitational constant* and r the distance to a gravitating body of mass M. The first derivative of U in any direction gives the gravitational field in that direction.

gravitational pressure See *gravitational load*.

gravity (acceleration due to gravity, g) The acceleration experienced by an object under the influence of the Earth's mass according to *Newton's law of gravitation*. The measurement of small variations in gravity forms the basis of a *geophysical exploration* method.

gravity anomaly The variation in the Earth's gravitational field caused by a geological feature of anomalous *density*.

gravity collapse structure A *structure* produced by movement down a slope under the influence of *gravity*.

Gravity Formula *Clairault's formula* with internationally agreed constants, which describes how *gravity* varies around the reference *spheroid* and used to compute the *latitude correction* in the *reduction* of *gravity* data.

gravity gliding (gravity sliding) The movement of a body of rock down an inclined plane in response to *gravity*. The plane is usually a *detachment horizon* or *décollement* parallel to the *bedding* with a *listric fault* at the *proximal* end to allow separation of the body.

gravity log A *geophysical borehole log* in which a specialized *gravimeter* is read remotely down a borehole in order to provide *density* information on the formations it penetrates.

gravity meter See *gravimeter*.

gravity settling The *gravity*-driven settling of heavy *minerals* to the floor of a *magma* chamber.

gravity sliding See *gravity gliding*.

gravity survey A survey in which variation in *gravity* is measured in order to locate and define subsurface bodies of anomalous *density*.

gravity tectonics *Tectonics* in which the primary driving mechanism is downslope sliding under the effect of *gravity*.

gravity unit (gu) The *SI* subunit used for *gravity* measurements; 1 gu = 1 μm s^{-2} = 0.1 *milligal*.

gravity wave An aqueous or airborne flow in response to *gravity*.

gray wethers See *sarsen*.

graywacke See *greywacke*.

greasy lustre A *lustre* in which a *mineral* appears to have an oily coating, caused by the scatter of light on a *microscopically* rough surface.

great circle A circle corresponding to a circumference of a sphere. cf. *small circle*.

Green-Ampt equation An empirical equation which describes the *infiltration* process.

greenalite ($(Fe,Mg)_3Si_2O_5(OH)_4$) A hydrated iron *serpentine mineral* exploited as an *ore*.

greenhouse effect The heat retention of the Earth by the absorption and reradiation of radiated heat back to its surface by clouds and gases, such as carbon dioxide, water vapour, methane and chlorofluorocarbons.

Greenland spar See *cryolite*.

greenockite (CdS) A yellow *ore mineral* of cadmium.

Greensand A *stratigraphic* term. See *Upper Greensand, Lower Greensand*.

greensand A *sandstone* with a green colour because of the presence of *glauconite*.

greenschist A green, *schistose*, *metamorphic rock* coloured green due to the presence of *chlorite*, *epidote* or *actinolite*.

greenschist facies A *metamorphic facies* of low-*grade regional metamorphism* of temperature range 300–500°C in which the *mineral assemblage* of *basic rocks* is *albite* + *epidote* + *chlorite* + *actinolite*, which replace *primary minerals*.

greenstone 1. A general term for a dark-green, altered, low- to medium-*grade*, metamorphosed *basic igneous rock* such as *spilite* or *dolerite*, the green colour reflecting the *greenschist facies mineral assemblage*. 2. A loose term used by archaeologists for a group of *basic igneous* and *metamorphic rocks* such as *amphibolite*, *schist*, *gabbro* and *epidiorite*, sometimes including *serpentinite* and *jade*.

greenstone belt A linear to irregular volcano-sedimentary belt, 10–15 km thick, occurring within *granite-greenstone terrain* and containing *mafic-ultramafic* volcanics with subordinate *greywacke*, *banded iron formation*, *chert* and carbonate. Probably originated in a *Precambrian back-arc basin*.

greenstone-granite terrain See *granite-greenstone terrain*.

greisen An *aggregate* of *quartz* and white *mica* (usually *muscovite* or *lepidolite*) with *accessory cassiterite*, *fluorite*, *rutile*, *topaz* and *tourmaline*, formed by the *alteration* of *granite* by *hydrothermal solutions*.

greisen deposit A *greisen* with associated tin-tungsten mineralization.

greisenization 1. The process of *greisen* formation. 2. Fluorine *metasomatism*.

Grenville orogeny See *Grenvillian orogeny*.

Grenvillian orogeny (Grenville orogeny) An *orogeny* during the Proterozoic at ~1000 Ma which affected the eastern part of the Canadian *shield*, equivalent to the *Daslandian orogeny* in Europe.

grey copper ore See *tetrahedrite*.

grey gneiss *Bimodal trondhjemite*.

grey wethers See *sarsen*.

greywacke (graywacke) Old term for an immature *sandstone* with >15% *clay minerals*.

grèze litée A bedded *scree* of angular rock debris whose *dip* is parallel to the *bedrock* slope, probably formed by *nivation* and *downwash* processes.

Griesbachian The lowest *stage* of the *Triassic*, 245.0–243.4 Ma.

Griffith crack (Griffith flaw) A *microscopic* to submicroscopic crack, flaw or inclusion around which *tensile stresses* concentrate, leading to crack propagation and *failure*.

Griffith failure criterion A relationship between *principal stresses* which gives the condition for *failure*, which assumes that *failure* occurs when the potential energy of a *Griffith crack* remains constant or decreases as it increases in length. $\sigma \geq \sqrt{(2E\lambda/\pi C)}$, where σ = *tensile stress*, E = *Young's modulus*, λ = surface energy of crack, C = crack half-length.

Griffith flaw See *Griffith crack*.

Griffith-Murrell failure criterion See *extended Griffith failure criterion*.

grike (gryke, kluftkarren) A trough separating *clints* in a *limestone pavement*, developing in response to *solutional* widening.

grit A hard, coarse-grained *sandstone*.

groove See *tool track*.

groove cast (groove mark) A millimetric- to decimetric-scale, elongated, sublinear scour or *sole mark*, formed when *clasts* (e.g. *pebbles*, shells etc.) or tools (e.g. icebergs) moving in a current erode a track into soft, *cohesive* sediment.

groove mark See *groove cast*.

grossular ($Ca_3Al_2Si_3O_{12}$) A *garnet*, frequently found in metamorphosed impure calcareous rocks.

ground mica See *flake mica*.

ground-penetrating radar (GPR) A *geophysical exploration* method using radar waves reflected back from subsurface discontinuities to depths of a few tens of metres to produce an output similar to a *seismic reflection seismogram*, but resulting from changes in the dielectric constant of subsurface layers rather than their *acoustic impedance*.

ground roll The large amplitude oscillation of the ground in the vicinity of an explosion, caused by *Rayleigh waves*.

ground truth The confirmation, or otherwise, of an interpretation of *remote* *sensing* data by actual examination of the ground area under consideration.

groundmass (matrix) The fine-grained material of a rock in which larger bodies may be set.

groundwater The water in porous rocks beneath the *water table*.

group A grouping of *formations* used in *lithostratigrapy*. See also *bed, member, formation, supergroup*.

group velocity The velocity of the energy in a signal suffering *dispersion*.

growan A partly *weathered* or decomposed *igneous rock*, particularly *granite*, probably forming by *alteration* during *metamorphism* or *deep weathering*.

growing ripple See *vortex ripple*.

growth fabric A *fabric* characteristic of the manner in which the rock formed, rather than the effects of *deformation*.

growth fault (synsedimentary fault) A *fault* along which movement is contemporaneous with sedimentation.

growth fibre A fibrous crystal growing in response to *strain* by a *crack-seal* mechanism.

growth twinning *Twinning* taking place during crystal growth by the continual addition of clusters of atoms.

grunerite ($Fe_7Si_8O_{22}(OH)_2$) A light brown *amphibole*, commonly found in metamorphosed, iron-rich sediments.

grus An accumulation of poorly sorted, angular rock fragments formed during the *weathering* of *crystalline rocks*, particularly *granite*.

gryke See *grike*.

Guadelupian A *stage* of the *Permian* in the Delaware Basin, USA equivalent to the *Wordian* and *Capitanean*.

Guan Ling A *Triassic* succession in China equivalent to the *Anisian*.

guano A deposit of calcium phosphate formed by the reaction of bird or bat excreta with *limestone*, once heavily worked for fertilizer.

gudmundite (FeSbS) A *sulphide mineral* of the *disulphide group* with an *arsenopyrite* structure.

Gulf 1. The younger *epoch* of the *Cretaceous*, 97.0–65.0 Ma. 2. A *Cretaceous* succession on the Gulf Coast of the USA comprising the *Woodbine*, *Eagle Ford*, *Austin*, *Taylor* and *Navarro*.

gull A fissure or crack, possibly sediment-filled, found on *escarpments* and caused by *cambering*.

gully erosion (gullying) The *erosion* of steep-sided channels and small ravines in poorly consolidated superficial material or *bedrock* by *ephemeral streams*, often prompted by the reduction or removal of vegetation cover.

gullying See *gully erosion*.

gumbo A certain type of soil which yields a sticky mud when wet.

Günz Glaciation The first of the *glacial* periods of the *Quaternary* of the Alps.

Gushan A *Cambrian* succession in China covering parts of the *Menevian* and *Maentwrogian*.

Gutenberg discontinuity A *seismic discontinuity* between the *core* and *mantle* at 2900 km depth.

gutter A linear depression in a sediment surface caused by the *erosion* associated with vortices in the flow.

gutter cast A type of *sole mark* in the form of a linear to sinuous, 'U'-shaped depression <10 cm wide, formed by fluid scouring parallel to the flow direction.

guyot A flat-topped, submarine peak, originating as a *volcano* above sea level and subjected to subaerial and marine planation, subsequently sinking below the surface. Origin related to the passage of an oceanic *plate* over a *hotspot.*

Gymnocodiaceae A group of *calcareous algae* with vegetative structure similar to green *algae*, but with sporangia resembling red *algae*. Range *Permian–Cretaceous.*

Gymnospermopsida/gymnosperms A class of kingdom *Plantae*, the 'naked seeded' plants. Range *Carboniferous–Recent.*

gypcrete A *duricrust* composed of *gypsum.*

gypsite An earthy variety of *gypsum* found coating *gypsum* outcrops in arid regions.

Gypsum Springs A *Triassic/Jurassic* succession in Utah/Idaho, USA, covering the U. *Triassic*, *Hettangian*, *Sinemurian*, *Pliensbachian*, *Toarcian*, *Aalenian* and part of the *Bajocian.*

gypsum (alabaster) ($CaSO_4.2H_2O$) An important *evaporite mineral.*

gyroconic Descriptive of a *gastropod* shell which coils loosely.

gyrolite ($Ca_2Si_3O_7(OH)_2.H_2O$) A hydrated calcium silicate, often found in *amygdales* with *apophyllite.*

gyromagnetic remanent magnetization A *remanent magnetization* produced by rotating a sample in an alternating *magnetic field.*

gyttja (nekron mud) A rapidly accumulating, organic *mud* formed in an *eutrophic lake.*

Gzelian The youngest *epoch* of the *Carboniferous*, 295.1–290.0 Ma.

H

habit 1. The relative development of individual crystal forms and faces. 2. The typical appearance of an organism.

Hackberryfrio An *Oligocene* succession on the Gulf Coast of the USA covering part of the *Rupelian* and the *Chattian.*

hackle mark A type of *plume structure.*

hackly fracture A *fracture* with jagged, sharp-edged surfaces.

hackmanite A variety of *sodalite* which changes from deep red to pale green on exposure to light.

hadal Descriptive of the deepest levels of the ocean, at depths >6000 m.

hade The angle made with the vertical by a *fault plane.*

Hadean An *era* of the *Precambrian,* 4560–3800 Ma.

Hadrynian The upper part of the *Proterozoic* of Canada, ~1000–570 Ma.

haematite (hematite) (Fe_2O_3) A major *ore mineral* of iron, also found as an *accessory mineral* in many rocks.

haematization (hematization) The *replacement* of the hard parts of an organism by *haematite.* cf. *pyritization.*

hagatalite A variety of *zircon* containing significant quantities of the *rare earth elements.*

Hagedoorn method See *plus-minus method.*

hailstone-generated structure A sedimentary *structure* formed by hail impacting on unconsolidated sediment. See also *raindrop-generated structure.*

hairpin An abrupt change in the direction of an *apparent polar wandering path,* usually indicative of a *continent–continent collision.*

half-graben 1. A wedge-shaped zone comprising a dipping *fault* on one side of a block of tilted *strata.* 2. In common usage, a valley formed by movement on a single *normal fault.*

half-life The time taken for the mass of a radioactive element to decrease by a half. See also *decay constant.*

half-spreading rate The rate of movement of a *plate* on one side of an *ocean ridge,* i.e. half the *spreading rate.*

half-width The horizontal distance over which a spatial function decays to half its maximum value. Particularly used for *magnetic* and *gravity anomalies,* as it can be used to estimate the *limiting depth* of the causative body.

halite (rocksalt) (NaCl) A common *evaporite mineral.*

hälleflinta A metamorphosed *volcanic* or *pyroclastic rock* with a *flint*-like appearance.

halloysite ($Al_4Si_4O_{10}(OH)_8.8H_2O$) A *clay mineral* composed of an irregular series of *kaolinite* layers and interlayer water.

halmyrolysis An early stage of *diagenesis* or decomposition of sediment on the seafloor.

haloclasty The disintegration of rock by the action of salts, which cause disruption by *crystallization*, *hydration* or thermal expansion.

halocline The interface between dense seawater and lighter freshwater in an *estuary* or between waters of different *salinity* within a water column.

halokinesis The mobilization and flow of subsurface salt, which can give rise to *salt pillows* and *salt diapirs*.

halotrichite (iron alum) ($Fe^{2+}Al_2(SO_4)_4.22H_2O$) A rare hydrated iron-aluminium sulphate forming fibrous crystals.

Hamagian A *stage* of the *Ordovician*, 462.3–457.5 Ma.

hambergite ($Be_2(OH,F)BO_3$) A beryllium borate found in *pegmatites*.

hammada A flat to gently-dipping, bare rock surface in a desert, sometimes with a thin cover of *gravel* or a *boulder lag*, forming from the *weathering* of surface rock.

Hammer chart A graticule used in conjunction with a topographic map in the calculation of *topographic corrections* in *gravity reduction*.

hamra A red, sandy soil with a high *clay* content found in desert regions.

hanging valley A tributary valley whose floor is at a higher level than the main valley, caused by the latter's deepening by *glacial erosion*.

hangingwall The wall and body of rock above an inclined or horizontal *fault* or *orebody*.

hangingwall ramp A *ramp* in which *bedding* or *datum* surfaces are seen in the *hangingwall*.

harbour bar A *coastal bar* located across a river mouth or harbour.

hard coal A high *rank coal* with a *calorific value* >23.86 MJ kg^{-1}, including most *bituminous coal* and *anthracite*.

hard pan A layer of iron oxyhydroxides above the *water table* formed by the *reprecipitation* of *minerals* leached from the overlying *vadose zone*.

hardebank Resistant *kimberlite* that does not break up when exposed. cf. *yellow ground*.

hardground An indurated surface resulting from the synsedimentary lithification of seafloor sediment and representing a decrease in sedimentation rate.

hardness 1. A *mineral* property determined by reference to *Mohs' scale*. 2. A property of water which prevents the formation of a lather with soap, caused by the presence of dissolved alkaline earth salts and usually expressed in ppm. Temporary hardness is caused by calcium bicarbonate and is removed by boiling; permanent hardness is due to other calcium and magnesium salts such as sulphates and is removed by *ion-exchange* processes or detergents.

Harker diagram A *variation diagram* which shows the chemical relationships of a suite of rocks by plotting constituents against *silica* content.

harmotome ($Ba(Al_2Si_6O_{16}).6H_2O$) A *zeolite* found in mineralized *veins*.

hartite ($C_{20}H_{34}$) A naturally occurring *hydrocarbon.*

harzburgite An *ultramafic/ultrabasic rock* composed of >40% *olivine*, 5–60% *orthopyroxene* and <5% *clinopyroxene*. Forms the residue when *basaltic magma* is derived from *mantle lherzolite*, so is found in *ultramafic xenoliths* in *alkali basalts* and in the *mantle* part of *ophiolites.*

Hastarian The lowest *stage* of the *Carboniferous*, 362.5–353.8 Ma.

Hastings Beds A *Cretaceous* succession in England covering part of the *Berriasian* and the *Valanginian.*

hastingsite ($NaCa_2Fe_4Al,Fe)Al_2Si_6O_{22}(OH)_2$) A calcic *amphibole* found in *igneous* and *metamorphic rocks.*

Haumurian A *Cretaceous* succession in New Zealand covering the upper part of the *Campanian* and the *Maastrichtian.*

Hauptdolomit (Stinkschiefer) A *Permian* succession in NW Europe covering the lower part of the *Wordian.*

Hauterivian A *stage* of the *Cretaceous* 135.0–131.8 Ma.

haüyne ($(NaCa)_{4-8}(AlSiO_4)_6(SO_4)_{1-2}$) A *feldspathoid* found in *phonolite* and related rocks.

Hawaiian eruption A *volcanic eruption* characterized by *fire fountaining* and the quiet extrusion of *mafic lava*, typically in an *ocean island* setting.

hawaiite An *alkaline volcanic rock* containing *andesine plagioclase* (An_{30-50}), *pyroxene*, iron-titanium oxide ± minor *quartz* or *nepheline.*

hawleyite (CdS) A *sulphide mineral* of the *sphalerite group.*

haycockite ($Cu_4Fe_5S_8$) A *sulphide mineral* of the *sphalerite group.*

HDR See *hot dry rock.*

head A superficial deposit formed under *periglacial* conditions.

head difference The difference in height of the water surface in two bodies of water.

head loss A lowering of the water level, e.g. in an *aquifer.*

head of water (hydraulic head) The height of the water surface above a particular point, giving a measure of the water pressure at that point.

head wave (critically refracted wave, seismic head wave) The refracted *seismic wave* incident at the *critical angle* so that it propagates along the boundary between two media at the higher *seismic velocity.* Its measurement is the basis of the *seismic refraction* method.

headwall erosion *Erosion* of the headwall of a ravine after *gullying* has been initiated, which can occur at very rapid rates.

heat flow (heat flux) The heat flowing from the Earth's interior through a unit area at the surface, generally in units of mW m^{-2}.

heat flow unit (HFU) The *cgs* unit of 1 cal cm^{-2} s^{-1} = 41.8 mW m^{-2}.

heat flux See *heat flow.*

heave 1. A group of *mass movement* processes including *frost heave*, *freeze-thaw* and *cambering.* 2. The horizontal separation resulting from movement on a *fault.*

heavy metal pollution The artificial introduction of heavy metals (As, Cd, Cu, Pb, Ni, Ag) into the environment in quantities that will adversely affect it.

heavy mineral A *mineral* with a specific gravity greater than 2.85, i.e. which sinks in bromoform.

heavy oil Oil with an *API gravity* $<22°$.

heavy oil sand A *tar sand* which can be exploited by techniques such as cyclic high pressure steam injection.

heavy spar A colloquial term for *barytes*.

heazlewoodite (Ni_3S_2) A *sulphide mineral* of the *metal excess group*.

hectorite ($Na_{0.3}(Al,Mg,Li)_3Si_4O_{10}(OH)_2.4H_2O$) A lithium-bearing *clay mineral* of *montmorillonite* type.

hedenbergite ($CaFeSi_2O_6$) A *clinopyroxene* found chiefly in *contact metamorphosed limestones*, iron-rich *metamorphic rocks* and *igneous rocks*.

held water The water retained above the *water table* by capillary action.

helictic structure (helictitic structure) Helical trails of *mineral* inclusions in a *porphyroblast* formed during *metamorphic recrystallization*, which may represent the remnants of an existing *fold fabric* enclosed by the *porphyroblast* or the rotation of the *porphyroblast* in the *matrix* during *metamorphism*.

helictite A small calcium carbonate *speleothem* growing in curved or spiral shapes in *limestone caves*.

helictitic structure See *helictic structure*.

Helikian The middle part of the *Proterozoic* of Canada, ~1800–1000 Ma.

heliodor A clear yellow variety of *beryl* coloured by iron, used as a *gem*.

helioplacoids The most primitive of the *Echinodermata*, classified in the stem group of both the *Pelmatozoa* and *Eleutherozoa*. Range L. *Cambrian*.

heliotrope A green and red variety of *chalcedony*, sometimes of *gem* quality.

helluhraun An Icelandic *lava flow* of *pahoehoe*.

hematite See *haematite*.

hematization See *haematization*.

hemi- Half.

Hemichordata A phylum which includes the acorn worms, related to the *Chordata*, and originating in the M. *Cambrian*.

hemicrystalline Descriptive of a rock composed of both crystals and *glass* in approximately equal proportions.

hemihydrate See *bassanite*.

hemimorphism A trait of certain crystals in which no symmetry element is present to cause repetition of the upper hemisphere faces in the lower hemisphere.

hemimorphite (electric calamine) ($Zn_4Si_2O_7(OH)_2.H_2O$) A minor *ore mineral* of zinc, sometimes found in the oxidized zone of other zinc *ores*.

hemipelagic deposit See *hemipelagite*.

hemipelagite (hemipelagic deposit) A generally fine-grained, structureless sediment of decimetric thickness formed by the settling of grains from suspension and from low density, low *viscosity turbidity currents*, *nepheloid layers* and other ocean currents.

Herangi A *Jurassic* succession in New Zealand comprising the *Aratauran* and *Ururoan*.

Hercynian orogeny See *Variscan orogeny*.

hercynite ($FeAl_2O_4$) A *spinel*.

Heretaungan An *Eocene* succession in New Zealand covering parts of the *Ypresian* and *Lutetian*.

Herkimer diamond A *quartz*-filled cavity occurring in *dolomite* from Herkimer, New York State, USA.

Hermann-Maugin symbol A shorthand notation for describing the symmetry of a crystal.

hermatypic Descriptive of a *reef*-building organism.

hermatypic coral A *coral* with symbiotic *algae*, living in water <50 m deep. cf. *ahermatypic coral*.

herringbone cross-bedding A type of *cross-lamination* in which the *foresets* alternate in direction as a result of reversing currents.

herringbone structure See *plume structure*.

Hervy A succession in E Australia of late *Devonian* age.

hessite (Ag_2Te) A rare *ore mineral* of *silver*.

hessonite (cinnamon stone) A yellow-brown variety of *grossular* containing iron.

Heterian A *Jurassic* succession in New Zealand covering the lower part of the *Kimmeridgian*.

hetero- Different from.

heterochthon See *allochthon*.

Heterocorallia A subclass of class *Anthozoa*, phylum *Cnidaria*; a very small group of *corals* which may be related to the *Rugosa* or a separate group. Range late *Devonian*–late *Carboniferous*.

heterogeneous nucleation An *exsolution* phenomenon in which exsolved phases appear at internal *crystal defects*, e.g. grain boundaries, *dislocations*. cf. *homogeneous nucleation*.

heterogeneous simple shear *Simple shear* in which the relationship between *displacement* and distance changes abruptly. cf. *homogeneous simple shear*.

heterogeneous strain *Strain* in which the relationship between *displacement* and distance changes abruptly. cf. *homogeneous strain*.

heterolithic bedding A fine interbedding of *sand* and *mud* formed in an area of variable current flow. The three main types are *flaser bedding*, *wavy bedding* and *lenticular bedding*.

heterolithic unconformity See *nonconformity*.

heteromorphism The *crystallization* of almost identical *magmas* into different *mineral assemblages* by their possession of different cooling profiles.

Heterostrachi An order of subclass *Diplorhina*, class *Agnatha*, superclass *Pisces*; the oldest known jawless *fish*, with heavy armour. Range U. *Cambrian*–*Devonian*.

Hettangian The lowest *stage* of the *Jurassic*, 208.0–203.5 Ma.

heulandite ($CaAl_2Si_7O_{18}.6H_2O$) A *zeolite* found in cavities in *basic igneous rocks*.

Hexacorallia See *Scleractinia*.

Hexactinellida A class of phylum *Porifera* (*sponges*) with *opaline silica* spicules arranged in rays at 90°. Range *Cambrian–Recent*.

hexagonal pyrrhotite (Fe_9S_{10}, $Fe_{11}S_{12}$) A *sulphide mineral* of the *nickel arsenide group*.

hexagonal system A *crystal system* whose members have three lateral axes of equal length intersecting at an angle of 60° to each other and reaching the edges of the six vertical faces, and a vertical axis of different length perpendicular to the other three.

Hexapoda/hexapods A superclass of phylum *Uniramia*; the *insects*. Range *Devonian–Recent*.

hexastannite ($Cu_2Fe_2SnS_6$) A *sulphide mineral* of the *wurtzite group*.

HFU See ***heat flow unit***.

hiatus A *chronostratigraphic* gap caused by non-deposition. cf. *disconformity*.

hidden layer See *blind layer*.

hiddenite (lithia emerald) A green *gem* variety of *spodumene*.

high energy window A period during the mid-*Holocene* when wave energy was higher than at present.

high-grade gneiss terrain One of the two main types of *Archaean crust*. cf. *granite-greenstone terrain*.

high quartz See *beta quartz*.

high sulphur crude oil *Crude oil* with >1.7% by weight of sulphur (not in the form H_2S).

Highgate resin (copalite, copaline) A colloquial name for *amber* from Highgate, N London.

hillebrandite ($Ca_2SiO_4.H_2O$) Hydrated dicalcium silicate found as fibrous masses in *thermally metamorphosed* impure *limestones*.

Hilt's law 'The *rank* of *coal* increases with depth.'

Himalayan mountain belt A broad mountain chain formed by *continental collision*, either a single collision or a series involving several *suspect terranes*, possibly followed by *indentation tectonics*. cf. *Andean mountain belt*.

hinge fault A *fault* whose *displacement* decreases to zero at one end, allowing it to act as a hinge.

hinge line The locus of points of maximum curvature of a surface.

hinterland 1. The region from which the surface rocks have been translated in an *orogenic belt*. cf. *foreland*. 2. An area of land drained by a river system. 3. The area supplying sediment to a depositional system.

Hirnantian The highest *stage* of the *Ordovician*, 439.5–439.0 Ma.

Hjulström curve An empirical curve showing the critical condition for *erosion* of river channel bed deposits in terms of mean flow velocity.

hogback A long ridge with a sharp crest and steep slopes on both flanks, produced by the differential *erosion* of steeply inclined *strata*.

Holkerian A *stage* of the *Carboniferous*, 342.8–339.4 Ma.

hollandite ($Ba_2Mn_8O_{16}$) An *ore mineral*

of manganese found at or near the contacts between lithium-rich *pegmatites* and *basic country rocks*.

holmquistite ($Li_2(Mg,Fe)_3(Al,Fe^{2+})_2Si_8O_{22}(OH)_2$) An *amphibole* found in lithium-rich *pegmatites*.

holo- Complete.

holoblast A *mineral* formed entirely during *metamorphism*.

Holocene (Recent) The youngest *epoch* of the *Quaternary*, 0.01 Ma–present.

Holocephali A subclass of class *Chondrichthyes*, superclass *Pisces*; the 'rabbit fishes' or 'rat fishes'. Range L. *Carboniferous–Recent*.

holocrystalline Descriptive of a rock which is completely crystalline.

holohedral Descriptive of a completely developed crystal.

holohyaline Descriptive of a rock which is composed entirely of *glass*.

holokarst Any *limestone* landscape with a fully developed range of *karst* features.

hololeucocratic Descriptive of an *igneous rock* with a *colour index* <5.

holomictic The status of a lake which undergoes complete water circulation, destroying any seasonal stratification. cf. *meromictic*.

Holostei An order of subclass *Chondrostei*, class *Osteichthyes*, superclass *Pisces*; *fish* characterized by heavy rhombic scales, a swim bladder, reduced, unjointed fin rays, a short, mobile jaw and a symmetrical tail. Range U. *Permian–Recent*.

Holothuroidea A class of subphylum *Eleutherozoa*, phylum *Echinodermata*; worm-like invertebrates with a non-rigid *calcitic* skeleton, poorly preserved as *fossils*. Range *Silurian–Recent*.

holotype The single, type specimen of a species.

homeomorph 1. A crystal similar to another in *crystal form* and *habit*, but with a different composition. 2. An organism similar to another but deriving from different ancestors.

homeomorphism The relationship between crystal *homeomorphs*.

homeomorphy The general similarity between species which are different in detail.

hominids Man-like animals coming to dominance in the *Quaternary*.

homoclinal With uniform *dip*.

homoclinal ridge An *erosional* feature with a form intermediate between a *hogback* and *cuesta*.

homogeneous nucleation An *exsolution* mechanism of *nucleation* of the second phase in which the nuclei are uniformly distributed, which takes place within grains in the absence of *crystal defects*. cf. *heterogeneous nucleation*.

homogeneous simple shear *Simple shear* in which the relationship between *displacement* and distance is linear or continuously changing. cf. *heterogeneous simple shear*.

homogeneous strain *Strain* in which the relationship between *displacement* and distance is linear or continuously changing. cf. *heterogeneous strain*.

homology The similarity between

parts of different organisms as a result of evolutionary differentiation of the parts from the same ancestor. cf. *homoplasy*.

homoplasy The similarity between parts of different organisms as a result of *convergent evolution* rather than a common ancestor. cf. *homology*.

homopycnal flow The immediate, three-dimensional mixing of river and basinal water of equal *density*, accompanied by considerable sediment deposition. cf. *hyperpycnal flow*, *hypopycnal flow*.

homotaxis Indicative that *strata* in different areas with the same *fossil* assemblage are not necessarily of the same age due to migration taking a finite time.

honestone See *whetstone*.

honeycombs See *alveoles*.

hoodoo A pillar of rock or weakly consolidated sediment formed by differential *erosion*, found in arid to semi-arid regions.

Hooke's law 'The extension of an elastic spring is proportional to the *force* applied to it.'

hopper crystal A crystal with recessed faces caused by more rapid growth at its edges.

horizon A time-plane with distinctive characteristics.

horizonation The formation of horizons in a *soil profile* as the result of *pedogenetic* processes.

horn A *glacial erosion* feature formed when a mountain is eroded from all sides to leave a pyramidal peak.

horn lead See *phosgenite*.

horn silver See *cerargyrite*.

hornblende ($(Ca,Na)_{2-3}(Mg,Fe,Al)_5 Si_6(Si,Al)_2O_{22}(OH)_2$) An important, widespread, monoclinic calcic *amphibole*.

hornblendite A *plutonic igneous rock* composed almost completely of *hornblende*.

hornfels A fine- to medium-grained *pelitic* rock, possibly containing *porphyroblasts* of *biotite*, *andalusite* and *cordierite*, commonly with relict sedimentary/*tectonic* structures, forming as a result of *thermal metamorphism* and solid-state *recrystallization* in a *contact aureole*.

hornfels facies An imprecisely defined *metamorphic facies* produced by *contact metamorphism*.

hornito A small chimney or cone of *lava spatter* on the surface of *pahoehoe*.

hornstone A very fine-grained *pyroclastic rock*.

horse A package of rocks surrounded and isolated by *faults* in an *imbricate structure*.

horse-tailing A feature made up of a number of mineralized *fractures* with the general appearance of a horse's tail, which is sometimes found at, or near, the end of a *vein*.

horsetail fault See *listric fan*.

horst *A fault block* elevated by a series of parallel to subparallel, outward-dipping, *conjugate step faults* with the same sense of *displacement*. cf. *graben*.

Horton equation An empirical equation which describes the *infiltration* process.

Hortonian overland flow *Infiltration*-excess *overland flow* which occurs when rainfall exceeds the *infiltration capacity* of the soil surface.

hortonolite An *olivine* with the composition Fo_{50-30}.

host grain-controlled replacement *Replacement* that is selective or specific to sites of high surface free energy in the host *mineral* or grain, typical of solutions of low chemical reactivity. cf. *precipitate controlled replacement.*

host rock See *country rock.*

hot dry rock (HDR) A method of extracting *geothermal* energy from poorly-permeable rocks, such as high heat-production *granite*, by drilling twin boreholes, fracturing the rock between them by the *hydrofrac technique* and using the boreholes for the ingress of cold water and egress of heated water.

hot spring A surface seepage from a natural *geothermal system.*

hot-working *Dislocation climb* taking place at an elevated temperature.

hotspot A heat source of limited areal extent arising from a *mantle plume* persisting for tens of millions of years and remaining nearly stationary with respect to the *mantle.*

howieite ($Na(Fe,Mn)_{10}(Fe,Al)_2Si_{22}O_{31}(OH)_{13}$) A black, hydrated sodium-manganese-iron silicate found in metamorphosed *shale*, siliceous *ironstone* and impure *limestone.*

HREE Heavy *rare earth element.*

Hsu diagram A method of illustrating the shapes of *strain ellipsoids.* cf. *Flinn diagram.*

Huashiban A *Carboniferous* succession in China covering the lower part of the *Bashkirian.*

hübnerite See *huebnerite.*

Hudsonian orogeny An *orogeny* affecting the Canadian *shield* in the *Precambrian* at ~1750–1800 Ma.

huebnerite (hübnerite) ($MnWO_4$) An end member of the *wolframite mineral* series.

hum A residual *limestone* hill, often rising from a *polje.*

humic coal See *banded coal.*

humification A biological, *pedogenetic* process in which organic matter is converted into *humus.*

humite ($Mg_7(SiO_4)_3(F,OH)_2$) A *nesosilicate* similar to *olivine* found in thermally metamorphosed *limestones.*

hummocky cross-stratification A *bedding structure* consisting of gently curved, low-angle *cross-stratification*, possibly forming under storm conditions or by a *flow regime* with a unidirectional component.

humus The organic part of soil.

Hunsrück Slate An *exceptional fossil deposit* of *Devonian* age in the Rhineland, Germany where *arthropods, molluscs, echinoderms, fish* and *plants* are preserved in *pyrite.*

huntite ($CaMg_3(CO_3)_4$) An *evaporite mineral.*

Huobachong A *Triassic* succession in China equivalent to the *Norian.*

Huronian An *era* of the *Precambrian*, 2450–2200 Ma.

hush A stream valley artificially deepened for *mineral* exploration.

Huttenlocher intergrowth A sub-optical intergrowth between *albite* and *anorthite*.

Huygen's principle 'A point on an advancing *wavefront* can be regarded as the centre of a fresh disturbance.' Used to predict the path of a *wavefront* of e.g. a *seismic wave*.

hyacinth 1. **(jacinth)** A red-brown variety of *zircon* valued a *gem*. 2. A brown variety of *grossular*.

hyaline Glassy.

hyalite A clear, colourless variety of *opal* with a globular or *botryoidal* surface.

hyaloclastite A poorly-sorted, non-bedded, generally *basaltic*, *volcanic rock* composed of quench-fragmented *glass* with very blocky and angular non-*vesicular* fragments. Formed when shards of glassy *lava* crusts come into contact with water or ice.

hyalocrystalline (hypocrystalline, hypohyaline) Descriptive of material composed of both *glass* and crystals.

hyalophane ($(K,Ba)(Al,Si)_2Si_2O_8$) A rare barium *feldspar*.

hyalopilitic Descriptive of a form of *pilotaxitic texture* in which the crystals are embedded in a glassy *matrix*.

hyalosiderite An *olivine* with the composition Fo_{70-50}.

hybrid fracture See *mixed mode fracture*.

hybrid joint A type of *joint* intermediate between a *shear joint* and an *extension joint*.

hybrid rock A rock formed by the *assimilation* of *wall rock* or *xenoliths* into a *magma*.

hybridization The process whereby valence shell *atomic orbitals* are re-arranged to point in specific spatial directions to allow the construction of complex molecules involving poly-valent atoms.

hydrargillite See *gibbsite*.

hydration The *alteration* of a material by the addition of water, e.g. in *weathering*. cf. *hydrolysis*.

hydration layer dating A dating method for *obsidian* artefacts based on the phenomenon of *obsidian*'s absorbing water at a rate dependent on its chemical composition and the ambient temperature.

hydraulic conductivity A measure of the ease with which a fluid flows and the ease with which a porous rock allows its passage. See *Darcy equation*.

hydraulic equivalence A term which expresses the size of a grain of given *density* in the equivalent size of a *quartz* grain with an identical *settling velocity*, enabling the behaviour of diverse grains to be compared.

hydraulic fracture (hydrofracture) *Fracture* caused by *pore fluid pressure*, either natural or induced.

hydraulic geometry The study of channel form in relation to the external controls of *discharge* and sediment.

hydraulic gradient The loss of pressure with distance along the direction of flow when water flows along a horizontal, constant diameter, horizontal pipe, i.e. the ratio of the *head loss* between two points to their separation.

hydraulic head See *head of water*.

hydraulic jump An abrupt increase in flow depth arising where a fast flow with a *Froude number* > 1 changes rapidly to a slow flow with a *Froude number* < 1, as happens at abrupt changes in bed relief in alluvial channels.

hydraulic mining A mining technique in which material is blasted out of the ground by high-pressure water jets.

hydraulic radius The ratio of a stream's wetted perimeter length to the cross-sectional flow in a channel, a measure of the efficiency of a section in conveying flow.

hydraulically rough A term indicating that sediment particles or other irregularities on a bed have a diameter about five times the thickness of the *viscous sublayer* of a flow, encouraging *turbulence* and vertical mixing. cf. *hydraulically smooth*.

hydraulically smooth A term indicating that sediment particles or other irregularities on a bed have a diameter less than the thickness of the *viscous sublayer* of a flow, and *turbulence* is probably not generated. cf. *hydraulically rough*.

hydroboracite ($CaMgB_6O_8(OH)_6 \cdot 3H_2O$) A borate *mineral* forming prismatic or *acicular* crystals.

hydrocarbon A solid, liquid or gas made up of compounds of carbon and hydrogen in varying proportions.

hydrocarbon saturation The condition of a *reservoir* in which the *pore fluid* has been reduced to the minimum possible level by *hydrocarbons*. This can be determined using *resistivity logging*.

hydrocarbon trap. See *gas trap, oil trap*.

hydrocerrussite ($Pb_3(CO_3)_2(OH)_2$) A colourless, hydrated lead carbonate occurring as an encrustation on lead *minerals*.

hydrofrac technique A method whereby artificial *hydraulic fracture* is induced in rocks penetrated by a borehole by sealing off part of it and pumping in fluid until fracturing occurs.

hydrofracture See *hydraulic fracture*.

hydrogeology The study of subsurface water (*groundwater*).

hydrogeomorphology (fluvial geomorphology) The study of landforms produced by *fluvial* processes.

hydrograph The graph of the *discharge* of a river or stream plotted against time for a particular point.

hydrogrossular ($Ca_3Al_2O_8(SiO_4)_{1-x}(OH)_{4x}$) A hydrous *garnet*.

hydrohaematite (hydrohematite, turgite) A natural mixture of *haematite* and *goethite*.

hydrohematite See *hydrohaematite*.

hydroisostasy An *isostatic* reaction to the removal or loading of a mass of water.

hydrolith A rock formed by *precipitation* from water.

hydrological cycle All processes encompassing the evaporation of water from the sea, its fall as precipitation on land and thence its flow through *aquifers* or watercourses back to the sea.

hydrology The study of the distribution, conservation and use of water in the Earth and its atmosphere.

hydrolysis The *alteration* of a *mineral* by the addition of hydroxyl (OH^+) ions, e.g. during *weathering*. cf. *hydration*.

hydrolyzate A sediment concentrating in fine-grained *alteration* products which contain elements (Al, K, Si, Na) which are easily hydrolyzed. Found in *clays*, *shales* and *bauxites*.

hydromagnesite ($Mg_5(CO_3)_4(OH)_2 \cdot 4H_2O$) Hydrated magnesium hydroxide and carbonate found in *amorphous* masses and rarely as crystals in *serpentinite*.

hydromica See *illite*.

hydromuscovite See *illite*.

hydrophane A variety of *opal*; *opaque* with a *pearly lustre* when dry and *transparent* when in water.

hydrophone An instrument used to detect seismic energy in water by converting pressure variations into a varying voltage with a piezoelectric material. cf. *geophone*.

hydrostatic The condition in which the *pressures* or *stresses* on a body are the same in all three dimensions.

hydrostatic gradient The systematic increase in *pore fluid* pressure with burial depth cause by the weight of the overlying water column up to sea level or the *water table*.

hydrostatic pressure See *pressure*.

hydrostatic stress See *pressure*.

hydrothermal alteration The *alteration*, ranging from minor colour changes to complete *recrystallization*, produced by *hydrothermal solutions*, often found alongside *veins* or *orebodies* from zones a few centimetres thick to several times the thickness of the *orebody*.

hydrothermal deposit A *mineral* deposit formed by *precipitation* from a *hydrothermal solution*, not necessarily of economic importance.

hydrothermal solution A hot, aqueous solution of high *salinity* responsible for many types of *mineral* and *ore* deposit, e.g. *veins*, *stockworks*, *massive sulphide deposits*. Originates from surface water, *groundwater*, sea water, *meteoric water*, formation water, *metamorphic* water and *magmatic* water.

hydrothermal vent The location of the exhalation of *hydrothermal solutions*, such as are found along *ocean ridge* systems, including *black smokers* and *white smokers*. Probably the source of many *massive sulphide deposits*.

hydrotroilite *Colloidal* hydrous ferrous sulphide formed by the reaction of *sulphur*, produced by the bacterial reduction of sulphates, with iron during the *diagenesis* of organic matter, which slowly converts to *pyrite*.

hydrovolcanic eruption See *phreatic eruption*.

hydroxyapatite ($Ca_5(PO_4)_3(OH)$) A phosphate *mineral* akin to *apatite*.

hydrozincite (zinc bloom) ($Zn_3(CO_3)_2(OH)_6$) A *secondary mineral* produced by the *alteration* of zinc *ore*.

hygromatophile element See *incompatible element*.

hyolithelminthid A small, conical or tubular *fossil* composed of phosphate, of uncertain taxonomic status, found in the *Tommotian.*

hyp- See *hypo-.*

hypabyssal Descriptive of an *igneous rock* crystallizing nearer the surface than a *plutonic* rock and further from the surface than a *volcanic rock.*

hyper- Exceeding.

hypermelanic Descriptive of a rock with 90–100% dark *minerals.*

hyperpycnal flow The water entering a *basin* on the floor of a lake or ocean which, because of its relatively high *density*, flows as a bottom-hugging current and causes sediment to travel a large distance from its point of entry. cf. *homopycnal flow, hypopycnal flow.*

hypersolvus granite A *granite* which crystallized above the *solvus* temperature and thus contains only one type of *alkali feldspar.* cf. *subsolvus granite.*

hypersthene ($(Mg,Fe)SiO_3$) A common *orthopyroxene.*

hypersthenite A coarse-grained *ultrabasic* rock mainly composed of *hypersthene.*

hypichnia *Trace fossils* found on the base of the preserving *bed.*

hypidioblastic texture See *crystalloblastic texture.*

hypidiomorphic See *subhedral.*

hypidiotopic fabric A *fabric* characterized by the presence of some *minerals* which exhibit their *crystal form.*

hypo- (hyp-) Below.

hypocentre See *earthquake focus.*

hypocrystalline See *hyalocrystalline.*

hypogene Caused by an ascending *hydrothermal solution.* cf. *epigenetic.*

hypohyaline See *hyalocrystalline.*

hypolimnion The lower, cold layer of a stratified lake undisturbed by diurnal or seasonal mixing. Tends to *anoxia*, allowing the preservation of organic material. cf. *epilimnion.*

hypopycnal flow The water entering a *basin* on the floor of a lake or ocean which, because of its relatively low *density*, takes the form of a buoyant surface plume or jet. cf. *homopycnal flow, hyperpycnal flow.*

hypothermal deposit An *epigenetic deposit* formed at temperatures in the range 300–600°C at depths of 3–15 km. Usually found as *fracture*-fill and *replacement* bodies near deep-seated *acid plutons* in deeply eroded *Precambrian* or *Palaeozoic* terrain. Important *ores* of *gold, tin* and tungsten are of this type.

hypsographic curve A graph of the distribution of elevation and depth with reference to *sea level.* cf. *hypsometric curve.*

hypsometric curve A graph of the percentage elevation and depth distribution on the continents and oceans. cf. *hypsographic curve.*

hypsometry The measurement of the land surface or seafloor with respect to a given *datum*, normally mean sea level.

hysteresis The difference between the paths followed with time during loading and unloading, e.g. of elastic loading, magnetization.

I

I In *earthquake seismology*, a *P wave* that has travelled through the inner *core*.

Iapetus The ocean assumed to have lain between North America and Europe/Africa ~500 Ma ago, whose closure was responsible for the *Caledonian orogeny*.

Ibexian (Canadian) A *Cambrian/Ordovician* succession in the E USA covering part of the *Dolgellian*, the *Tremadoc* and the early *Arenig*.

ice age 1. A long period of *glaciation*, e.g. the *Permo-Triassic* ice age. 2. The *geochronological* equivalent of a *chronostratigraphic stage*, e.g. the *Cenozoic* ice ages.

ice cap A *glacier* composed of a small *ice sheet* of $<50\,000$ km^2 which buries the landscape.

ice mound A *permafrost*-related landform, such as a *pingo* or *palsa*.

ice rafting The transport of material by floating ice.

ice sheet A *glacier* of $>50\,000$ km^2, composed of a flattened dome which buries the landscape.

Iceland spar A variety of clear, colourless *calcite*.

Ichang See *Yichang*.

ichnite A *trace fossil* comprising a footprint.

ichnofacies A model for a palaeoenvironment based on *trace fossils*.

ichnofauna The organisms responsible for *trace fossils*.

ichnofossil See *trace fossil*.

ichnology The study of *trace fossils* or their recent counterparts.

Ichthyopterygia A class of the *Reptilia* whose sole order is the *Ichthyosauria*. Range M. *Triassic*–U. *Cretaceous*.

Ichthyosauria/icthyosaurs An order of subclass *Ichthyopterygia*, class *Reptilia*; long-snouted marine *reptiles*. Range M. *Triassic*–U. *Cretaceous*.

Ichthyostegalia An order of subclass *Labyrinthodontia*, class *Amphibia*; one of the earliest body *fossils* of the *amphibians*, very similar to their *Crossopterygii fish* ancestors. Range L. *Carboniferous*–U. *Devonian*.

ICPES See ***i****nductively-****c****oupled* ***p****lasma* ***e****mission* ***s****pectrometry*.

idaite ($\sim Cu_3FeS_4$) A *sulphide mineral* with a *covellite* structure of the *layer sulphides group*.

Idamean A *Cambrian* succession in Australia covering parts of the *Maentwrogian* and *Dolgellian*.

iddingsite An *alteration* product of *olivine* comprising *goethite*, *quartz*, *montmorillonite* group *clay minerals* and *chlorite*.

idioblastic texture A *texture* of a *metamorphic rock* in which grains show fully-developed crystal forms.

idiomorphic See *euhedral*.

idiomorphic fabric See *idiotopic fabric*.

idiotopic fabric (idiomorphic fabric) A *fabric* in a *sedimentary rock* in which most constituent crystals are *euhedral*.

idocrase See *vesuvianite*.

IGC **I**nternational **G**eological **C**ongress.

IGCP **I**nternational **G**eological **C**orrelation **P**rogram.

IGF See ***I**nternational **G**ravity **F**ormula*.

igneous body A volume of *igneous rock* with discrete boundaries with the surrounding *country rock* into which it was emplaced.

igneous breccia A *rudite* formed of igneous material.

igneous foliation A *foliation* in an *igneous rock* defined by compositional variation or changes in crystal size or shape, possibly produced by *differentiation* or flow in a *magma*.

igneous rock A rock which has solidified from molten or partially molten material.

ignimbrite A poorly-sorted, *pyroclastic rock*, comprising mainly *pumice* and *ash*, possibly with broken *phenocrysts* and dismembered vent wall material, of large volume (1 km^3–2000 km^3).

ignimbrite plain See *ash-flow field*.

ignimbrite plateau See *ash-flow field*.

IGRF See ***I**nternational **G**eomagnetic **R**eference **F**ield*.

IGSN See ***I**nternational **G**ravity **S**tandardization **N**et*.

ijolite A *plutonic* rock with >90% *nepheline* and *mafic minerals*, usually *pyroxene*, and also *amphibole*, *sphene*, *apatite* and *melanite*. Normally has a normal igneous *texture*, particularly *subophitic* and *comb-structure*. Forms concentric intrusions and *dykes* in continental areas.

ikaite ($CaCO_3.6H_2O$) A chalk-like *carbonate mineral* forming submarine pillars.

Ikskiy A *Permian* succession on the eastern Russian Platform equivalent to the *Artinskian*.

Ilibeyskiy A *Permian* succession in the Timan area of the former USSR equivalent to the L. *Sakmarian*.

Illinoian glaciation The third of the *glaciations* of the *Quaternary* in North America.

illite (hydromica, hydromuscovite) ($(K_{1.5}Al_2(Al_{1.5}Si_{2.5}O_{10})(OH)_2)$) A common *clay mineral*.

illitization The *alteration* of a *mineral* to *authigenic illite*.

illuviation The accumulation of material in the lower soil zone by the leaching and *eluviation* of fine-grained material and water-soluble *minerals* from the upper soil zone and their downward transport.

ilmenite (titaniferous iron ore) ($FeTiO_3$) An iron-titanium *oxide ore mineral*, also occurring as an *accessory mineral* in many rocks.

ilvaite ($CaFe_3^{2+}Fe^{3+}O(Si_2O_7)(OH)$) A *sorosilicate* chiefly found as a product of contact *metasomatism*.

imaginary component (out-of-phase component, quadrature) The part of the secondary electromagnetic field that is 90° out of phase with the primary field in *electromagnetic induction methods*. It is large in the presence of a good conductor and at low EM frequencies.

imbibition The absorption of a fluid by a granular rock by capillary action.

imbricate Overlapping like tiles.

imbricate fan An *imbricate structure* in which *faults* branch from a *floor thrust* and terminate in *fold* or *strain* zones in the overlying *strata*.

imbricate fault One of a number of closely-spaced *faults* in an *imbricate structure*.

imbricate slice The rock between *imbricate faults* in an *imbricate structure*.

imbricate stack An *imbricate structure* in a *thrust belt*.

imbricate structure A set of subparallel, overlapping slices of rock bounded by closely-spaced *faults* which join and form one *fault* at depth, developed at all scales.

imbrication A *fabric* in a *clastic rock* resulting from the alignment of *clasts* in which the plane containing the long and intermediate axes dips at a small angle (<20°) upstream.

immature oil Heavy, low *API gravity*, *naphthene*- or *asphalt*-rich oil, which has not undergone sufficient cracking to convert it into a light oil.

immiscible Descriptive of two or more liquids that are incapable of mixing to form a single liquid and separate into two phases. cf. *miscible*.

impact hypothesis See *bolide impact hypothesis*.

impact ripple See *ballistic ripple*.

impactite A rock formed by *bolide* impact.

impactogen A continental *rift valley* system in the *distal* region of a *continent–continent collisional orogen* resulting from tensional *stresses* associated with *indentation tectonics*.

impsonite A member of the *asphaltite* group.

in-phase EM component See *real component*.

in-sequence thrust A *thrust* in which movement on the *fault* folds or tilts an existing *fault* surface of the same phase of *deformation*.

***in situ* combustion** A technique of recovering heavy oil from a *reservoir* by combustion, which breaks it down to *coke* and light oil, the latter being pushed towards producing *wells*. Used when primary methods of recovery have failed.

***in situ* mining** The extraction of *minerals* without physical mining, such as in the *Frasch process*.

***in situ* stress measurement** The direct measurement of stress *in situ* using a *borehole breakout log*, a *flatjack*, the *hydrofrac technique*, an *inclusion stress meter* or *overcoring*.

Inarticulata A class of phylum *Brachiopoda*, characterized by the absence of hinging between the *valves*. Range *Tommotian–Recent*.

inclination The angle between the total *magnetic field* and the horizontal.

inclined extinction See *oblique extinction.*

inclined fold (tilted fold) A *fold* whose *axial plane* is inclined.

inclusion trail A linear array of *microscopic* inclusions within a crystal, commonly fluids, *iron oxides* and iron hydroxides.

incompatible element (hygromatophile element) An element that is difficult to substitute into the crystal structure of a rock-forming *mineral* because of size, charge or valency requirements, which consequently is less likely to crystallize out of a *magma* and often concentrates in *pegmatitic* or *hydrothermal fluids.* The *mantle* is depleted in such elements and the *crust* enriched.

incompetent bed A layer which flows during *deformation.*

incompressibility See *bulk modulus.*

incongruent melting Melting in which there is dissociation or reaction with the melt so that one crystalline phase is converted to another plus a liquid of different composition. cf. *congruent melting.*

incongruent solution *Dissolution* yielding a solution with different proportions from those in the solid.

inconsequent stream A stream which does not follow major land surface or geological features. cf. *anteconsequent, consequent, insequent, obsequent, resequent* and *subsequent streams.*

indentation tectonics (escape tectonics, extrusion tectonics) *Deformation* affecting the *lithosphere* of the overriding *plate* in a *continent–continent collision,* in which it is dissected by a series of long *transcurrent faults,* whose geometry depends on the shape of the indenting continent and which allow further convergence. Indentation tectonics was probably the cause of the continued *deformation* of Asia and rotation of Indochina by the convergence of India with Asia during the formation of the Himalaya.

index fossil A *fossil* species whose abundance characterizes a specific *horizon.*

index mineral A *mineral* whose first occurrence marks the limit of a *metamorphic zone.*

indialite ($(MgFe)_2Al_4Si_5O_{18}.nH_2O$) A *cyclosilicate* that is the high-temperature *polymorph* of *cordierite.*

Indian topaz 1. See *citrine.* 2. A yellow variety of *corundum.*

indicated reserve See *ore reserve.*

indicator boulder An *erratic* whose origin can be used to determine the source area of *till.*

indicatrix A three-dimensional geometrical figure that represents geometrically the different vibration directions in a *mineral* in terms of an ellipsoid whose axes are proportional to the *refractive index* of a ray travelling parallel to it.

indigo copper See *covellite.*

indicolite (indigolite) A blue, sodium-rich variety of *tourmaline.*

indigolite See *indicolite.*

Indigskiy A *Permian* succession in the Timan area of the former USSR equivalent to the L. *Asselian.*

Induan See *Ust'kel'Terskaya.*

induced magnetization The phenomenon whereby a material is made to be magnetic by the presence of an external *magnetic field*, which is lost on removal of the field.

induced polarization A *geophysical exploration* method based on the phenomenon that metallic *minerals* can store electric charge, which is gradually released when the energizing current is removed. The only method capable of detecting *disseminated deposits*.

induction log A *geophysical borehole log* in which the *resistivity* structure of the wallrock is determined by inducing current to flow in it by *electromagnetic induction*, used when the *drilling mud* is insulating or the borehole is cased and *electric logs* cannot be made.

inductively-coupled plasma emission spectrometry (ICPES) An analytical technique for the determination of *major*, minor and *trace elements*. An aerosol of the sample is injected into a plasma of high-temperature, ionized argon. The wavelengths of the radiation emitted by the atoms of the sample are diagnostic of the elements present.

induration Hardening.

industrial diamond A *diamond* of less than *gem* quality, used for drilling, cutting, lapping etc.

industrial mineral A *mineral* of economic importance in itself rather than because of the element(s) it contains.

inertinite A group of *coal macerals* which have high *carbon* and low hydrogen contents and are hard with a high relief.

inertite A *microlithotype of coal* group comprising *inertinite macerals*, of which *fusite*, *fusinite* and *semifusite* are the most common.

inertodetrinite *A coal maceral* consisting of broken fragments of *inertite* and *sclerotine*.

infaunal Descriptive of an organism that burrows into the substrate. cf. *epifaunal*.

inferred reserve See *ore reserve*.

infiltrability See *infiltration capacity*.

infiltration The entry of water into the soil, usually by downward flow through the surface.

infiltration capacity (infiltrability) The maximum rate at which water can enter the soil by *infiltration*.

infiltration rate The volume flux of water flowing into the soil per unit area of surface.

infrared spectrometry A method used to study the hydroxyl groups of *clay minerals*.

infrastructure The part of an *orogenic belt* which was deformed and metamorphosed at deep levels of the *crust*. cf. *suprastructure*.

infusorial earth See *tripolite*.

injection complex An association of *igneous* or *metamorphic rocks* which are intimately intermixed.

injection gneiss A layered *gneiss*, usually of *granitic* or *granodioritic* composition, formed by *magma* injection along parallel *structures* in the *host rock*. A type of *migmatite*.

injection structure A *structure* in a sediment formed when one layer forces into another.

injection well A *well* in a gas field or *oilfield* through which water, gas or steam is injected into the *reservoir* to maintain its pressure and thus enhance recovery.

inland sabkha See *playa lake*.

inlier An area of older rocks surrounded by younger rocks. cf. *outlier*.

inosilicate (band silicate, chain silicate) A *crystal structure* classification in which the *coordination polyhedra* are Si tetrahedra and these form chains when each tetrahedron shares two corners with adjacent tetrahedra.

INPUT® **I**nduced **Pu**lse **T**ransient. An *electromagnetic induction method* using a transient primary field to produce a decaying secondary field when the primary is absent, so that it can be accurately monitored.

Insectivora An order of infraclass *Eutheria*, subclass *Theria*, class *Mammalia*; *mammals* which include insects in their diet, probably the basic stock from which the other *Eutheria* were derived. Range U. *Cretaceous–Recent*.

insects Members of superclass *Hexapoda*, phylum *Uniramia*. Range *Devonian–Recent*.

inselberg A large, steep-sided *outcrop* rising abruptly from a flat landscape, formed as a residual produced by the *parallel retreat* of *bedrock* slopes or as a remnant on a land surface affected by *deep weathering*.

insequent stream A stream developed as a result of indeterminate features. cf. *anteconsequent, inconsequent, obsequent, resequent* and *subsequent streams*.

insolation The heat received from the sun.

insolation weathering (thermoclastis) The shattering or disintegration of surface rock by the rapid expansion and contraction resulting from large temperature fluctuations, probably not a process of major importance.

inspissation The process of drying of oil that has reached the surface.

instantaneous rotation The relative rotation of a *plate* about its *Euler pole* at a given instant in time. cf. *finite rotation*.

instantaneous strength The short-term *strength* of a material. cf. *creep strength*.

intensity See *earthquake intensity*.

intensity of magnetization The strength of the magnetic behaviour of a material, which can be expressed as an *induced magnetization* and possibly a *remanent magnetization*.

inter- Between.

inter-arc basin An elongate *basin* between the outer and inner arcs of an *island arc*, usually filled with volcanic sediments.

interconnectedness An index of the proportion of individual sediment bodies of a specific type that are in touch with each other in a given succession.

interdistributary bay A bay between distributaries in a *delta*.

interdune The topographically-low, generally flat ground between *aeolian dunes*, often with a *gravel* surface.

interfacial angle The angle between faces of a crystal, measured with a *goniometer*.

interference figure The pattern of coloured curves and black areas seen in the *thin section* of a *mineral* through polarized light and crossed Nichols.

interference ripple A *bedding* surface sedimentary *structure* comprising *ripples* in at least two orientations at a high angle forming a polygonal pattern, produced by simultaneous or sequential currents in different directions.

interflow 1. **(subsurface stormflow)** Rapid subsurface flow within the soil layer. 2. Flow at an intermediate level within a water body, often at a *pycnocline*.

interfluve An area of high ground between two adjacent river valleys.

interglacial A phase of relatively warm temperatures between *glacials*, during which *ice sheets* retreated, forest replaced *tundra* and sea level rose.

intergranular displacement *Displacement* along an adjacent grain boundary during *deformation*. cf. *intragranular displacement*, *transgranular displacement*.

intergranular fracture *Fracture* taking place along adjacent grain boundaries. cf. *circumgranular fracture*, *intragranular fracture*.

intergranular texture See *granulitic texture*.

intergrowth An interlocking arrangement of crystals arising from simultaneous *crystallization* or *exsolution*.

interlayer slip *Deformation* accomplished by *displacement* along *bedding* or *foliation* surfaces.

intermediate argillic alteration A type of low grade *wall rock alteration* in which *plagioclase* is altered to *kaolin* and *montmorillonite* group *minerals*.

intermediate-focus earthquake An *earthquake* with a depth of *focus* between 70 km and 300 km. cf. *shallow-focus earthquake*, *deep-focus earthquake*.

intermediate rock A rock with $<10\%$ *quartz* plus either *plagioclase* in the range An_{10-50}, or an *alkali feldspar* or both *alkali* and *plagioclase feldspars*.

intermediate sulphur crude oil *Crude oil* with 0.6–1.7% sulphur.

internal boudinage A type of *boudinage* in which there is the development of sinusoidal thickening and thinning within an *anisotropic*, homogeneous material, resulting from layer-parallel *extension* in which an instability occurs.

internal wave A *wave* occurring along a density interface within a body of stratified fluid, e.g. a *thermocline*.

International Geomagnetic Reference Field (IGRF) A complex formula used to compute the *geomagnetic correction* in the *reduction* of *magnetic survey* measurements.

International Gravity Formula (IGF) A formula used for the *latitude correction* in the *reduction* of *gravity survey* data collected during the period 1930 to 1967. Now superseded by the *Gravity Formula*.

International Gravity Standardization Net (IGSN) A world-wide network of locations where the absolute value of *gravity* is known, used to convert the relative readings of *gravimeters* into absolute values.

International Program for Ocean Drilling (IPOD) A phase of the *Deep Sea Drilling Program*.

International Seabed Authority The authority responsible for the administration of the seabed and its resources beyond 200 miles of shorelines.

interpenetrant twin A *twinning* phenomenon in which two individuals appear to have grown through each other.

interpluvial A relatively dry phase between *pluvials* of the *Pleistocene* and *Holocene*.

intersection cleavage A *cleavage* which crosses another planar feature and thus creates a *lineation* with it.

intersection lineation A *lineation* produced by the intersection of two planes, e.g. *cleavage* and *bedding*.

intersection point The point on the longitudinal profile of an *alluvial fan* above which there is incision into older *fan* deposits and below which there is deposition.

intersertal texture The *texture* of a random network of *feldspar* laths whose gaps are filled with *glass* or very small crystals.

interstadial A phase of relative warmth during a major *glacial*, of insufficient magnitude and/or duration to be classed as an *interglacial*.

interstice entrapment A *placer* formation mechanism involving the trapping of particles between grains of the stream bed.

interstitial Within the pores or between the grains or crystals of a rock.

interstitial crystal defect A *crystal defect* in which an alien atom occupies a small gap between atoms in the regular arrangement.

intertidal bar A *sandwave* or *swash bar* which migrates rapidly onshore due to *swash* action and is destroyed by storm activity, in which it is welded to the upper foreshore to form a *berm*.

interval velocity The *seismic velocity* over a given vertical interval, which may comprise units with different velocities.

intra- Inside.

intraclast See ***intraformational clast***.

intracrystalline plasticity *Deformation* by the movement of *dislocations* through a *crystal lattice*.

intrafolial fold A *fold* consisting solely of the *hinge* of an *asymmetrical fold* which has been rotated and stretched by continued *simple shear*.

intraformational clast (intraclast) A *clast* formed by the *erosion* of sediment soon after deposition and incorporated into a slightly younger deposit. cf. *lithoclast*.

intraglacial See *englacial*.

intragranular displacement *Displacement* taking place within a grain during *deformation*. cf. *intergranular displacement*, *transgranular displacement*.

intragranular fracture A *fracture* in a granular material which takes place within the grains. cf. *circumgranular fracture*, *intergranular fracture*.

intragranular glide *Glide* in which the *crystal lattice* either side of the *slip plane* remains undistorted.

intramicrite A *limestone* with >10% *intraclasts* in a *micrite matrix*.

intrasparite A *limestone* with >10% *intraclasts* in a *sparite matrix*.

intratelluric crystal A crystal that has grown slowly at depth before more rapid secondary *crystallization*.

intrenched meander A meander in a river channel that is incised into the landscape as a result of *erosion*.

intrusive igneous body An *igneous body* emplaced at depth.

Inverian event An *orogenic* phase affecting NW Scotland at ~2400 Ma during the later stages of the *Scourian*.

inverse grading (negative grading, reverse grading) An upward increase in mean grain size in a sedimentary *bed*, implying the sedimenting grains were independently mobile and deposited under the influence of grain-grain interactions.

inverse problem A problem common to the interpretation of *potential fields* relating to the fact that the anomaly of any given body can be computed, but the derivation of the body responsible for any particular anomaly has no unique solution.

inversion 1. In geophysical interpretation, the production of a model whose anomaly simulates an observed anomaly. 2. In *structural geology*, the reversal of the sense of vertical movement of a *fault*, block or region, such as the transformation of a sedimentary *basin* into a mountain by *orogenesis*.

inversion twinning A *twinning* phenomenon resulting from a phase transition during which *crystal symmetry* is lowered and the *twin* orientations are in the higher symmetry.

inverted relief Topography in which areas of high relief, such as *anticlines*, have become depressions due to enhanced *erosion* of the uplands.

invisible gold *Gold* invisible even under *microscopic* examination because of its small grain size or inclusion in a *solid solution* in a *mineral*.

involute With edges that roll under or inwards.

inyoite ($CaB_3O_3(OH)_5.4H_2O$) A colourless, *transparent* borate *mineral*.

iodargyrite (iodyrite) (AgI) A rare *supergene ore mineral* of *silver*.

iodobromite ($Ag(Cl,Br,I)$) A rare *supergene ore mineral* of *silver*.

iodyrite See *iodargyrite*.

iolite A *gem* variety of *cordierite*.

ion microprobe An instrumental analytical method for mapping the distribution of elements and determining their concentrations, similar in its usage to the *electron microprobe*. A beam of ions bombards a polished section *in vacuo*. The ejected ionic species are identified and quantified in a *mass spectrometer*. The advantage of the technique is that isotopes, *trace elements* and light elements can be determined to a high accuracy.

ion milling A preparation method for *transmission electron microscopy* in which beams of argon ions or atoms are fired at a demounted, polished *thin-section*, eroding its surface to provide small, electron-transparent holes.

Iowan glaciation The fourth of the *glaciations* experienced by North America in the *Quaternary*.

IPOD See ***I**nternational **P**rogram for **O**cean **D**rilling*.

Iren'skiy 1. **(Fillipovskiy)** A *Permian*

succession on the eastern Russian Platform equivalent to the *Kungurian*. 2. **(Vyl'skiy)** A *Permian* succession in the Timan area of the former USSR equivalent to the *Kungurian*.

iridescence A shimmering effect exhibited by some *minerals* caused by the *diffraction* of light by fine-scale lamellar intergrowths.

iridium (Ir) A very rare, platinum-group, *native element*.

iridosmine (Ir,Os) A very rare natural alloy of iridium and osmium.

iris (rainbow quartz) A variety of *quartz* exhibiting a chromatic reflection of light from *fractures*.

IRM See *isothermal remanent magnetization*.

iron (Fe) A rare *native element*, also found in *meteorites*.

iron alum See *halotrichite*.

iron glance A specular variety of *haematite*.

iron hat See *gossan*.

iron meteorite (siderite) A *meteorite* comprising iron plus < 20% nickel.

iron pan See *carstone*.

iron pyrites See *pyrite*.

ironstone A *sedimentary rock* containing iron-rich *nodules* or layers, of actual or potential economic importance.

irregular fracture See *uneven fracture*.

irregular mullion A *mullion structure* comprising cylinders with irregular cross-sections.

Ischnacanthiformes An order of subclass *Acanthodii*, class *Osteiichthyes*, superclass *Pisces*; bony *fish* of uncertain affinity. Range U. *Silurian*–U. *Devonian*.

island arc A *destructive plate boundary* at which there is a *subduction zone* with oceanic *lithosphere* subducting beneath oceanic *lithosphere*. The subducted slab is manifest in a *Benioff zone*, and *volcanic rocks* associated with the *subduction* are erupted at the surface.

island silicate See *sorosilicate*.

iso- Equal.

isoash map A *coal quality map* showing trends in the *ash* content of a *coal*.

isobar 1. A line joining points of equal *pressure*. 2. A line joining points of equal *principal stress* magnitude.

isobath A line joining points of equal water depth.

isochore A line of equal drilled thickness of a rock unit. cf. *isopachyte*.

isochromatic A line joining points of equal maximum *shear stress*, obtained by photoelastic methods.

isochron 1. A line joining points of equal age. 2. A line joining points of constant ratio of radioactive isotopes used in *radiometric age dating*.

isoclinal fold A *fold* with an *interlimb angle* of 0°–10°, i.e. parallel or nearly parallel limbs.

isocline A line joining points of equal magnetic *inclination*.

isoclinic A line joining points of similar *principal stress* orientations.

isofacies map A map showing the limits of *facies*, which may include several rock types.

isogal A line joining points of equal *gravity anomaly*.

isogam A line joining points of equal *magnetic anomaly*.

isogon A line joining points of equal magnetic *declination*.

isograd A line joining locations of the same *metamorphic grade*.

isogyre A black pattern in an *interference figure*.

isoline See *isopleth*.

isolith map A map showing the total thickness of beds of one lithology in a *stratigraphic* succession composed of several lithologies.

isometric system See *cubic system*.

isomorphism The phenomenon of two or more *minerals* of similar chemistry crystallizing with the same *crystal symmetry*.

isomorphous Descriptive of substances which form a series of *solid solutions*.

isopach 1. See *isopachyte*. 2. A line joining points of equal mean *stress* magnitude.

isopachyte (isopach) A line joining points of equal *stratigraphic* thickness. cf. *isochore*.

isopleth (isoline) A line joining points of equal abundance.

isopycnal A line joining points of equal pressure in water.

isoseismal A line joining points of equal *earthquake intensity*.

isostasy The study of the response of the Earth to the removal and imposition of large loads. Isostatic theory states that at the *depth of compensation* all *pressures* exerted by the rocks above are equal. This condition is satisfied by the *Airy* and *Pratt hypotheses*, which propose different geometrical arrangements for the subsurface compensation of major Earth features.

isostatic anomaly The *Bouguer anomaly* minus the effect of the compensation predicted by the *Airy* or *Pratt hypotheses*. Non-zero values indicate *isostatic equilibrium* is not extant.

isostatic equilibrium The condition when a surface load, or lack of mass, is perfectly balanced by a mass deficiency, or mass excess, at depth.

isostatic rebound The recovery of *isostatic equilibrium* after the removal of a load, e.g. after the removal of an *ice sheet*, the *lithosphere* has to rise to regain equilibrium.

isostructural Descriptive of *minerals* with similar chemical, physical and *crystallographic* characteristics.

isosulph map A *coal quality map* showing trends in the *sulphur* content of a *coal*.

isotherm A line joining points of equal temperature.

isothermal Having the same temperature.

isothermal remanent magnetization (IRM) A very complex *natural remanent magnetization* acquired when a rock is struck by lightning.

isotopic abundance The ratio of the quantity of an isotope to the total amount of the element.

isotopic fractionation The phenomenon whereby the isotopic composition of a *mineral* is different from that of the fluid from which it *precipitated*.

isotopic stratigraphy *Stratigraphy* based on the isotopic composition of *minerals*.

isotropic Having no order or *preferred orientation*. Does not affect doubly-polarized light.

isotropic fabric A *fabric* of a *clastic rock* which has *equant clasts* or no *preferred orientation* of non-*equant clasts*.

isotropic point The point where the *principal stresses* are equal.

isotypic Descriptive of a pair of *isomorphs* whose relative sizes of ions are the same, but whose absolute sizes are different, so that no *solid solution* is possible.

isovol A line joining points of equal volatile content in *coal*.

ISSC International Subcommission on Stratigraphic Correlation.

Isuan An *era* of the *Precambrian*, 3800–3500 Ma.

itabirite See *banded iron formation*.

itacolumnite A micaceous *sandstone* with interlocking *clasts* so that the rock bends when in the form of a thin slab.

italite A rare, coarse-grained, *plutonic igneous rock* comprising *leucite* and a small amount of *glass*.

IUGG International Union of Geophysics and Geodesy.

IUGS International Union of Geological Sciences.

IUGS classification An internationally adopted scheme for the classification of *igneous rocks* proposed by the *IUGS* in 1989.

Iuosuchanskaya A *Triassic* succession in Siberia equivalent to the *Rhaetian*.

Ivorian A *stage* of the *Carboniferous*, 353.8–349.5 Ma.

J

J In *earthquake seismology*, an *S wave* generated by conversion from a *P wave* at the boundary between the inner and outer *core* and reconverted on exit.

J-type lead (Joplin-type lead) An *anomalous lead* which provides a negative age, probably due to its acquiring extra radiogenic lead during its history.

jacinth See *hyacinth.*

Jackson An *Eocene* succession on the Gulf Coast of the USA equivalent to the *Priabonian.*

jacobsite ($MnFe_2O_4$) A rare black *spinel.*

jacupirangite An *igneous rock* comprising *titanaugite*, *aegirine-augite* and *accessory nepheline.*

jade A precious *gem* consisting of *jadeite* or *nephrite.*

jadeite ($NaAlSi_2O_6$) A green *pyroxene* found in *metamorphic rocks.*

jamesonite ($Pb_4FeSb_6S_{14}$) A minor *ore mineral* of lead.

jargoon A yellow or smoky variety of *zircon.*

jarosite ($KFe_3(SO_4)_2(OH)_6$) A *secondary mineral* found coating iron *ores.*

jasper A *granular, microcrystalline* variety of *quartz*, usually coloured red by *haematite.*

jasperoid A form of *wall rock alteration*, common in some *epithermal deposits*, in which fine-grained, *haematite*-stained *silica* is developed.

jaspillite See *banded iron formation.*

Jeffreys-Bullen model A model for the *seismic velocity* structure of the Earth, in which velocity varies only radially.

jeppeite ($(K,Ba)_2(Ti,Fe)_6O_{13}$) An *oxide mineral* found as an *accessory mineral* in *lamproites.*

jet A lustrous variety of *lignite* found as isolated masses in some bituminous *shales*, used in jewellery.

Jiangtangjiang (Chientangkiang) An *Ordovician* succession in China covering part of the *Harnagian*, the *Soudleyan*, *Longvillian*, *Marshbrookian*, *Actonian*, *Onnian* and *Ashgill.*

jimthompsonite ($(Mg,Fe)_{10}Si_{12}O_{32}(OH)_4$) A *biopyribole.*

JMA scale A scale of *earthquake intensity* (0–VII) used by the **J**apanese **M**eteorological **A**gency, suitable for local building styles.

joesmithite ($(Ca,Pb)Ca_2(Mg,Fe^{2+},Fe^{3+})_5(Si_6Be_2O_{22})(OH)_2$) An *amphibole* containing beryllium in the tetrahedral chain.

johannsenite ($CaMnSi_2O_6$) A *pyroxene* found in association with *rhodonite* and *bustamite* in *metasomatized limestones.*

JOIDES (Joint Oceanographic Institutions for Deep Earth Sampling) An organization established to undertake deep-sea drilling, subsequently evolving into the *Deep Sea Drilling Program* and the *International Program for Ocean Drilling.*

Joint Oceanographic Institutions for Deep Earth Sampling See *JOIDES.*

joint A *fracture* on which any *shear displacement* is too small to be visible to the unaided eye.

joint set A group of *joints* with a common orientation.

jökull A small *ice cap.*

jökullhlaup A catastrophic flood caused by a *volcanic eruption* beneath an *ice sheet.*

Joplin-type lead See *J-type lead.*

Jotnian orogeny An *orogeny* affecting the Baltic *shield* in the *Proterozoic.*

Jurassic The middle *period* of the *Mesozoic*, 208.0–145.6 Ma.

juvenile bomb A *pyroclastic* fragment of congealed *magma* over 64 mm in size.

juvenile gas Gas originating in the interior of the Earth, not previously having been at the surface.

juvenile water Water derived from *magma.*

K

K In *earthquake seismology*, a *P wave* that has travelled through the *core*.

K-cycle A concept of land evolution involving the cyclic *erosion* of soils on upper slopes during unstable climatic phases and soil development during stable phases.

K-feldspar See *potassium feldspar*.

K–T boundary See *Cretaceous–Tertiary boundary*.

Kachian A *Palaeocene* succession in the former USSR covering the upper part of the *Thanetian*.

kaersutite (titanium hornblende) ($Ca_2(Na,K)(Mg,Fe^{2+},Fe^{3+})_4TiSi_6Al_2O_{22}(O,OH,F)_2$) A titanium *amphibole*.

Kaiatan An *Eocene* succession in New Zealand covering the upper part of the *Bartonian*.

Kaihikuan A *Triassic* succession in New Zealand covering the lower part of the *Ladinian*.

kainite ($KMgSO_4Cl.3H_2O$) An *evaporite mineral*.

Kainozoic See *Cenozoic*.

kaiwekite A *volcanic rock* containing *phenocrysts* of *olivine*, *titanaugite*, *barkevikite* and *anorthoclase*.

kaliophilite A *polymorph* of *kalsilite*.

kalsilite ($KAlSiO_4$) A rare *feldspathoid* found in complex *phenocrysts* and in the *matrix* of certain *lavas*.

kamacite (Fe,Ni) An iron-nickel alloy found in *meteorites*.

kame An irregular, undulating mound of bedded *sands* and *gravels* deposited unevenly along the front of a stationary or decaying *ice sheet*.

kame terrace A terrace between a hillside and *glacier* formed by *glaciofluvial* activity.

kamenitza A solution *pan* or *basin* formed on a *limestone* surface by the *dissolution* of calcium carbonate.

Kamyshinian A *Palaeocene* succession in the former USSR covering part of the *Thanetian*.

kandite A group of *clay minerals* including *kaolinite*, *dickite*, *nacrite*, *halloysite* and meta-*halloysite*.

Kanev An *Eocene* succession in the former USSR covering the *Ypresian* and part of the *Lutetian*.

Kansan glaciation The second of the *glaciations* of North America in the *Quaternary*.

kaolin (china clay) An important *mineral* product composed principally of *kaolinite*.

kaolinite ($Al_2Si_2O_5(OH)_4$) A common *clay mineral* formed by the *weathering* or *hydrothermal alteration* of *feldspars* and other aluminous silicate *minerals*.

kaolinization (kaolinitization) The *alteration* of a *mineral* (commonly *muscovite, biotite* or *feldspar*) by a *dissolution-precipitation* mechanism to *authigenic kaolinite*.

kaolinitization See *kaolinization*.

Kapitean A *Neogene* succession in New Zealand covering parts of the *Messian* and *Zanclian*.

Karaganian A *Miocene* succession on the Russian Platform covering the lower part of the *Tortonian*.

Karakan An *Ordovician* succession in Kazakhstan covering the upper part of the *Llanvirn*.

karat See *carat*.

Karatau The youngest *period* of the *Riphean*, 1050–800 Ma.

Karelian orogeny An *orogeny* affecting the Baltic *shield* in the *Archaean* from ~2000–1900 Ma.

Karman-Prandtl velocity law A law expressing the velocity profile within a turbulent *boundary layer* as a logarithmic function of the distance from the bed: $U = \kappa^{-1}\sqrt{(\tau/\rho)}.\ \ln(y/k)$, where U = velocity, κ = *von Karman's constant*, τ = *shear stress*, ρ = fluid *density*, y = height of point at which U is measured and k = *coefficient of roughness*.

karren (lapiés) Minor solutional features developed on carbonate rocks, formed mostly by *dissolution*.

karst A terrain with distinctive landforms and drainage (often underground), mainly originating from *solutional erosion* and commonly developed on carbonate rocks or *evaporites*.

Kashirskian A *stage* of the *Carboniferous*, 309.2–307.1 Ma.

Kasimovian An *epoch* of the *Carboniferous*, 299.9–295.1 Ma.

kataphorite See *katophorite*.

katatectic layer A layer of *solution* residue, commonly of *gypsum* and/or *anhydrite*, at the top of a *salt dome*.

katazone A *depth zone* of high temperature and very high *hydrostatic pressure*. See also *epizone, mesozone*.

katophorite (cataphorite, kataphorite) ($Na_2Ca(Fe^{3+},Al)_5(AlSi_7)O_{22}(OH)_2$) An *amphibole* found in *basic, alkaline igneous rocks*.

kavir A *playa*, continental *sabkha* or other saline desert *basin* which may be flooded periodically.

Kawhia A *Jurassic* succession in New Zealand comprising the *Temaikan, Heterian* and *Ohauan*.

Kazakhstania A continent situated between Siberia and *Gondwanaland* from *Cambrian* to *Devonian* times.

Kazanskiy A *Permian* succession on the eastern Russian Platform covering part of the *Ufimian*, the *Wordian* and part of the *Capitanian*.

keatite (SiO_2) A synthetic tetragonal form of *silica*.

kegel A conical hill on a *limestone* landscape.

kegelkarst (cockpit karst, cone karst) A *limestone* landscape characterized by *kegels* interspersed by closed depressions, typical of *karst*.

keilhauite A variety of *sphene* containing >10% *rare earth elements*.

Kelvin model See *viscoelasticity*.

Kelvin-Voight model See *visco-elasticity*.

kelyphitic rim A rim of one *mineral* around another in an *igneous rock* resulting from reaction of the enclosed *mineral* with other constituents of the rock.

Kenoran orogeny (Algoman orogeny) An *orogeny* of *Precambrian* age at ~2400 Ma which affected the Canadian *shield*.

kentallenite A coarse-grained, *basic igneous* rock comprising *olivine*, *augite* and *biotite* with minor *plagioclase* and *orthoclase* in equal amounts.

kenyite A fine-grained *igneous rock*; an *olivine*-bearing *phonolite* with *phenocrysts* of *anorthoclase* ± *augite* in a *glassy groundmass*.

keratophyre A rock of the *spilite* suite made up of *albite*, *chlorite*, *epidote* and iron-titanium oxides, formed by the *alteration* of intermediate *volcanic rocks*.

kermesite (pyrostibnite) (Sb_2S_2O) Antimony oxysulphide, a *secondary mineral* forming by the *alteration* of *stibnite*.

kernite ($Na_2B_4O_6(OH)_2.3H_2O$) A *mineral* important as a source of boron compounds.

kerogen A bituminous material, found in *oil shales* and other *sedimentary rocks*, which is composed of organic matter and can yield oil on distillation.

kersantite A *calc-alkaline lamprophyre* containing *biotite*, *plagioclase* (usually An_{10-50}) and *augite*, ± *diopside* and *olivine*.

kesterite (Cu_2ZnSnS_4) A *sulphide mineral* derived from the *sphalerite* structure by ordered substitution.

Ketilidian orogeny An *orogeny* of *Proterozoic* age affecting Greenland at ~1800–1600 Ma.

kettlehole A topographic depression left when *drift*-covered ice melts.

Keuper 1. A *Triassic* succession in Germany covering part of the *Ladinian* and the *Carnian*, *Norian* and *Rhaetian*. 2. A traditional name for the U. *Triassic* in Europe.

key-hole vugs Bubble-like *fenestrae* found in *sands*.

Kharakovian An *Oligocene* succession in the former USSR equivalent to the *Rupelian*.

Khedalichenskaya A *Triassic* succession in Siberia equivalent to the *Carnian* and *Norian*.

Kibalian orogeny See *Bugando-Toro-Kibalian orogeny*.

kidney ore A form of *haematite* with a fibrous, radiating, internal structure and a *reniform*, red, external surface.

kidney stone 1. A *reniform limestone pebble* or *nodule*. 2. See *nephrite*.

kieselguhr See *diatomite*.

kieserite ($MgSO_4.H_2O$) An *evaporite mineral*.

Kieslager deposit See *Besshi-type deposit*.

killas A term for *Devonian-Carboniferous*, low-*grade phyllites* in SW England.

kilobar The *cgs* unit of *pressure*, equivalent to 100 M*Pa*.

kimberlite A *serpentinized*, carbonated, commonly brecciated, *porphyritic mica-peridotite* made up of *phenocrysts* of *olivine* and *phlogopite* in a fine-grained *groundmass* of *olivine*, *phlogopite*, *pyrope*, iron-titanium oxide, *perovskite* plus *serpentinite*, *chlorite* and carbonates. Found in *volcanic pipes* and characteristically containing a wide range of *xenoliths* of *crustal* and *mantle* origin. The main source of *diamonds*.

Kimmeridgian A *stage* of the *Jurassic*, 154.7–152.1 Ma.

Kinderhookian A *Carboniferous* succession in the USA covering the lower part of the *Tournaisian*.

Kinderscoutian A *stage* of the *Carboniferous*, 322.8–321.5 Ma.

kinematic Referring to motion; used to describe phenomena related to the relative motion of an object.

kinematic indicator A *structure* that can provide the direction and sense of *displacement* of a *fault* or *shear zone*.

kinematic sieving A process suggested as a partial explanation for *inverse grading* in which 'vibration-strain' during flow promotes the downward filtration of small grains between large ones.

kinematic symmetry axes The direction of *flow*, the plane of *flow* and the plane normal to these, assuming *plane strain*.

kinematic viscosity The ratio of *viscosity* to *density*, allowing comparison of the resistance to shape change between materials of different *densities*.

kink A type of *fold* with an angular *profile*.

kink band A *microscopic*- to *mesoscopic*-scale localized band where the orientation of a *structure* changes abruptly.

kink band boundary See *kink plane*.

kink plane (kink band boundary) The edge of a *kink band*.

kinking The *fold* process forming a *kink*.

kipuka (dagala) An 'island' of land surrounded by *lava*.

kirschsteinite ($CaFeSiO_4$) An *olivine*.

Kiruna-type deposit See *volcanic-associated massive oxide deposit*.

Klabava An *Ordovician* succession in Bohemia equivalent to the *Arenig*.

Klazminskian A *stage* of the *Carboniferous*, 295.1–293.6 Ma.

klippe An *outlier* formed by *thrusting*.

kluftkarren See *grike*.

knebelite ($(Mn,Fe)_2SiO_4$) An *olivine* usually formed as a *metamorphic* product in iron-manganese *ore* deposits.

knee twin See *geniculate twin*.

knoch and lochan Topography dominated by a mixture of eroded rock ridges and small *basins*, formed by *glacial erosion*.

Knott's equations Equations which define how *seismic wave* energy is partitioned between reflected *P* and *S waves* when it impinges on a discontinuity; similar to *Zoeppritz' equations*.

kobellite ($Pb_2(Bi,Sb)_2S_5$) Lead-bismuth-antimony sulphide, found in *granite pegmatite veins*.

Koenigsberger ratio (Königsberger ratio) The ratio of intensity of *natural remanent magnetization* to magnetization induced by the local *geomagnetic field*.

Kogashyk An *Ordovician* succession in

Kazakhstan covering the upper part of the *Arenig*.

Kokhanskiy A *Permian* succession on the eastern Russian Platform equivalent to the U. *Asselian*.

Koln stone (Cullen stone) A *millstone* of *Mayen lava* traded widely in NW Europe from the Iron Age.

komatiite An *ultramafic volcanic rock* with >18% MgO composed of *olivine* and *pyroxene* ± *chromite* in a *glassy* or *devitrified groundmass*. It has the morphological features of subaerial and submarine *basaltic lava* flows (i.e. *pillow* and *hydroclastite structure*) and a distinctive *spinifex texture*. Characteristic of *Archaean* terrains.

Komichanskiy A *Permian* succession in the Timan area of the former USSR equivalent to the U. *Artinskian*.

Königsberger ratio See *Koenigsberger ratio*.

Konkian A *Miocene* succession on the Russian Platform covering parts of the *Tortonian* and *Messian*.

Kopaly An *Ordovician* succession in Kazakhstan covering the lower part of the *Llanvirn*.

Kopanina-Schichten A *Silurian* succession in Bohemia covering part of the *Gorstian* and the *Ludfordian*.

kopje (koppie) A rocky hill probably formed by the exhumation of relatively unweathered rock and *corestones* from within *deep weathering* profiles.

koppie See *kopje*.

Korangian A *Cretaceous* succession in New Zealand covering parts of the *Aptian* and *Albian*.

kornerupine ($Mg_3Al_6(Si,Al,B)_5O_{21}(OH)$) A rare magnesium-aluminium borosilicate, sometimes used as a *gem*.

Kosov An *Ordovician* succession in Bohemia equivalent to the *Hirnantian*.

koum See *erg*.

Kraluv An *Ordovician* succession in Bohemia approximately equivalent to the *Rawtheyan*.

krennerite ($AuTe_2$) A rare *gold ore mineral*.

Krevyakinskian A *stage* of the *Carboniferous*, 303.0–299.9 Ma.

krotovina (crotovina) An animal burrow filled by later material.

kulaite An *amphibole*-bearing *nepheline basalt*.

Kungurian A *stage* of the *Permian*, 259.7–256.1 Ma.

kunzite A *gem* variety of *spodumene* with a clear lilac colour.

kupfernickel See *niccolite*.

Kupferschiefer 1. A *Permian* succession in NW Europe covering the upper part of the *Kungurian*. 2. A *copper*-rich *shale* of *Permian* age found in Germany, Poland, Holland and England.

Kuroko-type deposit A type of volcanic-associated *massive sulphide deposit*, usually an *ore* of *copper*, zinc and lead ± *gold* and *silver*, formed in a *back-arc basin* environment.

kurtosis A measure of the 'peakedness' of a frequency distribution.

kutnahorite ($CaMn(CO_3)_2$) A manganese-rich *dolomite*.

Kuyalnitskian (Cimmeridian) A *Pliocene* succession on the Russian Platform covering parts of the *Zanclian* and *Piacenzian.*

kyanite (disthene) (Al_2SiO_5) A *nesosilicate* commonly formed by the *regional metamorphism* of an *argillaceous* rock.

kyanite group An industrial name for the *sillimanite minerals.*

kylite See *theralite.*

L

L Letter used to indicate a phase of *lineation* formation, subscripted to denote each separate phase.

L-S tectonite A rock with both a *lineation* and a *foliation.*

L-tectonite A rock with well-developed *lineations.*

labile Unstable.

labradorescence The brilliant play of colours shown by *labradorite.*

labradorite A *plagioclase feldspar,* An_{50-70}.

Labyrinthodontia A subclass of class *Amphibia*; *amphibians* with a large head and limbs splayed laterally from a relatively stubby body. Range M. *Devonian*–U. *Triassic.*

laccolite See *laccolith.*

laccolith (laccolite) A *concordant* minor intrusion with a flat floor and convex upper surface. Generally with a diameter up to ~8 km and a thickness from a few metres to hundreds of metres.

lacuna See *disconformity.*

lacustral 1. See *lacustrine.* 2. See *pluvial.*

lacustrine (lacustral) Referring to a lake.

ladder ripples (ladder-back ripples) *Interference ripples* comprising a set of long wavelength *wave ripples* and a second set of shorter wavelength orthogonal to them, observed to form during a fall in water level.

ladder-back ripples See *ladder ripples.*

Ladinian A *stage* of the *Triassic,* 239.5–235.0 Ma.

lag breccia (co-ignimbrite breccia) A coarse-grained deposit rich in *lithic* fragments which accumulated at the same time as an *ignimbrite* near the *volcanic vent* by the accumulation of *clasts* too large to be transported away.

lag deposit See *lag gravel.*

lag fault (low-angle fault) A *normal fault* with a *dip* of $<45°$.

lag gravel (lag deposit) A residual accumulation of coarse particles from which the fine material has been *winnowed* away.

lagerstätte See *exceptional fossil deposit.*

Lagomorpha An order of infraclass *Eutheria,* subclass *Theria,* class *Mammalia*; the rabbits and hares. Range U. *Eocene–Recent.*

lagoon A body of water enclosed by a barrier or between a barrier and its associated coastline.

lahar A flow of volcanic debris and water, travelling at great speed, deposited as a poorly-sorted mass.

Lahn-Dill iron deposit A *syngenetic*, *conformable*, iron-rich layer or lens, dominantly siliceous at the base and calcareous at the top. The principal *minerals* are *haematite*, *magnetite*, *siderite* and *limonite*. Probably of *volcanic-exhalative* origin.

Lamé constant (λ) An *elastic constant* equal to the *bulk modulus* less two thirds the *shear modulus*.

lamellibranchs See *Bivalvia*.

laminar flow A non-*turbulent flow* in which the mean flow velocity and instantaneous velocities at any point are exactly the same, characterized by the dominance of viscous forces over inertial forces, and *Reynolds numbers* <500. cf. *turbulent flow*.

laminar twinning See *polysynthetic twinning*.

lamination A fine, discrete layer of rock 0.005–1.00 mm thick.

laminite A sediment with millimetric-scale *lamination*, common in lake environments and useful in determining *palaeoclimate*.

lamp shell A colloquial name for the *Brachiopoda*.

lamprobolite See *oxy-hornblende*.

lamproite A potassium- and magnesium-rich *mafic* to *ultramafic alkaline lamprophyre*-type rock of volcanic or *hypabyssal* origin composed of Ti-rich *phlogopite*, *clinopyroxene*, alkali *amphibole*, *olivine*, *leucite* and *sanidine* with *accessory chrome spinel*, *priderite*, *wadeite*, *nepheline*, *ilmenite*, *sherbakovite*, *jeppeite*, *apatite*, *perovskite*, *sphene* and *amphibole*.

lamprophyllite ($Na_3Sr_2Ti_3(Si_2O_7)_2(O,OH,F)_2$) A rare titanium *nesosilicate* found mainly in *nepheline syenites* and their *pegmatites*.

lamprophyre A minor intrusion of *mesocratic* and *melanocratic mineral* composition containing *biotite* or *phlogopite* ± *amphibole*, with *clinopyroxene*, *olivine* and occasionally *melilite* in a *groundmass* of *feldspars* or *feldspathoids*.

lanarkite (Pb_2SO_5) A rare lead sulphate which occurs with *anglesite* and *leadhillite* in the oxidized zone of lead deposits.

Lancefieldian An *Ordovician* succession in Australia covering parts of the *Tremadoc* and early *Arenig*.

land mammal age A unit based on the rich *Tertiary mammal* faunas of the Great Plains of North America, used to correlate continental deposits over long distances.

land plant A plant spending most of its life on land and commonly with a fluid-conducting vascular system. Such plants appear in the *fossil* record from at least as early as the *Silurian*.

landfill See *sanitary landfill*.

Landon An *Oligocene* succession in New Zealand comprising the *Whaingaroan* and *Duntroonian*.

Landsat satellites A series of satellites used in *remote sensing* to provide images particularly useful for geological studies. Previously known as the *ERTS satellites*.

landscape marble A *limestone* showing a pattern on a cut and polished surface similar to a landscape. Believed to arise from biogenic activity, the mixing of different coloured sediment by injection or the action of gas produced during the decay of bituminous sediments.

landslide (landslip) The rapid movement of a mass of soil downslope along a curved or planar *failure* surface, without *deformation* of the *soil structure*.

landslip See *landslide*.

Lang topographic method An *X-ray diffraction* method for the direct imaging of microstructural *crystal defects*.

langbeinite ($K_2Mg_2(SO_4)_3$) An *evaporite mineral*.

Langhian A *stage* of the *Miocene*, 16.3–14.2 Ma.

langite ($Cu_4SO_4(OH)_6.2H_2O$) A blue to green-blue, hydrated copper sulphate found as a *secondary mineral* in copper deposits.

lansfordite ($MgCO_3.5H_2O$) An unstable, hydrated magnesium carbonate.

lapiés See *karren*.

lapilli *Pyroclastic* fragments between 2 mm and 64 mm in size.

lapis lazuli An ornamental stone comprising a mixture of *lazurite*, *calcite*, *pyroxenes* and other silicates.

Laplace's equation A relationship obeyed by all *potential fields* which states that in Cartesian coordinates the sum of the second derivatives of the field in three orthogonal directions is zero.

Laramide orogeny An *orogeny* responsible for the formation of the Rocky Mountains of America in late *Cretaceous* and *Palaeocene* times.

lardalite See *laurdelite*.

large-ion lithophile (LIL) An element of large ionic radius and valency of 1 or 2 which becomes concentrated mainly in the potassium silicates of silicic melts during igneous *fractionation*.

larnite (Ca_2SiO_4) A rare *mineral* formed by the *contact metamorphism* of *limestone*.

larsenite ($PbZnSiO_4$) A rare *olivine*.

larvikite (laurvikite) A variety of *syenite* in which the *feldspar* shows blue *iridescence*.

lateral ramp A *ramp* that trends parallel to the transport direction of a *thrust system*.

lateral resistivity log An *electric geophysical borehole log* which measures the *resistivity* of the formations to a considerable distance from the borehole.

lateral secretion The derivation of *mineral*-forming materials from the *wall rocks* around a *vein* or other *mineral* deposit.

laterite A *residual deposit* of iron and aluminium hydroxides formed by the *weathering* of rocks in humid, tropical conditions.

lateritization The process whereby rock is converted to *laterite* by the extraction of *silica*.

laterolog An *electric geophysical borehole log* which measures the *resistivity* of the formations within a thin circular disc around the borehole.

latite A *volcanic rock* similar to *calc-alkaline andesite* in its *quartz* content but with a higher K_2O content and K_2O/Na_2O ratio, so that *phenocrysts* of *plagioclase*, *biotite* ± *sanidine* are more abundant than *pyroxene* and *hornblende*.

latitude correction The correction applied to *gravity* data for the variation of *gravity* with latitude using the *Gravity Formula*. Magnetic data require a *geomagnetic correction* for latitude and longitude.

Laue back-reflection method An *X-ray diffraction* method in which the incident beam is directed at the sample through the centre of a flat piece of film. Back-diffracted radiation causes a pattern of spots on the film which can be interpreted with the aid of a net.

laumontite ($(CaAl_2Si_4O_{12}).4H_2O$) A *zeolite* found in the *veins* and cavities of many rock types, especially *basaltic*.

Laurasia The northern *supercontinent* prior to the *continental splitting* that formed the Atlantic Ocean, comprising North America, Greenland, Europe and Asia excepting Siberia and other blocks which joined during and after the *Triassic*.

laurdelite (lardalite) A coarse-grained *syenite* resembling *larvikite* but containing rhomb-shaped *alkali feldspar* crystals and large *nephelines*.

Laurentia A *supercontinent* present during and subsequent to the *Variscan orogeny* comprising the Canadian *shield*, Greenland, parts of NW Europe and some other parts of North America.

laurvikite See *larvikite*.

lautarite ($Ca(IO_3)_2$) Calcium iodate, found rarely in *caliche*.

lava Molten rock material at the surface.

lava cave A *cave* formed as molten rock solidified, possibly termed a *pseudokarst* feature.

lava flow A dense mass of molten or partially molten rock moving as a stream on the surface.

lava levée A retaining wall of *scoria* at the side of a *lava flow*.

lava tube A hollow subsurface passage, up to 30 m in width, 15 m in height and up to tens of kilometres long, in a solidified *lava flow* formed when *lava* withdrew from a distributary tunnel.

Law of accordant junctions (Playfair's law) 'Tributary rivers enter a main river at the same level as that river without any sudden drop.' Not always correct, as channel sizes may affect the form of the junction.

Law of Bravais 'The favoured low-energy crystal faces are those parallel to lattice planes with a high density of lattice points, as those planes cut the smallest area of each *unit cell* and so break the smallest number of bonds in the structure.' A law governing *crystal morphology*.

Law of constancy of angles 'In a given compound the angles between corresponding faces (*interfacial angles*) are always the same.'

Law of divides 'The nearer the divide, the steeper the slope, with all points on a single slope being interdependent.' A law explaining the smooth slopes in *badland* areas.

Law of effective stress '*Pore fluid pressure* reduces the *normal stress* across a plane by the magnitude of the *pressure*.'

Law of equal declivities 'Slopes on either side of a *divide* are interdependent.' An extension of the *Law of divides*, which explains why a *badland* ridge always stands midway between two streams of equal elevation.

Law of mineral stability '*Minerals* are in thermodynamic equilibrium only in the environment in which they form.' An explanation for many *eogenetic* reactions.

Law of the Sea Convention An agreement that all deep-sea mining of the seafloor beyond 200 miles of coastal states would be under the jurisdiction of an *International Seabed Authority*, which might undertake such mining itself.

lawsonite ($CaAl_2(Si_2O_7)(OH)_2.H_2O$) A *sorosilicate* commonly found in *blueschists*.

Laxfordian orogeny An *orogeny* affecting NW Scotland in the *Precambrian* from 1800–1600 Ma.

layer-parallel shear *Shear* parallel to the layering in the folding of a layered bed.

layer-parallel shortening *Shortening* parallel to the layering in the formation of *chevron folds* from *kink bands* and *box folds*.

layer silicate See *phyllosilicate*.

layer sulphides group A group of *sulphide minerals* characterized by a *structure* derived from the *nickel arsenide group* by the omission of a complete cation layer, with metal atoms in octahedral coordination and octahedra in the same layer sharing edges.

layered anorthosite complex A very extensive intrusion, often with a retained *cumulate texture* and rich in *chromite*, with much more abundant *anorthosite* and *leucogabbro* than normal for a basic intrusion. Unique to the *granulite-gneiss belts* of the *Archaean*.

layered igneous rock An *igneous rock* which displays *mineralogical* and/or chemical layering.

lazulite ($(Mg,Fe^{2+})Al_2(PO_4)(OH)_2$) A deep blue, strongly *pleochroic*, hydrated aluminium-magnesium-iron-(calcium) phosphate found in high-*grade*, aluminous *metamorphic rocks* and *granite pegmatites*.

lazurite ($((Na,Ca)_8(AlSiO_4)_6(SO_4,S,Cl)_2)$) A rare *tectosilicate* found in *contact metamorphosed limestones*.

leaching A process of *pedogenesis* in which soluble *minerals* are removed from the soil.

lead-glance See *galena*.

lead-lead dating A *radiometric dating* method based on the proportion of radiogenic ^{207}Pb and ^{206}Pb, the former of which accumulates six times more rapidly than the latter.

leader A thin mineralized *vein* related to the main *ore*-carrying *vein* and which aids its discovery.

leadhillite ($Pb_4SO_4(CO_3)_2(OH)_2$) A hydrated lead carbonate and sulphate found in association with lead *ores*.

leaky transform fault A *transform fault* across which there is a component of *extension* so that movement along it forms a gap up which *mantle* material may penetrate.

lean ore A low *grade ore*.

lebensspur See *trace d'activité animale*.

lechatelierite (SiO_2) Natural fused *silica* or *silica glass*, the main constituent of *fulgurites*.

lectotype See *type specimen.*

lee side The downstream side of a body sheltered from the dominant flow direction. cf. *stoss side*.

lee slope The downstream slope of a body sheltered from the dominant flow direction. cf. *stoss slope*.

Leighton-Pendexter classification A largely outmoded classification scheme for *limestones* and *dolostones*.

Leiner A *Permian* succession in NW Europe covering the upper part of the *Wordian*.

lenad A term for the *feldspathoids leucite* and *nepheline*.

Lenan A *Cambrian* succession in Siberia covering part of the *Atdabanian*, the *Lenian* and part of the *Solvan*.

Lenian A *stage* of the *Cambrian*, 554–536.0 Ma.

lenticular bedding A type of *heterolithic bedding* characterized by lenses and ripples of *sand* in a *mud matrix*.

Leonard A *Permian* succession in the Delaware Basin, USA covering part of the *Sakmarian*, the *Artinskian*, *Kungurian* and *Ufimian*.

lepidoblastic texture A *metamorphic texture* comprising *foliation* or *schistosity*.

lepidocrocite (γFeO.OH) A *secondary mineral* of iron, a *polymorph* of *goethite*.

lepidolite (lithia mica) ($K(Li,Al)_{2-3}(AlSi_3O_{10})(O,OH,F)_2$) A pink or lilac, lithium-bearing *mica* found in *granitic* rocks and *pegmatites*.

lepidomelane An iron-rich variety of *biotite* found in *igneous rocks*.

Lepidosauria A subclass of class *Reptilia*; the 'scaly' *reptiles*, including the lizards and snakes. Range U. *Permian–Recent*.

lepisphere A spherical, *microcrystalline aggregate* of bladed crystals formed during the transformation of *opal* to *quartz*.

Lepodocystoidea A class of phylum *Echinodermata* close to the common ancestry of the *Crinoidea* and *Cystoidea*. Range L. *Cambrian*.

Lepospondyli A subclass of class *Amphibia*; a group of extinct forms whose relationships are uncertain. Range *Triassic–Permian*.

leptite (leptynite) An *equigranular metamorphic rock* comprising *quartz* and *feldspars*.

Leptostraca An order of subclass *Phyllocarida*, class *Malacostraca*, subphylum *Crustacea*, phylum *Arthropoda*; small, marine *crustaceans* with a *bivalved* carapace and an abdomen of seven somites. Range *Recent*.

leptynite See *leptite*.

Letna An *Ordovician* succession in Bohemia covering part of the *Costonian* and the *Harnagian*.

leucite ($KAlSi_2O_6$) A *feldspathoid* common in potassium-rich *lavas*.

leucitite A fine-grained, often *porphyritic lava* or minor intrusion composed of *leucite* (30–50%) and *clinopyroxene* with *accessory nepheline*, iron-titanium oxides and *apatite*. Occurs in continental *rifts*, *island arcs* and other complex, post-*tectonic*, continental settings.

leucitophyre A fine-grained *igneous rock*, commonly a *lava*, comprising *phenocrysts* of *leucite* and other *minerals* in a *trachytic groundmass*.

leuco- Of lighter colour.

leucocratic Descriptive of a rock with 0–30% dark *minerals*.

leucogabbro A *gabbro* with a predominance of *felsic minerals*.

leucosapphire See *white sapphire*.

leucosome A centimetric- to metric-scale, coarse-grained, quartzofeldspathic *vein* found in *pelitic* and *psammitic metamorphic rocks*, which may represent a *migmatitic* product of a low melting point liquid segregated from the sediment during high-*grade metamorphism*.

leucoxene A fine-grained, yellow-brown *alteration* product of titanium-rich *minerals*, made up principally of *rutile*.

levée A raised bank beside a terrestrial or subaqueous channel formed by the rapid deposition of sediment from water escaping the channel, which fines and thins away from it. Levées allow water to rise above the level of the *floodplain*, and when breached major flooding or channel *avulsion* occurs.

Levy-Mises equations A fundamental law of *plasticity* in which *strain rates* are proportional to *deviatoric stress*.

levyne See *levynite*.

levynite (levyne) ($(Na,Ca)_2(Al,Si)_9O_{18}.8H_2O$) A *zeolite* found in cavities in *basalt*.

Lewisian A division of the *Precambrian* in Scotland overlain by the *Torridonian* and affected by the *Scourian* and *Laxfordian* events.

Lg wave An *S wave* of high amplitude trapped in the *crust*, which acts as a waveguide.

lherzolite A *phaneritic* rock similar to *peridotite* which contains 40–90% *olivine*, >5% *orthopyroxene* and >5% *clinopyroxene* with *accessory plagioclase*, *spinel* ± *garnet*. It yields *basaltic magma* on *partial melting*, so is taken as a model for the composition of the *upper mantle*. Occurs in some *ophiolites* and as *xenoliths* in *alkali basalts*, *lamproites* and *kimberlites*.

Liangshan A *Permian* succession in China equivalent to the *Asselian*.

Lias A term for the L. *Jurassic*.

Liben An *Ordovician* succession in Bohemia covering the *Mid* and *Late Llandeilo* and part of the *Costonian*.

libethenite (Cu_2PO_4OH) Hydrated copper phosphate, found rarely in the oxide zones of metalliferous *veins*.

libolite A *pitch*-like member of the *asphaltite* group.

lichenometry A relative or absolute dating method for the exposure of a surface using the concentric growths of long-lived lichen, often used in the dating of *glacial* deposits.

liddicoatite ($Ca(Li,Al)_3Al_6B_3Si_6O_{27}(O,OH)_3(OH,F)$) A calcium-rich form of *tourmaline*.

life assemblage See *biocenosis*.

lift force The force experienced by a sediment grain, in a direction at right angles to the flow direction, generated by *pressure* differences over its surface.

light oil Oil with an *API gravity* >30°.

light red silver ore See *proustite*.

light ruby silver See *proustite*.

lignite A soft, low *rank*, earthy, brown-black *coal*, sometimes with a massive *sapropelic* form but more commonly composed of humic material with wood and plant remains in a finer-grained, organic *groundmass*.

LIL See *large-ion lithophile*.

Lillburnian A *Miocene* succession in New Zealand equivalent to the *Serravallian*.

limburgite An alkali-rich and/or *silica-undersaturated, volcanic rock* or minor intrusion made up of *olivine* and *pyroxene* crystals in *basaltic glass*. Originally, the name given to alkali-poor *komatiites* from southern Africa.

lime (CaO) A substance produced by the calcining of high purity *limestone*, with a wide variety of industrial uses.

limestone A rock comprising >50% calcium carbonate, since the *Cambrian* partly or wholly of biogenic origin.

limestone pavement A glacially stripped platform of *limestone* dissected into blocks and runnels by solutional *weathering*.

limiting depth The maximum depth at which the top of an anomalous body could lie and still give rise to an observed *geophysical anomaly*.

limnic Descriptive of the environment of a freshwater lake.

limnic basin A freshwater *coal basin* formed in a river *delta* in an intra*cratonic* lake. cf. *paralic basin*.

limnology The study of lakes.

limonite (brown iron ore) ($FeO.OH.nH_2O$) A general term for a hydrated *iron oxide mineral*.

linarite ($PbCuSO_4(OH)_2$) Deep blue, hydrated lead-copper sulphate found as a *secondary mineral* in the oxidized zone of copper and lead deposits.

line crystal defect A *dislocation* in a *crystal structure*. See also *edge dislocation, screw dislocation, stacking fault*.

line mapping Detailed geological mapping at scales of 1:2500 or more, used for complex areas, in which offsets are taken to an *exposure* from a surveyed base line.

lineage zone A *biostratigraphic zone* based on an evolutionary succession of species of a particular genus, so that *stratigraphic* gaps are precluded.

lineament A major, linear, topographic feature of regional extent of structural or volcanic origin, most easily appreciated from *remote sensing* data, e.g. a *fault* system.

lineation A repeated or *penetrative* linear *structure* in a rock mass. Commonly used for a *metamorphic fabric* of non-specific genesis, but also applied to features of *fault planes, folds*, elongate crystal alignments in *igneous rocks* and *current lineations*, etc. in *sedimentary rocks*.

linguoid Shaped like a tongue.

linnaeite (Co_3S_4) A cobalt *ore mineral* with *spinel* structure.

linsen A *sand* lens of *lenticular bedding* in *mud*.

liparite See *rhyolite*.

Lipopterna An order of infraclass *Eutheria*, subclass *Theria*, class *Mammalia*;

South American ungulates from rabbit- to camel-size, whose nostrils sit well back in the skull, suggesting the presence of a trunk. Range L. *Eocene–Recent.*

liptite A rare *microlithotype of coal* composed of *exinite macerals.*

liptobiolith A *fossil* gum or resin, e.g. *amber.*

liptodetrinite A *coal maceral* of the *exinite* group composed of fragments of *alginite, cutinite, resinite* and *sporinite.*

liquation deposit An oxide-sulphide deposit forming as the result of *liquid immiscibility* in a mixed sulphide-silicate *magma.* The dense sulphides form globules that sink and accumulate at the base of the *magma.*

liquefaction The sudden loss of *shear resistance* associated with the collapse of the grain-supported framework in an *underconsolidated sediment.* This collapse causes a temporary increase in the *pore fluid pressure* as the suspension resediments from the base upwards until a more tightly-packed, grain-supported state is achieved. *Pore fluids* escape during this process, which may be initiated by *earthquakes* or other types of shock.

liquefied flow A type of *sediment gravity flow* which is kept in motion by the buoyancy imparted to the particles by the escape of *pore fluid.*

liquefied natural gas (LNG) *Methane* liquefied at $-160°C$ and one atmosphere, reducing its volume by a factor of over 600.

liquefied petroleum gas (LPG) Liquefied propane and *butane* extracted from *wet gas.*

liquid immiscibility The separation of a homogeneous liquid into two contrasting liquids. Responsible for the diversification of some iron-rich *basaltic* and highly *alkaline magmas* and in the formation of *liquation deposits.*

liquid limit An *engineering geology* term for the minimum amount of water required to be mixed with sediment so that it will flow under standard conditions. cf. *plastic limit.*

liquidus The temperature above which a *mineral assemblage* is entirely liquid. cf. *solidus.*

Lissamphibia A subclass of class *Amphibia*; the frogs, toads and *Apoda.* Range *Triassic–Recent.*

listric Smoothly curving.

listric fan A set of *synthetic, horsetail listric faults* produced in an *extensional fault system* as the *sole fault* migrates into the *footwall.*

listric fault A spoon-shaped, *rotational fault* in which the *hangingwall* is rotated towards the *fault* in the same sense as the movement on the *fault.*

lit-par-lit injection The injection of *magma* or fluid along *bedding, cleavage* or *schistosity* planes to produce *migmatite.*

Litenschichten A *Silurian* succession in Bohemia covering the *Rhuddanian, Aeronian, Telychian, Sheinwoodian* and part of the *Homerian.*

litharge (γPbO) A red *oxide mineral* of lead.

lithia emerald See *hiddenite.*

lithia mica See *lepidolite.*

lithic Formed of rock.

lithic block A *pyroclastic* fragment of initially solid rock over 64 mm in size.

lithic tuff A *tuff* in which rock fragments are predominant.

lithiophilite ($Li(Mn,Fe)PO_4$) A phosphate *mineral* occurring in *pegmatites.*

litho- Pertaining to rock.

lithoclast A *clast* of an origin not associated with the main depositional system. cf. *intraclast.*

lithofacies A body of *sedimentary rock* characterized by specific and distinctive physical and chemical characteristics. cf. *biofacies.*

lithographic stone An ultra-fine-grained *limestone* suitable for printing plates.

lithographic texture A *texture* resembling a *lithographic stone.*

lithology A description of the *macroscopic* features of a rock type.

lithophile element (oxyphile element) An element with an affinity for oxygen, which thus occurs as an oxide or silicate rather than a sulphide or *native element.*

lithophysa A centimetric-scale, rounded mass found in *glassy* and partly crystalline, *felsic volcanic rocks* comprising concentric shells separated by voids. Possibly formed during rapid *crystallization* alternating with gas expansion or by the chemical *alteration* of *spherulites,* with which they are found.

lithophysae Concentric shells of *aphanitic* material encircling hollow cores which make up centimetric-scale masses in *glassy rocks.*

lithosome A sediment body deposited under uniform physical and chemical conditions.

lithosphere The upper shell of the solid Earth comprising the *crust* and *upper mantle* (*viscosity* $> 10^{21}$ Pa s) which deforms in a *brittle* fashion when subjected to a *stress* of ~ 100 MPa. Its base is at a depth of 2–3 km under *ocean ridges*, increasing to up to 180 km beneath old *oceanic crust.* Beneath *cratonic* areas it is at least 250 km thick and possibly as much as 500 km.

lithostatic pressure (lithostatic stress, load pressure, load stress) The vertical *stress* due to the weight of overlying rocks, i.e. $\rho g z$, where ρ = bulk *density* of overlying rock and any contained fluids, g = *gravity* and z = depth.

lithostatic stress See *lithostatic pressure.*

lithostratigraphy The subdivision and correlation of sequences of *strata* by means of rock type. Much less reliable than *biostratigraphy* as rocks rarely persist laterally for great distances, but can be used in the absence of *fossils.*

lithotypes of banded coal The four basic constituents of *banded coal* that can be used for identification in hand specimen and classification: *clarain, durain, fusain* and *vitrain.*

Little Ice Age The period from 1550–1850 AD with extended cold seasons and *glacier* expansion.

littoral The *beach* environment.

liver opal (menilite) *Opal* with a colour resembling liver.

lizardite ($Mg_3Si_2O_5(OH)_4$) A *serpentine mineral.*

Lizzie A *Carboniferous/Permian* succession in Queensland, Australia, covering parts of the U. *Carboniferous* and *Asselian.*

Llandeilo An *epoch* of the *Ordovician,* 468.6–463.9 Ma.

Llandovery The lowest *epoch* of the *Silurian,* 439.0–430.4 Ma.

Llanvirn An *epoch* of the *Ordovician,* 476.1–468.6 Ma.

LNG See *liquefied natural gas.*

load cast (load structure) A type of *sole mark* formed during *wet-sediment deformation* when, for example, *sand* is deposited on *mud,* into which it presses to leave bulbous projections.

load pressure See *lithostatic pressure.*

load stress See *lithostatic pressure.*

load structure See *load cast.*

loam A soil containing approximately equal proportions of *sand, silt* and *clay.*

local magnitude An *earthquake magnitude* scale for crustal events within 600 km of the *seismograph,* based on wave amplitudes and a simple correction for distance.

Lochkovian 1. A *stage* of the *Devonian,* 408.5–396.3 Ma. 2. A *Devonian* succession in Czechoslovakia covering the *Lochkovian* and part of the *Pragian.*

Lochkovium A *Devonian* succession in Bohemia equivalent to the *Lochkovian.*

Lockportian A *Silurian* succession in North America covering part of the *Sheinwoodian,* the *Homerian* and part of the *Gorstian.*

lode A mineralized body resulting from the extensive *replacement* of pre-existing *host rock. Vein* is now the preferred term for all such bodies, regardless of their genesis.

lodestone A stone rich in *magnetite* which aligns in the *geomagnetic field,* used in the past for determining the *magnetic north* direction.

loellingite (löllingite) ($FeAs_2$) A *sulphide mineral* of the *disulphide group* found mainly in *mesothermal vein deposits* or *pegmatites.*

loess *Silt* of *aeolian* derivation, often forming extensive, thick deposits.

Logan Canyon A *Cretaceous* succession on the Scotian shelf of Canada covering the *Aptian, Albian* and part of the *Cenomanian.*

logan stone A large, exposed *boulder,* balanced so that it is easily rocked.

löllingite See *loellingite.*

Lomnitz law A law controlling *creep,* $\varepsilon_1(t) = A[(1 + at)^a - 1]$, where $\varepsilon_1(t)$ = *transient creep,* t = time and A and a are constants for a particular *rheology.*

long river profile The elevation of a river channel plotted against distance downstream.

longitudinal coast See *Pacific-type coast.*

longitudinal joint (bc joint) A *joint* parallel to a *fold axis* and perpendicular to the folded layer at the *fold hinge.*

longshore bar See *swash bar.*

longshore drift The shore-parallel transport of sediment in one direction as the result of oblique *wave* action.

Longtanian A *stage* of the *Permian,* 250.0–247.5 Ma.

Longvillian A *stage* of the *Ordovician*, 449.7–447.1 Ma.

Longwangmiao A *Cambrian* succession in China covering parts of the *Atdabanian* and *Lenian*.

lonsdaleite (C) A hexagonal *polymorph* of *diamond* found in *iron meteorites*.

looping A method of transferring the absolute value of *gravity* from a location where it is known to a new *base station* prior to a *gravity survey*. Consecutive alternate readings are taken at the two locations in order to account for the *drift* of the *gravimeter* taking the measurements.

loparite A niobium-bearing variety of *perovskite*.

lopolith A saucer-shaped igneous intrusion with upper and lower surfaces that are concave upwards, usually of *mafic* composition.

losing stream A stream which loses water when flowing across permeable rocks. cf. *gaining stream*.

loughlinite ($Na_2Mg_3Si_6O_{16}.8H_2O$) A poorly-known *mineral* found in thin *veins*.

Louisiana A *Devonian* succession in the USA equivalent to the *Famenian*.

Love wave A *seismic surface wave* in which particle motion is horizontal and perpendicular to the direction of propagation. The wave travels by *multiple reflection* in the surface low velocity layer.

low angle fault See *lag fault*.

low-grade metamorphism *Metamorphism* at low to moderate temperature and pressure.

low quartz See *alpha quartz*.

low velocity zone (LVZ) A zone at the top of the *asthenosphere* in oceanic areas in which *P waves* are slowed by ~10% compared with waves at higher levels. The zone is interpreted as a region where the melting point of *mantle minerals* is most closely approached so that there is about 0.1% *partial melting*.

löweite ($Na_{12}Mg_7(SO_4)_{13}.15H_2O$) A hydrated sodium-magnesium sulphate; an *evaporite mineral*.

Lowell-Guilbert Model A type of *porphyry copper deposit* in which a cylindrical zone of *potassic alteration* of the host *stock* is surrounded in turn by a phyllic zone of *sericitization*, a zone of *intermediate argillic alteration* and a zone of *propylitic alteration*. The *ore* may be in the host *stock*, *country rocks* or both. cf. *Diorite Model*.

lower continental crust The *continental crust* between 10–12 km depth or the *Conrad discontinuity* and the *Mohorovicic discontinuity*, formed of *granodioritic* rocks in the *granulite facies* with a *density* of ~3.0 Mg m^{-3}. cf. *upper continental crust*.

lower flow regime A *flow regime* characterized by increasing *shear stress* on the bed as flow velocity increases. It is accompanied by *ripple* formation and then *dunes*, causing *flow separation* and thus an increased resistance to flow. cf. *upper flow regime*.

Lower Greensand A *Cretaceous* succession in England covering the *Aptian* and part of the *Albian*.

lower mantle The *mantle* below 400 km or 700 km down to the *Guten-*

berg discontinuity, characterized by very uniform physical properties. Probably composed of close-packed oxides with *perovskite* structure. cf. *upper mantle*.

lower-stage plane bed A flat sediment bed forming at low *shear stresses* from *sand* coarser than 0.6 mm diameter and replacing the *ripples* of finer sediment.

löwigite See *alunite*.

LPG See *liquefied petroleum gas*.

LREE Light *rare earth element*.

Ludfordian A *stage* of the *Silurian*, 415.1–410.7 Ma.

Ludlow An *epoch* of the *Silurian*, 424.0–410.7 Ma.

Lugeon test See *packer test*.

Luisan A *Miocene* succession on the west coast of the USA covering parts of the *Langhian* and *Serravallian*.

luminescence The emission of light by a material when irradiated by electromagnetic radiation of different wavelength.

lump ore See *direct-shipping ore*.

lunate Crescent shaped.

lunette An arcuate *dune* ~20 m in height formed on the *lee side* of a *deflated lagoon*, lake *basin* or river bed in semi-arid areas.

Luning A *Triassic* succession in Nevada, USA equivalent to the *Carnian*.

lustre A *mineral* property caused by the interference of light with the *mineral* surface. See *adamantine lustre*, *greasy lustre*, *pearly lustre*, *resinous lustre*, *silky lustre*, *vitreous lustre*.

lutaceous Formed from mud.

Lutetian A *stage* of the *Eocene*, 50.0–42.1 Ma.

lutite A fine-grained *sedimentary rock* in which *silt* makes up one third to two thirds of the total.

luxullianite A rock formed by the *tourmalinization* of *granite* with the addition of boron, comprising *schorl*, *quartz* and corroded, reddened *feldspar*.

luzonite (Cu_3AsS_4) A rare *ore mineral* of copper found in low- to medium-*grade* copper deposits.

LVZ See *low velocity zone*.

Lycopsida A class of division *Trachaeophyta*, kingdom *Plantae*; the club mosses and related plants. Range *Devonian–Recent*.

lysocline The level in the ocean, above the *carbonate compensation depth*, separating well-preserved (above) from poorly-preserved (below) assemblages of a given calcareous, micro*fossil* group.

M

M-discontinuity See *Mohorovicic discontinuity*.

Ma Okou A *Permian* succession in China equivalent to the *Kungurian* and *Zechstein*.

maar A type of *tuff ring* in which the centre of the *crater* has been affected by down-faulting or sagging, so that it lies below the surrounding ground surface.

Maastrichtian The highest *stage* of the *Cretaceous*, 74.0–65.0 Ma.

maceral The basic organic constituent of *coal*, recognizable at *microscopic* scale and made up of the remains of plant materials existing at the time of *peat* formation. Three maceral groups are recognized on the basis of their physical appearance: *vitrinite*, *exinite* and *inertinite*.

machair A coastal area of calcareous, sandy soils covered by rich grassland.

mackinawite $(Fe,Co,Ni,Cr,Cu)_{1+x}S$) A *sulphide mineral* of the *layer sulphides group* with tetragonal PbO structure.

macrinite A *coal maceral* of the *inertinite* group resembling *fusinite* but without cell structure.

macrocrystalline Descriptive of a material in which crystals are visible to the unaided eye. cf. *microcrystalline*.

macrophyric Descriptive of the *texture* of a medium- to fine-grained *igneous rock* with *phenocrysts* >2 mm long. cf. *microphyric*.

macropore A structural void in a soil.

macroscopic Referring to a *structure* or feature of a scale of kilometres to hundreds of kilometres. cf. *mesoscopic*, *microscopic*.

maculose Spotted.

Madagascar aquamarine A strongly *dichroic*, blue variety of *beryl*, valued as a *gem*.

Madagascar topaz See *citrine*.

made ground An area of land that has been constructed by man, often using a *landfill* of natural materials.

Madeira topaz A brown, heated *amethyst*.

madupite A *silica-saturated lamproite* with *sanidine* rather than *leucite*, often carrying *diamonds*.

Maentwrogian A *stage* of the *Cambrian*, 517.2–514.1 Ma.

mafic A general term for *ferromagnesian minerals*. cf. *felsic*.

maghemite (γ-Fe_2O_3) An *oxide mineral* of the *magnetite* series.

magma A melt, generally containing suspended crystals and dissolved gases or volatiles, formed by total or *partial melting* of solid *crustal* or *mantle* rocks. Comprises polymers of interconnected, disordered Si-O tetrahedra with cations such as magnesium, iron, calcium, sodium and potassium in loose coordination with the oxygens and may range in composition from *ultramafic* through *basaltic*, *andesitic*, *dacitic* and *rhyolitic*. Magma diversification can take place through the processes of *magmatic differentiation* (including *fractional crystallization*, *liquid immiscibility* and *vapour transport*), *assimilation* or *magma mixing*.

magma chamber A subsurface accumulation of *magma*.

magma mixing A mechanism of *magma* diversification, generally on a local scale, involving the mixing of two *magmas* of contrasting composition to form a hybrid intermediate in composition.

magma pressure The *hydrostatic pressure* created by *magma*.

magmatic differentiation The separation of an initially homogeneous *magma* into two or more *magmas* of contrasting composition by the processes of *fractional crystallization*, *liquid immiscibility* or *vapour transport*.

magmatic segregation deposit See *orthomagmatic segregation deposit*.

magnesia (MgO) An industrial product derived from *magnesite*.

magnesia alum See *pickeringite*.

magnesian limestone A *limestone* with a small proportion of *dolomite*.

magnesiochromite ($MgCr_2O_4$) An *oxide mineral* with *spinel crystal structure*.

magnesioferrite ($MgFe_2O_4$) An *oxide mineral* with *spinel crystal structure*.

magnesite ($MgCO_3$) A *carbonate mineral*, the source of *magnesia*.

magnetic anisotropy The phenomenon whereby a *ferromagnetic* particle is more easily magnetized in one direction than others, so deflecting the original direction in the direction of *anisotropy*.

magnetic anomaly The variation in the *geomagnetic field* arising from variation in the magnetic properties of underlying rocks.

magnetic coercivity The *magnetic field* required to reduced the magnetization of a *ferromagnetic material* to zero.

magnetic dating A relative or absolute dating method making use of the *secular variation* of the *geomagnetic field*, *polarity reversals* or *apparent polar wander*, which cause *remanent magnetization* directions to vary with time. Comparison of directions for different regions at the same time will demonstrate if they have moved relative to each other since the magnetization was acquired.

magnetic domain A small (~1 μm) volume of a *ferromagnetic* material within which electron spins are coupled to produce a unidirectional *magnetic field* within it.

magnetic elements See *geomagnetic elements*.

magnetic epoch A time interval of constant *geomagnetic field* polarity, normally longer than 10 000 years.

magnetic equator (aclinic line) A line joining the locations of zero magnetic *inclination*, which approximately follows the geographic equator.

magnetic field (magnetic flux, magnetic induction) The *force* experienced by a unit positive *magnetic pole* at the point of measurement.

magnetic flux See *magnetic field.*

magnetic gradiometer An instrument comprising a pair of *proton, cesium vapour* or *fluxgate magnetometer* sensors separated by a short distance in the vertical or horizontal planes, used to measure directly the vertical or horizontal gradient of the *geomagnetic field.*

magnetic induction See *magnetic field.*

magnetic lineations See *oceanic magnetic anomalies.*

magnetic log A *geophysical borehole log* in which the *geomagnetic field* is measured down a borehole to provide information on the presence of magnetic rocks.

magnetic meridian See *magnetic north.*

magnetic moment For a dipole, the product of the *magnetic pole* strength and the distance between the poles.

magnetic north (magnetic meridian) The direction to the Earth's magnetic north pole. cf. *true north.*

magnetic observatory A fixed installation which continuously monitors all the *magnetic elements* and thus records *diurnal* and *secular variation* of the *geomagnetic field.*

magnetic permeability A constant describing the magnetic properties of the medium separating the causative body of a *magnetic anomaly* and its point of observation.

magnetic pole The points at which the lines of magnetic force surrounding a dipole converge (single poles very rarely exist in isolation). The Earth's magnetic poles are displaced from the geographic poles and are not exactly antipodal.

magnetic potential The scalar quantity which provides the *magnetic field* of a source in any direction when differentiated in that direction.

magnetic pyrites See *pyrrhotite.*

magnetic quiet zone A region with few or no *magnetic anomalies*, particularly an area of *oceanic crust* with no *magnetic lineations* due to an absence of *geomagnetic polarity reversals* or the loss of original magnetizations.

magnetic reversal (polarity reversal) A change in the orientation of the *geomagnetic field* in which the south *magnetic pole* becomes the north pole and vice versa.

magnetic spectrometer An instrument that can be attached to a *transmission electron microscope* which allows the energy lost by inelastically-scattered electrons to be measured. This is characteristic of the elements present and allows microanalyses to be made.

magnetic storm A severe type of *diurnal variation* in the *geomagnetic field* in which the field varies greatly in amplitude over short time periods. Caused by the arrival in the ionosphere of charged particles generated during sunspot activity.

magnetic survey A survey undertaken with a *magnetometer* on land, at sea or in the air to search for magnetic rocks and *minerals*, archaeological artefacts or buried ferrous metals.

magnetic susceptibility A measure of the magnetic behaviour of a material in a *magnetic field*; a dimensionless constant of proportionality in the relationship between *intensity of induced magnetization* and the *magnetizing force* of the inducing *magnetic field*.

magnetic variometer (variometer) An early form of *magnetometer* based on a small dipole suspended in the *geomagnetic field*.

magnetite (Fe_3O_4) An *oxide mineral* with the *spinel crystal structure*; the most common *ferrimagnetic mineral*.

magnetizing force The phenomenon responsible for the creation of a *magnetic field*.

magnetochronology A *geochronological* sequence based on *geomagnetic polarity reversals*.

magnetogram A recording of temporal variations of the *geomagnetic elements*.

magnetohydrodynamics The generation of a *magnetic field* by the motion of electrically charged particles; such an origin has been proposed for the *geomagnetic field*.

magnetometer An instrument for measuring a *magnetic field*. *Fluxgate*, *proton* and *cesium vapour magnetometers* are used in *magnetic surveys* to measure the *geomagnetic field*. *Spinner* and *cryogenic magnetometers* measure the fields associated with rock samples in *palaeomagnetic* studies.

magnetostratigraphy The use of geomagnetic *polarity reversals* recorded in stratal sequences for the purposes of correlation, particularly useful in the absence of *fossils* and for correlating between marine and terrestrial sequences.

magnetotelluric survey (MT survey) An *electromagnetic induction method* of *geophysical exploration* making use of naturally occurring electromagnetic fields to investigate the distribution of *electrical conductivity* in the subsurface. The *depth of penetration* extends to some 7 km, so that the method has application in *hydrocarbon* surveys.

magnitude See *earthquake magnitude*.

magnitude-frequency relationship The relationship between *earthquake magnitude* and the number of events of a given *magnitude*, which shows that the number increases by a factor of ten for each *magnitude* unit less than the one considered.

major element An element present in a rock in high concentration, so that it controls the presence of *minerals* such as *feldspar* and *pyroxene*. cf. *trace element*.

major fold The larger, *macroscopic fold* in a complexly folded region, generally with a wavelength of kilometric scale. cf. *minor fold*.

malachite ($Cu_2CO_3(OH)_2$) A bright green *carbonate mineral*, often found in the oxidized parts of copper *ores*.

Malacostraca A class of subphylum *Crustacea*, phylum *Arthropoda* characterized by a carapace covering the head and part of the trunk, which comprises a distinct thorax and abdomen, and terminates in a telson. Range *Cambrian–Recent*.

Malakovian A *Triassic* succession in New Zealand equivalent to the *Nammalian*.

malignite A *mesocratic* variety of *nepheline syenite*.

malleable A *tenacity* descriptor indicating a *mineral* that can be hammered into a thin sheet.

Malm The youngest *epoch* of the *Jurassic*, 157.1–145.6 Ma.

Malvinokaffric Province A biogeographic province extant in *Silurian* times, including much of South America and adjacent parts of south and north Africa.

mamillated (mamilliary) With the form of portions of spheres.

Mammalia/mammals A class of vertebrate animals that suckle their young, the *fossils* of which are classified by their teeth. Range late *Triassic–Recent*.

mamilliary See *mamillated*.

man-made earthquake An *earthquake* induced by human activity as a result of mining, the construction of reservoirs, explosion *aftershocks* and fluid injection into deep boreholes. All these factors change the regional *stress* pattern or add/redistribute *pore fluids* and thus can trigger events.

manganepidote See *piemontite*.

manganese nodule A *concretion* of ferromanganese oxides on the ocean floor, forming by the extraction of metals from seawater and the *pore fluids* of seafloor *muds*. Generally 5–200 mm in size and spheroidal, ellipsoidal or *botryoidal* in form, sometimes with a rock core or no core at all and usually with internal *mineral* zoning. Not presently of significant economic importance.

manganese spar See *rhodochrosite*.

manganite (γ-MnO(OH)) An *ore mineral* of manganese found mainly in low-temperature *hydrothermal veins*.

manganoan cummingtonite See *tirodite*.

manganophyllite A variety of *phlogopite* or *biotite* rich in manganese.

manganosite (MnO) A green *oxide mineral* of manganese, which becomes black on exposure.

manganotantalite ($(Mn,Fe)Ta_2O_6$) An *ore mineral* of tantalum found mainly in *granite pegmatites*.

Mangaorapan An *Eocene* succession in New Zealand covering the middle part of the *Ypresian*.

Mangaotanian A *Cretaceous* succession in New Zealand covering parts of the *Cenomanian* and *Turonian*.

Mangapanian A *Pliocene* succession in New Zealand covering part of the *Piacenzian*.

mangerite A *charnockitic* rock equivalent to *hypersthene monzonite*.

mangrove swamp An intertidal region of mudflats and mangrove vegetation found along sheltered, low-energy shorelines in tropical areas.

manjiroite ($(Na,K)Mn_8O_{16}.nH_2O$) A *secondary ore mineral* of manganese.

Manning equation A formula for estimating stream velocity (v) from *channel roughness*, slope (s) and *hydraulic radius* (R): $v = kRsn$, where k = constant (1 in *SI* units), n = Manning roughness coefficient of the stream bed.

mantle The inner shell of the Earth between the *Mohorovicic* and *Gutenberg discontinuities*, with a silicate *mineralogy* distinct from the *crust* above and *core* below. Its composition is probably equivalent to a mixture of ~75% *dunite* and ~25% *basalt*. The uppermost mantle beneath oceanic areas is the location of the *low velocity zone*. The *mineralogy* of the mantle changes with depth to denser *mineral* phases as the pressure increases (e.g. *olivine* → *spinel* → *perovskite*) and this is responsible for the *transition zone* between 370 and 700 km.

mantle bedding A uniform thickness of *pyroclastic* material over all but the steepest topography.

mantle drag The *force* exerted on the base of the *lithosphere* by movement of the *asthenosphere*. If the *asthenosphere* moves at a higher velocity than the *lithosphere*, the latter's velocity is enhanced and vice versa. This is unlikely to be an important mechanism of *plate* movement as the *low velocity zone* at the top of the *asthenosphere* probably would not allow efficient coupling between *lithosphere* and *asthenosphere*.

mantle plume A persistent column of hot material, in the form of a vertical cylinder with a radius of ~150 km, rising to the *crust* from the *mantle*, possibly originating by localized streaming from the *core-mantle* boundary. Responsible for *hotspots*.

mantle transition zone A layer in the *mantle* between 400 and 700 km across which there is an increase in *seismic velocity* due to phase changes of the *minerals* present to more closely-packed, denser forms.

mantled gneiss dome A *structure* comprising a variably-oriented, dome-shaped core of *granitic gneiss* overlain by supracrustal metasedimentary and metavolcanic rocks. May originate by the *deformation* of an *unconformity*, by intrusion of the core material into the cover or from the buoyant ascent of low *density granitic basement* into denser cover rocks.

manto A horizontal to subhorizontal, tubular *orebody*, more rarely a *tabular orebody*.

Maozhuang A *Cambrian* succession in China covering parts of the *Lenian* and *Solvan*.

Maping A *Carboniferous*/*Permian* succession in China equivalent to the *Kasimovian*, *Gzelian* and *Asselian*.

marble A metamorphosed *limestone* formed by *recrystallization* during *thermal* or *regional metamorphism*. It may form an attractive *building stone*, although usage of the term marble by stonemasons also encompasses unmetamorphosed *limestones*.

marcasite (white iron pyrites) (FeS_2) A relatively common *sulphide mineral*, occasionally used as a *gem*.

March analysis The analysis of the behaviour of passive markers in a homogeneous body that is deforming by *viscous flow*. Elongate markers originally parallel to the axis of *shortening* rotate symmetrically from this direction by an amount dependent on the *strain* to form a *bimodal preferred orientation*.

marekanite A *rhyolitic perlite* which has been broken down to rounded *pebbles*.

margarite ($CaAl_2(Al_2Si_2O_{10})(OH)_2$) A *brittle mica*.

marginal basin See *back-arc basin*.

marginal ore *Ore* which just repays the cost of exploitation.

marginal sea See *back-arc basin.*

marialite ($Na_4(AlSi_3O_8)_3(Cl_2,CO_3,SO_4)$) A *metamorphic mineral* of the *scapolite* series.

marine abrasion platform See *shore platform.*

marine band A horizon of marine origin within a succession of non-marine *strata* and representative of a brief *transgression.* Constitutes a useful *marker bed.*

marine snow The discarded mucus feeding sheets or strands of gelatinous zooplankton such as of *pteropods, salps* and appendicularians. This traps suspended particulate matter and transports it to the seabed, although it itself is not preserved.

marker bed A distinctive *bed* useful in constructing a *lithostratigraphy.*

markfieldite A variety of *diorite* with *porphyritic texture* and a *granophyric groundmass.*

Markowitz wobble A systematic change in position of the Earth's axis of rotation of unknown origin with a periodicity of 30 years and an amplitude of 25 marcs.

marl A friable, calcareous *mudstone.*

Marl Slate The extensive basal deposit of the *Permian Zechstein Sea* in NE England, equivalent to the *Kupferschiefer* in continental Europe.

marlstone Indurated *marl.*

marmorization The thermal *recrystallization* of *limestone* to produce *marble.*

Marsdenian A *stage* of the *Carboniferous,* 321.5–320.6 Ma.

marsh gas *Natural gas* (*methane*), of no commercial importance, produced by bacterial alteration of organic matter near the Earth's surface.

Marshbrookian A *stage* of the *Ordovician,* 447.1–444.5 Ma.

Marsupiala/marsupials An order of infraclass *Eutheria,* subclass *Theria,* class *Mammalia; mammals* in which development of the young takes place in an external pouch. Range U. *Cretaceous–Recent.*

martite *Haematite* or an intergrowth of *haematite* and *magnetite* which replaces *magnetite* along *cleavage* planes.

maskelynite A *glass* of *plagioclase* composition occurring in colourless, *isotropic* grains in *meteorites,* probably representing re-fused *feldspar.*

mass deficiency The difference in mass between a body of relatively low *density* (e.g. a sedimentary *basin*) and the relatively high *density country rock* which would otherwise occupy its space. It can be estimated by a *gravity survey.*

mass excess The difference in mass between a body of relatively high *density* (e.g. an *orebody*) and the relatively low *density country rock* which would otherwise occupy its space. It can be estimated by a *gravity survey.*

mass extinction The *extinction* of a large number of *fossil* groups in a limited period of time, such as occurred at the *K-T boundary,* possibly as a result of *bolide impact.*

mass flow A *slide* of sediment downslope under the force of *gravity.*

mass movement (mass wasting) The

movement of mass takes place by *slide* or flow processes under the influence of *gravity*. Flow processes include *solifluction, soil creep, debris avalanches, earth flows* and *mud flows*. *Slide* processes include *rock falls, rock slides, planar slumps* and *rotational slumps*. Additionally there are *frost heave, freeze-thaw* and *cambering* movements.

mass solute transfer A *diagenetic* process of transport of dissolved species from reaction sites in donor sediments where *dissolution* occurs to reaction sites in receptor sediments where *precipitation* takes place.

mass spectrometry The determination of the ratios and/or concentrations of isotopes in rocks or other materials. Ionized atoms of the sample are accelerated by a voltage through a *magnetic field* which curves their path by an amount dependent on ionic mass. Variation of the voltage allows ions of different masses to be focused on a detector. Particularly important in *rubidium-strontium, potassium-argon* and *accelerator radiocarbon dating*.

mass wasting See *mass movement*.

massicot (PbO) A rare yellow lead oxide, found as a *secondary mineral* associated with *galena*.

massive sulphide deposit (volcanic-associated massive sulphide deposit) A large, usually *stratiform, conformable orebody* composed mainly of iron sulphide, usually *pyrite* ± *pyrrhotite*, along interfaces between volcanic units or volcanic units and sediments. Usually underlain by a *stockwork* which acted as the feeder for the mineralizing fluids. Probably *syngenetic* in origin, forming by *volcanic-exhalative* processes, such as at *black smokers*.

master joint A *joint* whose extent is considerably greater than others in the set and against which less prominent *joints* terminate.

Mata A *Cretaceous* succession in New Zealand comprising the *Haumurian* and *Piripauan*.

matrix 1. See *groundmass*. 2. The fine-grained material separating *clasts* in a *sedimentary rock*.

matrix-supported conglomerate A *rudite* with 30–85% *clasts* which are commonly not in contact. cf. *clast-supported conglomerate*.

maturation The processes whereby organic matter is transformed into oil and gas.

Matuyama A *magnetostratigraphic epoch* of *reversed polarity* in the *Pleistocene*, 2.42–0.71 Ma.

Mauretanian Orogeny An *orogeny* affecting NW Africa in *Devonian* times.

Maxillopoda A class of subphylum *Crustacea*, phylum *Arthropoda* including the barnacles. Range U. *Silurian–Recent*.

maximum octahedral shear failure criterion A *failure criterion* in which *failure* occurs when the octahedral *shear stress* reaches a constant value dependent on the material.

maximum projection sphericity The ratio of the maximum and minimum cross-sectional areas of a particle, used in the description of the *clasts* of a *sedimentary rock*.

maximum strain energy of distortion failure criterion A *failure criterion* in which *failure* occurs when the *strain* energy of distortion reaches a constant value dependent on the material.

Maxwell model See *elastoviscous deformation.*

Maxwell substance A substance in which the *strain rate* is equal to the ratio of *shear stress* to three times the *viscosity*.

Mayan A *Cambrian* succession in Siberia covering part of the *Menevian.*

Mayen lava (Andernach lava, Niedermendig lava) A highly *vesicular*, grey *nepheline tephrite* from the Mayen region of western Germany, extensively quarried for *millstones* from the Bronze Age to the 19th century.

Mazon Creek An *exceptional fossil deposit* of U. *Carboniferous* age in Ilinois, USA, containing many species of plants and animals preserved in *siderite nodules.*

meander belt An area of land occupied by a meandering channel between *avulsions.* Submarine channels may also meander.

meander scroll The topography, comprising low, curved ridges of relatively coarse sediment parallel to a river channel, of exhumed *point bars* resulting from differential *erosion* of *beds* in a truncated *epsilon cross-stratified* complex.

meandering stream A stream with planform *sinuosity* >1.3, occurring in alluvial, submarine and tidal environments.

measured reserve See *ore reserve.*

mechanical infiltration of fines The introduction of *clay* and *silt* grade sediment into coarse-grained, porous, permeable sediment where surface *seepage* accompanies *alluviation* in areas of low *water table*, thus giving rise to a secondary *matrix.*

median valley See *axial rift.*

medium oil Oil with an *API gravity* of 22°–30°.

meerschaum See *sepiolite.*

mega- Very large.

megablast A large *porphyroblast* in a coarse-grained *igneous rock.*

megabreccia A very coarse *breccia* in which the *clasts* may exceed 1 km in length, which may have been formed by a *landslide.*

megacryst A crystal in an *igneous rock* which is large compared to those in the *matrix.*

megaripple A *ripple* with a wavelength >1 m.

megaturbidite (seismoturbidite) A very thick-bedded *turbidite* >1 m in thickness, which is commonly laterally continuous over an entire depositional *basin.* Internal *cross-stratification* or *ripple-lamination* may be developed, indicative of flow reflection from the *basin* margins. Believed to originate from very large volumes of sediment produced by *failure* of the *basin* margin, perhaps in response to *earthquake* activity.

meimechite An *ultramafic volcanic rock* with *phenocrysts* of *olivine.*

meionite $(Ca_4(Al_2Si_2O_8)_3(Cl_2CO_3SO_4))$ A *metamorphic mineral* of the *scapolite* series.

mela- Prefix attached to an *igneous rock* name signifying that it is of darker colour than usual.

melaconite A massive variety of *tenorite* in the form of a black, earthy material formed by oxidation in copper *veins*.

mélange A metric- to kilometric-scale body of rock composed of chaotic blocks of *competent strata* in a finer-grained *matrix*. Sedimentary mélanges (*olistostromes*) originate by avalanching, *gravity* slumping or sliding; *tectonic* mélanges are the result of *tectonic deformation* in which *sedimentary rock* is dismembered.

melanite A black to dark brown, titanium-bearing *andradite*.

melanocratic Descriptive of a rock with 60–90% dark *minerals*.

melanosome A dark-coloured band rich in *mafic* and aluminous *minerals*, found between coarse-grained, quartzofeldspathic *veins* in *regionally metamorphosed pelites* and *psammites*. Represents a layer which has been highly shortened by *dissolution* or melting as the *quartz* and *feldspar* of the parent rock were removed along non-*penetrative cleavages* caused by high *shear stress* during *metamorphism*, leaving the *mafic minerals* behind.

melanterite ($FeSO_4.7H_2O$) A green-blue, *secondary mineral* of iron, formed by the *weathering* of *pyrite*, *marcasite* and cupriferous *pyrite ores*.

Melekesskian A *stage* of the *Carboniferous*, 313.4–311.3 Ma.

melilite A group of *feldspathoid minerals* including *gehlenite*, *akermanite* and *soda melilite*.

melilitite An *ultramafic volcanic rock* comprising *melilite* and *pyroxene*.

melitolite An *ultramafic plutonic* rock comprising *melilite*, *pyroxene* and *olivine*.

melteigite An *ijolite* with 70–90% *mafic minerals*.

member A grouping of *beds* used in *lithostratigraphy*. See also *bed*, *formation*, *group*, *supergroup*.

membrane polarization (electrolytic polarization) A mechanism of *induced polarization* caused by the varying mobility of ions in *pore fluid* travelling through small pores in a rock. Charge build-up on either side of pores gradually disperses when the polarizing current is removed, creating a decaying voltage.

membrane tectonics The *deformation* of a *plate* resulting from its movement from regions of different radii of curvature on the Earth's surface. Radial tensional *stresses* are generated when moving from small to large radii of curvature regions and compressional *stresses* in the opposite direction. Suggested as a possible, but unlikely, mechanism behind *continental splitting*.

meneghinite ($CuPb_{13}Sb_7S_{24}$) A rare *ore mineral* of copper and lead.

Menevian A *stage* of the *Cambrian*, 530.2–517.2 Ma.

menilite See *liver opal*.

Meotic (Sarmatian) A *Miocene* succession on the Russian Platform covering the upper part of the *Messinian*.

Meramec A *Carboniferous* succession in the USA covering the lower part of the *Viséan*.

Mercalli scale A twelve point scale of *earthquake intensity* devised in 1902. See also *modified Mercalli scale*.

mercury (Hg) A naturally occurring liquid *native element.*

mere A small lake of uncertain origin on *glacial* outwash deposits and other superficial materials, possibly the result of *thermokarst* or *solution* processes.

Merioneth The youngest *epoch* of the *Cambrian*, 517.2–510.0 Ma.

Merions A *Devonian* succession in E Australia equivalent to the *Pragian.*

merocrystalline Descriptive of a rock containing both crystals and *glass.*

merokarst A *limestone* landscape with only partially developed *karst* landforms.

meromictic The status of a lake which is permanently stratified with a well-developed *epilimnion* and *hypolimnion* as a result of atmospheric disturbances being insufficient to break down the layering. cf. *holomictic.*

Merostomata A class of phylum *Arthropoda* including the king crabs and water scorpions. Range *Ordovician–Recent.*

mesa A steep-sided, flat-topped *plateau* or promontory surrounded by a flat *erosional* plain, forming as a result of *parallel retreat* or protection from *erosion* by a hard capping such as a *cuirasse.*

mesocratic Descriptive of a rock with 30–60% dark *minerals.*

mesogenesis See *burial diagenesis.*

mesolite ($Na_{16}Ca_{16}(Al_{48}Si_{72}O_{240}).64H_2O$) A *zeolite* found in cavities of *volcanic rocks.*

Mesosauria An order of subclass *Anapsida*, class *Reptilia*; an aquatic group of *reptiles* with long, needle-like teeth for catching small *crustaceans.* Range late *Carboniferous*–early *Permian.*

mesoscopic Referring to a *structure* or feature of a scale of metres. cf. *macroscopic, microscopic.*

mesosiderites A heterogeneous group of *stony iron meteorites* with a metal content possibly exceeding 40%, but not forming a continuous network.

mesosphere A largely outdated term for the mechanically strong layer beneath the *asthenosphere.*

mesostasis The final fraction of a *magma* to crystallize, in the spaces between existing crystals.

mesothermal deposit An *epigenetic mineral* deposit, intermediate between *epithermal* and *hypothermal deposits*, formed at 200–300° and depths of 1200–4500 m and generally found associated with near-surface *intrusive igneous rocks.* Comprises *fracture* fills and *replacement bodies*, often with well developed *zoning.*

mesotype An *igneous rock* with a *colour index* of 30–60.

Mesozoic An *era* comprising the *Triassic, Jurassic* and *Cretaceous*, 245.0–65.0 Ma.

mesozone A *depth zone* of medium temperature and high *hydrostatic pressure.* See also *epizone, katazone.*

Messel Oil Shales An *exceptional fossil deposit* of *Eocene* age near Darmstadt, Germany, containing a *lacustrine* assemblage of articulated vertebrates, *insects* and *plants.*

Messinian The highest *stage* of the *Miocene*, 6.7–5.2 Ma, during which many *evaporite basins* were developed in the Mediterranean region.

meta- Prefix indicating a metamorphosed variety.

meta-anthracite The highest *rank* of *anthracite*, with at least 98% *fixed carbon*.

metabasite Any metamorphosed *basic igneous rock*.

metacinnabar ($Hg_{1-x}S$) A high temperature form of *cinnabar*.

metacryst See *porphyroblast*.

metal factor parameter A measure used in the quantification of frequency domain *induced polarization* data, defined as $2\pi 10^5(\rho_0 - \rho_\infty)/\rho_\infty \rho_0$, where ρ_0 and ρ_∞ correspond to *apparent resistivities* measured at low and high alternating current frequencies respectively.

metal excess group A group of *sulphide minerals* characterized by the presence of greater numbers of metal atoms than sulphur atoms.

metalimnion The top of the *hypolimnion*, a zone of rapid temperature change.

metallic lustre A *lustre* similar to polished steel seen in some *opaque minerals*.

metallogenic epoch A period of time during which there was abundant mineralization of the same type.

metallogenic province A region of the *crust* in which there is abundant mineralization of the same type.

metalloid A substance with both metallic and non-metallic properties, e.g. arsenic.

metaluminous Descriptive of an *igneous rock* in which Al_2O_3 exceeds $(CaO + Na_2O + K_2O)$.

metamorphic aureole (contact aureole) The zone of *metamorphism* of the *country rocks* around an intrusion, which may involve *metasomatism* as a result of heating *groundwater* as well as *recrystallization* and the development of new *minerals*.

metamorphic banding Banding caused by the *recrystallization* of original *fabrics* to produce new planar *fabrics* such as *foliation* and *gneissosity*.

metamorphic differentiation (metamorphic segregation) The separation of components or phases in a rock by *metamorphic* processes.

metamorphic facies A group of rocks that reached chemical equilibrium at the same pressure and temperature range of *metamorphism* and characterized by particular *mineral assemblages*.

metamorphic grade The intensity of *metamorphism*, an indicator of the temperature and pressure conditions extant.

metamorphic rock A rock which results from the partial or complete *recrystallization* in the solid state under temperature and pressure conditions elevated with respect to the surface.

metamorphic segregation See *metamorphic differentiation*.

metamorphic zone An area of *metamorphic rocks* defined by the appearance of certain *mineral assemblages*.

metamorphism The processes by which rocks are changed by the solid-state application of heat, pressure and fluids, excluding *weathering* and *diagenesis*. See also *autometamorphism*, *dynamic metamorphism*, *regional metamorphism*, *retrograde metamorphism*, *thermal metamorphism*.

metapedogenesis The human alteration of the properties of a soil, both deliberate and unintentional.

Metaphyta The plants. Range *Precambrian–Recent*.

metaquartzite A *quartzite* formed by the *metamorphism* of *sandstone*. cf. *orthoquartzite*.

metasilicate See *cyclosilicate*.

metasomatism A *metamorphic* process in which the chemical composition of a rock is changed significantly, usually as a result of fluid flow.

metastable Existing under conditions outside the normal range of stability.

Metatheria An infraclass of subclass *Theria*, class *Mammalia*; the *marsupials*. Range late *Cretaceous–Recent*.

Metazoa The animals. Range *Precambrian–Recent*.

meteoric diagenesis *Diagenesis* caused by rainfall-derived *groundwater*. Important in the formation of *limestones* as they are prone to subaerial exposure. *Aragonite* and magnesium *calcite* are unstable in *meteoric waters* and are replaced by low-magnesium *calcite*.

meteoric water *Groundwater* derived from rainfall or *infiltration*.

meteorite An extraterrestrial body, derived from the Asteroid belt of the solar system, impacting the Earth's surface. Three main classes exist: *iron meteorites*, *stony iron meteorites* and *stony meteorites*. These are subdivided into *chondrites* and *achondrites* on the basis of the presence or absence respectively of *chondrules*.

methane (CH_4) The lightest component of *crude oil*.

methane series (paraffin series) Straight-chain *hydrocarbons* with the general formula C_nH_{2n+2}, found in *crude oil*.

methanogenesis The production of *methane* during *fermentation*.

meulière A *sarsen*-like stone found in the *Tertiary* of France.

Mexican onyx A *translucent*, veined, partly-coloured variety of *aragonite*.

mgal See *milligal*.

miarolitic cavity A small, crystal-lined cavity in an *intrusive igneous rock* resulting from the segregation of small gas pockets into an irregular cavity defined by surrounding crystals.

miarolitic fabric A *fabric* in an *intrusive igneous rock* formed by the alignment of *miarolitic cavities*.

miaskite A *leucocratic*, *biotite nepheline monzosyenite*.

mica schist A *schist* rich in *mica*, commonly *muscovite*.

micaceous Containing or resembling *mica*.

micaceous iron ore A *specular* variety of *haematite* with a flaky *habit* reminiscent of the *micas*.

micas Sheet silicates characterized by a platy morphology and perfect basal *cleavage* in consequence of their atomic structure. The general formula is

$X_2Y_{4-6}Z_8O_{20}(OH,F)$ where X is K or Na, Y is Al, Mg, Fe, Mn, Cr, Ti, Li etc. and Z is Si, Al or Fe^{3+}. Common in *igneous* and *metamorphic rocks*, and also found in *sedimentary rocks*.

Michel-Lévy chart A chart of standard colours used in measuring *birefringence*.

micrinite A *coal maceral* of the *inertinite* group made up of very small rounded grains ~ 1 μm across.

micrite An abbreviation of **micr**ocrystalline *calcite*; very fine-grained (< 4 μm) carbonate making up the *matrix* in *limestones*.

micritic limestone A *limestone* with a *matrix* of *micrite* rather than a sparry cement.

micritization The degradation of coarse calcareous material by *replacement* or reduction in grain size to *micrite*, often by biological activity.

micro- Extremely small; × 10^{-3} when attached to a unit.

microatoll (patch reef) An individual massive *coral* colony in a *lagoon*.

microboring A submillimetric diameter boring up to 1 mm in length made by a microorganism. The *dissolution* process used is important in the *bioerosion* of carbonates.

microcline ($KAlSi_3O_8$) The low-temperature form of *potassium feldspar*.

microcontinent A small fragment or remnant of *continental crust* up to about the size of Malagasy.

microcrack A *microfracture* on which there has been no visible *displacement*. cf. *microfault*.

microcrystalline Descriptive of material that is so fine-grained that its crystals can be viewed only microscopically. cf. *macrocrystalline*.

microdiorite A medium-grained *igneous rock* with the *mineral assemblage* and chemistry of *diorite*.

microfabric A *microscopic* scale *fabric*.

microfalaise A small cliff.

microfault A *microfracture* on which there has been visible *displacement*. cf. *microcrack*.

microfelsitic texture A *cryptocrystalline texture* in the *groundmass* of *felsic igneous rocks* formed by the *devitrification* of an originally *glassy matrix*.

microfracture A *microscopic* discontinuity across which there has been separation. Observed prior to faulting with a *preferred orientation* orthogonal to the least *principal stress*. Significant in *dilatancy-diffusion theory*.

microgal (μgal) The *cgs* unit of *gravity anomalies* in *microgravity* surveys, equal to 10^{-2} g.u. or 10^{-6} cm s^{-2}.

microgranite A medium-grained, *microcrystalline igneous rock* with the composition and *texture* of a *granite*.

microgranodiorite A medium-grained, *microcrystalline igneous rock* with the composition and *texture* of a *granodiorite*.

micrographic texture (micropegmatite texture) A *microscopic* form of *graphic texture*.

microgravity A technique of measuring *gravity* to *microgal* accuracy using specialized *gravimeters*. Used in the search for subsurface voids, monitoring underground water movement and in measuring rates of *neotectonic* movement.

microlite 1. ($Ca_2Ta_2O_6(O,OH,F)$) A *mineral* found in *pegmatites*. 2. A very small crystal, usually in the *glassy groundmass* of a rapidly chilled *lava*, representing an initial stage of crystal *nucleation* and growth.

microlithon A *tabular* body of rock defined by *cleavage* surfaces, formed during the *buckling* of a layered sequence along the hinges of buckles. See also *gleitbretter*.

microlithotype of coal A *maceral* association visible at a *microscopic* scale, given the suffix -ite for distinction from a *maceral* and commonly containing several *maceral* types. The seven microlithotypes are *clarite*, *durite*, *inertite*, *liptite*, *trimacerite*, *vitrinertite* and *vitrite*.

microlog A *geophysical borehole electric log* in which small electrodes mounted on a pad are pressed against the wall-rock. Provides information particularly on the *mudcake*.

micropegmatite texture See *micrographic texture*.

microperthite A *microscopic* intergrowth of *albite* and *potassium feldspar*.

microphyric Descriptive of the *texture* of a medium- to fine-grained *igneous rock* with *phenocrysts* <2 mm long. cf. *macrophyric*.

micropiracy A type of *river capture* in which small *rills* and gullies migrate back and forth across a hillslope, giving rise to even *erosion*.

microplate A small *plate* with identifiable margins, which may subsequently become a *displaced terrane*.

microporosity The *porosity* arising from the presence of pores <0.5 μm in diameter, generally within the *argillaceous matrix* of a *sedimentary rock*.

microprobe See *electron microprobe*.

Microsauria An order of subclass *Lepospondyli*, class *Amphibia*; mainly terrestrial *amphibians* with close similarities to the *Reptilia*. Range late *Carboniferous*–early *Permian*.

microscopic Descriptive of features visible under an optical microscope, with a size in the range 5 μm to 2 mm. cf. *macroscopic*, *mesoscopic*.

microseism The dominant, naturally-occurring seismic *noise*, taking the form of long-duration *Rayleigh waves* of 5–20 s period and amplitude 0.1–10 000 nm. Mainly generated by sea *waves*.

microspar A recrystallized *micrite* with crystals 4–20 μm in size. cf. *pseudospar*.

microspherulitic texture The *texture* of an *igneous rock* in which *microscopic spherulites* are distributed through the *groundmass*.

microstalactitic structure A gravitationally-driven *cementation fabric* developed in the *vadose zone* on the lower surface of *allochemical* grains.

microstructure The *microscopic* features of a rock.

microsyenite A medium-grained, *microcrystalline igneous rock* with the composition and *texture* of a *syenite*.

microtonalite A medium-grained, *microcrystalline igneous rock* with the composition and *texture* of a *tonalite*.

mictite (mixtite) A *clastic rock* with a very wide range of grain sizes.

mid-ocean ridge See *ocean ridge*.

mid-ocean ridge basalt See *MORB*.

middle ground bar A *bar* dividing a river channel in a *delta*, formed where the water is shallow and the flow is dominated by friction.

Midway An *Eocene* succession on the Gulf Coast of the USA equivalent to the *Danian* and *Thanetian*.

migmatite A *metamorphic rock* injected with *igneous material*.

migmatization The process of forming a *migmatite* by *partial melting* of the parent rock under extreme *metamorphism*.

migration 1. See *seismic migration*. 2. Of oil and gas, the movement from the *source rock* into the *reservoir rock* beneath the *oil* or *gas trap*, largely controlled by their *buoyancy*. 3. The lateral or down-current movement of a *bedform* or channel.

Milankovich cycle Periodic perturbations in the Earth's orbit round the Sun caused by the effects of the other solar system planets. These affect the tilt of the Earth's spin axis with respect to the Sun and thus cause changes in the *insolation* experienced. This has been related to the periodicity of *glaciations*.

milarite ($KCa_2AlBe_2(Si_{12}O_{30}).H_2O$) A hydrated aluminium-beryllium-calcium-potassium silicate found in *veins* in *granitic* rocks and *pegmatites*.

Miller index An index used in *crystallography* to specify the orientation of a crystal face.

Miller-Bravais index A modified *Miller index*.

millerite (capillary pyrite) (NiS) A rare *ore mineral* of nickel found as a low-temperature *mineral* in *limestone*, *dolomite*, *haematite*, *serpentine* and carbonate *ore veins*.

milligal (mgal) The *cgs* unit of *gravity* equal to 10^{-3} cm s^{-2} or 10^{-5} m s^{-2}.

millstone One of a pair of stones used for grinding, chosen for its roughness, hardness and low degree of contamination of the ground material. Rocks which have been used include *limestone*, *travertine*, *sandstone*, *granite*, *basalt*, *rhyolite* and *leucitite*.

Millstone grit A coarse *sandstone* division in the British *Carboniferous*, approximately corresponding to the *Namurian*.

mima mound An earth mound up to 2 m in height and 20–50 m in diameter, generally found at a density of 50–100 ha^{-1}. May be erosional remnants or formed from deposition around vegetation, by frost sorting or by communal rodents.

mimetic growth The growth of a *metamorphic mineral* which mimics pre-existing features in shape or orientation, e.g. *mica* growing in *slate* in an orientation controlled by the *mineral* arrangement in the *argillite*.

mimetic twinning *Twinning* which closely imitates a higher symmetry.

mimetite ($Pb_5(AsO_4)_3Cl$) A *secondary ore mineral* of lead occurring in the oxidized zone of lead deposits, forming a *solid solution* with *pyromorphite*.

Mindel glaciation The second of the *glaciations* affecting the Alps in *Quaternary* times.

Mindyallen A *Cambrian* succession in Australia covering the lower part of the *Maentwrogian.*

mineral A naturally occurring, homogeneous solid with a defined chemical composition and highly ordered atomic arrangement. cf. *mineraloid.*

mineral assemblage (mineral paragenesis) *Minerals* coexisting in equilibrium.

mineral lineation A *lineation* formed by *minerals* of elongated crystal *habit.*

mineral paragenesis See *mineral assemblage.*

mineral wax See *ozocerite.*

mineralogical limit The concentration of a *resource* below which an element is no longer recoverable as a distinct *mineral* phase.

mineralogy The study of *minerals.*

mineraloid A naturally occurring substance which does not conform to the definition of a *mineral,* e.g. *native mercury.*

minette A *calc-alkaline lamprophyre* with *phenocrysts* of *biotite* and *clinopyroxene* ± *hornblende,* with *alkali feldspar* > *plagioclase.*

minette ironstone A rock containing *siderite, berthierine* (often *oolitic*) and *calcite* (causing the *ore* to be self-fluxing) with *silica* >20%. Widespread in the *Mesozoic* of Europe.

miniripple A *wave ripple* with a wavelength <10 mm.

minium (Pb_3O_4) A brown-red *oxide mineral* of lead found as an oxidation product of *galena* and other lead *minerals.*

minnesotaite ($(Fe,Mg)_3Si_4O_{10}(OH)_2$) A rare, iron-rich analogue of *talc* found in *Precambrian Iron Formations.*

minor fold A *fold* distinguishable at *outcrop* scale. cf. *major fold.*

minor structure A *structure* distinguishable at *outcrop* scale.

minverite A *dolerite* containing a brown, soda-rich *hornblende.*

mio- Less.

Miocene An *epoch* of the *Neogene,* 23.3–5.2 Ma.

miogeosyncline A *geosyncline* with no related magmatism.

mirabilite (Glauber salt) ($Na_2SO_4.10H_2O$) An *evaporite mineral* formed by the *hydration* of *thernardite.*

mire See *bog.*

Mirnyy A *Silurian/Devonian* succession in the Mirnyy Creek area of NE Siberia equivalent to the *Pridoli* and *Lochkovian.*

miscible Capable of mixing to form a single liquid. cf. *immiscible.*

mise-à-la-masse A *resistivity method* in which one current electrode is sited within the conducting body. The second is placed at effectively infinity and *equipotential* lines are mapped with a pair of potential electrodes, providing more information about the extent of the conductor than standard surface methods.

mispickel An old name for *arsenopyrite.*

Missisauga A *Cretaceous* succession on the Scotian shelf of Canada covering the *Berriasian, Valanginian, Hauterivian* and part of the *Barremian.*

Mississippi Valley-type deposit A type of *carbonate-hosted base metal deposit.*

Mississippian Sub-*period* of the *Carboniferous*, 362.5–349.5 Ma.

Missourian A *Carboniferous* succession in the USA covering the lower part of the *Kasimovian.*

missourite A *melanocratic plutonic igneous rock* comprising *clinopyroxene*, *olivine* and *leucite.*

mixed mode fracture A combination of tensile-, sliding- or tearing-type *fractures.*

mixing length The mean size of an eddy. See *Von Karman's constant.*

mixtite See *mictite.*

mizzonite ($m Ca_4(Al_6Si_6O_{24})CO_3 + n Na_4(Al_3Si_9O_{24})Cl$) A *scapolite* group *mineral* comprising a mixture of *meionite* and *marialite*, found in metamorphosed *limestone* and altered *basic igneous rocks.*

MKSA The **M**etre-**K**ilogram-**S**econd-**A**mpere system of units.

MO See *molecular orbital.*

Moberg An Icelandic name for *hyaloclastite* formed by subglacial cooling of *basaltic magma* and altered to *palagonite.*

mobile belt (orogenic belt) A large-scale, linear belt of *continental crust* affected by *tectonic* activity over a given period of geological time. cf. *craton.*

Mocha stone See *moss agate.*

modal composition (mode) The composition of a rock sample in terms of the volumetric proportions of the *minerals* in it.

modderite (CoAs) A *mineral* with a distorted derivative of the nickel arsenide structure.

mode 1. See *modal composition.* 2. The most common value in a set of numbers.

modified Griffith failure criterion An extension of the *Griffith failure criterion* which introduces frictional sliding between crack surfaces.

modified Mercalli scale The most widely used scale of *earthquake intensity*, modified to allow for changes in design and construction standards since being devised in 1902.

modulus of rigidity See *rigidity.*

mofette 1. An opening of volcanic origin emitting carbon dioxide, nitrogen and oxygen. 2. *Solfatara* rich in carbon dioxide.

Mogi doughnut See *seismic gap.*

mogote A generally steep-sided, residual *limestone* hill in tropical *karst* terrain.

Mohnian A *Miocene* succession on the west coast of the USA covering part of the *Serravalian*, the *Tortonian* and part of the *Messinian.*

Moho See *Mohorovicic discontinuity.*

Mohorovičić discontinuity (M-discontinuity, Moho) The *seismic discontinuity* between the *crust* and *mantle.*

Mohr circle A circle used in a *Mohr diagram.*

Mohr diagram A graph in which a state of *stress* or *strain* is represented by circles.

Mohr envelope A line showing the relationship between *shear* and *normal stress* at *failure* on a *Mohr diagram*.

Mohr locus A closed line representing the *strain* on a general section of the *strain ellipsoid*.

Mohs' scale A ten point scale of *hardness* to which *minerals* are compared.

molasse The copious sediment derived from a newly-elevated mountain range, i.e. post-orogenic continental sediment. cf. *flysch*.

molecular orbital (MO) A wave function which describes the behaviour of an electron in the presence of many nuclei.

Mollusca/molluscs A phylum characterized by a fleshy mantle and often a calcareous shell of from one to eight parts in the form of a coiled, hollow cone in which the soft parts can be viewed as various modifications of a hypothetical 'archimollusc' body plan. The 'archimollusc' possesses a ventral muscular foot beneath the visceral organs, which contain a gut running from a mouth with a rasping plate to a mantle cavity with paired gills. Range *Cambrian–Recent*.

molybdenite (MoS_2) The major *ore mineral* of molybdenum.

monadnock An upstanding rock, hill or mountain on an otherwise flat plain.

monalbite ($NaAlSi_3O_8$) A *monoclinic*, high-temperature form of *albite*.

monazite ($(Ce,La,Y,Th)PO_4$) A *rare earth* phosphate occurring as an *accessory mineral* in *granite* and concentrated as an *ore mineral* in *beach sand*.

monchiquite An alkaline variety of *lamprophyre* comprising Al-*titanaugite*, *barkevikite* ± *kaesurtite*, *biotite/phlogopite*, ± *olivine*, in a *glassy groundmass* containing *nepheline* or *analcime*.

mono- Single.

monocline An asymmetric *fold* with one limb dipping at a lower angle than the other.

monoclinic system A *crystal system* whose members have three unequal axes, two of which intersect at an oblique angle and the third perpendicular to the plane of the other two.

monomict (oligomict) Descriptive of a *clastic sedimentary rock* composed of a single *mineral* type. cf. *polymict*.

monomineralic Composed of a single *mineral*.

monophyletic Descended from a common ancestor. cf. *polyphyletic*.

Monoplacophora A class of phylum *Mollusca* with multiple paired gills, kidneys, gonads and shell attachment muscles, usually with only a slightly coiled shell. Range *Cambrian–Recent*.

Monorhina A subclass of class *Agnatha*, superclass *Pisces* including lamprey-type *fish*. Range U. *Silurian–Recent*.

Monoskaya A *Triassic* succession in Siberia covering the upper part of the *Nammalian*.

Monotremata An order of subclass *Prototheria*, class *Mammalia*; egg-laying *mammals*. Range *Miocene–Recent*.

monotropy The relationship between two *polymorphs* in which only one is stable and the change to this form is irreversible. cf. *enantiotrophy*.

montebrasite ((Li,Na)Al(PO_4)(OH,F)) A variety of *amblygonite* found in *granite pegmatites.*

monticellite ($CaMgSiO_4$) An *olivine* found mainly as a *metamorphic* or *metasomatic mineral.*

montmorillonite ((Al,Mg)$_8$(Si_4O_{10})$_3$(OH)$_{10}$.$12H_2O$) A *clay mineral*; the principal component of *bentonite clays.*

monzodiorite An *igneous rock* intermediate in composition between *monzonite* and *diorite.*

monzogabbro A *plutonic* rock intermediate in composition between *gabbro* and *monzonite.*

monzogranite A *granite* with equal proportions of *alkali* and *plagioclase feldspars.*

monzonite (syenodiorite) A medium- to coarse-grained *intrusive igneous rock* containing *plagioclase* ($Ab_{>50}$), $<20\%$ *quartz*, *amphibole*, *alkali feldspar* ($>10\%$) and/or *pyroxene*. It grades into *tonalite* with increased *quartz* and *diorite* with $<10\%$ *alkali feldspar.*

monzonorite A *norite* containing *orthoclase.*

mooihoekite ($Cu_9Fe_9S_{16}$) A rare *sulphide mineral.*

moonmilk A white, *cryptocrystalline* substance which forms *speleothems*, normally composed of *carbonate minerals*, which feels like cream cheese when wet and is a very fine powder when dry.

moonstone A variety of *albite* or *oligoclase* with an *opalescent* play of colours, used as a semiprecious *gem.*

moraine A depositional landform generated directly by a *glacier.*

Morarian orogeny An *orogeny* of *Precambrian* age affecting northern Scotland from ~1050–730 Ma.

MORB (mid-ocean ridge basalt) A type of *tholeiitic basalt* found in *oceanic ridges*, characterized by very low K_2O and TiO_2; low Fe, P_2O_5, Ba, Rb, Sr, Pb, Th, U and Zr; and high CaO. Depleted in light *rare-earth elements* with respect to heavy *rare-earth elements*. Has not been contaminated and so retains the chemical signature of its *mantle* source.

mordenite ($Na_3KCa_2(Al_8Si_{40}O_{96})$.$28H_2O$) A *zeolite* found in *veins* and cavities of *igneous rocks.*

morganite A pink *gem* variety of *beryl.*

morion A black variety of *smoky quartz.*

Morrison A *Jurassic* succession in Colorado, Idaho, Utah and Wyoming, USA, covering part of the *Oxfordian*, the *Kimmeridgian* and part of the *Tithonian.*

Morrowan A *Carboniferous* succession in the USA equivalent to the *Bashkirian.*

mortar structure (mortar texture) An optical *microstructure* of large strained grains surrounded by smaller, *recrystallized strain*-free grains. Typical of *mylonites* in *monomineralic* rocks.

mortar texture See *mortar structure* .

Mortensnes A *stage* of the *Vendian*, 600–590 Ma.

mortlake See *oxbow lake.*

morvan The intersection of two *erosional* surfaces.

mosaic evolution Evolution in which not all changes are seen in all representatives at the same time.

mosaic texture See *saccharoidal texture*.

Moscovian An *epoch* of the *Carboniferous*, 311.3–303.0 Ma.

moss agate (Mocha stone) A semi-precious *gem* variety of *agate* with a moss-like patterning.

Mössbauer spectroscopy The recoilless emission and resonant absorption of gamma rays by the nuclei of solids, principally iron, which provide information on the nature and environment of the atom.

mottramite ($Pb(Cu,Zn)VO_4OH$) *Descloizite* in which most of the zinc is replaced by copper.

Motuan A *Cretaceous* succession in New Zealand covering part of the upper *Albian*.

mouldic fabric A type of grain *dissolution porosity* caused by the *leaching* of grains or replacive cements whose characteristic morphology is preserved in the new void.

moulin A vertical cylindrical shaft, 0.5–1.0 m wide and up to 25 m deep, through which surface meltwater flows into a *glacier*.

mound dune See *nebkha*.

mound spring An *artesian spring* occurring preferentially along a *fault* and which gives rise to a small mound.

mouth bar A *bar* at the mouth of a river dividing the flow into channels.

moveout (stepout) The increase in *arrival time* with distance from the detector of a *seismic wave* reflected from a discontinuity. See also *dip moveout*, *normal moveout*.

MSK scale A twelve point *earthquake intensity* scale similar to the *modified Mercalli scale*, used in the former USSR.

MT survey See *magnetotelluric survey*.

mud Sediment whose particles have a size $< 62\ \mu m$.

mud ball See *armoured mud ball*.

mud cake The solid part of a *drilling mud* which is left on the wall of a borehole when the *mud filtrate* has penetrated the adjacent formations.

mud clast See *rip-up clast*.

mud crack A vertical to subvertical shrinkage crack formed by the contraction of cohesive muddy sediment, which may be preserved if infilled by a different sediment.

mud filtrate The liquid part of a *drilling mud* which can penetrate adjacent formations and displace *groundwater* or *hydrocarbons*.

mud pebble See *armoured mud ball*.

mud ripple A surface *bedding structure* on a *mudstone*, and its *sandstone* cast, similar in form and scale to the straighter-crested *current ripple*. Grades into *flute mark*, and also probably of erosional origin.

mud volcano 1. A *hot spring* which produces boiling mud. 2. A conical mound formed when liquid mud is forced to the surface as a result of *compaction* or *earthquake* activity. cf. *sand volcano*.

mudflow A *mass flow* of debris mixed with water with a high proportion of *mud*. Very similar to a *lahar*, but richer in fine-grained material.

mudlump A small-scale landform in a *delta*, probably formed by the *diapiric* intrusion of plastic *clays* through *sands*.

mudrock (mudstone) A *sedimentary rock* of *mud* grade.

mudstone See *mudrock*.

mugearite A *silica*-poor sodic *trachyandesite*.

mullion structure A linear, cylindrical *structure* comprising elongate rods or columns 20 mm–2 m across and up to 100 m long which are either complete or incomplete in section. Surfaces may be smooth or corrugated and define a *lineation* parallel to *fold axes*. Most common in strongly deformed *metamorphic rocks*. May form by *buckling* of the surface between *competent* and *incompetent beds*, by the intersection of *bedding* and *cleavage*, by *fold hinges* or by *boudinage*.

mullite ($Al_6Si_3O_{15}$) A rare aluminosilicate formed by the intense heating of *andalusite*, *kyanite* or *sillimanite*.

multiple See *multiple reflection*.

multiple bars A set of up to 10 shore-parallel *coastal bars*.

multiple faults (fault set) A group of *faults* with similar orientations and *displacement vectors*, probably forming simultaneously in a common *stress* field.

multiple intrusion An igneous intrusion made up of successively emplaced *magmas* of distinctive, similar composition distinguished by internal contacts and/or *phenocryst* content and *crystallinity*.

multiple reflection (multiple, seismic multiple) A *seismic wave* that has been reflected more than once before being recorded, and which can obscure the arrival of primary reflections.

multiple twinning See *polysynthetic twinning*.

multiplexing The process of interleaving multiple input information channels into a single output channel, used, for example, in sending the output of several *geophones* to a recorder. cf. *demultiplexing*.

multispectral scanner A *remote sensing* device carried by the first *Landsat satellites* which imaged four spectral bands in the region 500–1100 nm with a *pixel* size of ~80 m.

multistorey sandbody A vertical succession of *sandstone beds* deposited by the infilling of river channels with little intervening *mud*, formed by the rapid migration of a channel network over an *alluvial plain* whose sediment has little opportunity of preservation.

Multituberculata An order of subclass *Allotheria*, class *Mammalia*; small rodent-like *mammals*. Range late *Jurassic–Eocene*.

Muschelkalk 1. A *Triassic* succession in Germany covering the *Anisian* and part of the *Ladinian*. 2. The traditional German name for the Middle *Triassic*, during which the *Tethys* Ocean transgressed over the region from Europe and N Africa to China.

muscovite (white mica) ($KAl_2(AlSi_3O_{10})(OH)_2$) A very common *mica*.

muskeg 1. The waterlogged marshland of NW Canada. 2. A *peat*-filled *basin* with sphagnum moss.

Muskingum method An empirical method of *flood routing* using the continuity equation and a relationship between flow rate and temporary storage of water in the channel during flooding.

Myachkovskian A *stage* of the *Carboniferous*, 305.0–303.0 Ma.

mylonite A fine-grained, *foliated fault rock* with a *recrystallized texture* with 50–90% *matrix* and a strong *lineation* caused by *shear* in a major *ductile fault* or *shear zone*. See also *protomylonite, ultramylonite*.

mylonite gneiss A *mylonite* with a well-developed *gneissosity*.

mylonite schist A *mylonite* with a well-developed *schistosity*.

mylonite zone A belt of *mylonite*, up to hundreds of kilometres long and kilometres thick, along a major *fault zone*.

mylonitization The process of forming a *mylonite*, generally by *crystal plastic strain* by *dislocation climb* and *recovery*, with new *mineral* grains forming by *dynamic recrystallization*.

Myriapoda/myriapods A superclass of phylum *Uniramia*. Terrestrial animals breathing through trachea or the body wall with a head bearing a pair of antennae and mandibles and two pairs of feeding maxillae, and many trunk somites. Range *Silurian–Recent*.

myrmekite (myrmekitic texture) A *vermicular* intergrowth of *quartz* and sodic *feldspar* adjacent to a crystal of *alkali feldspar*, probably of secondary origin when *quartz* is released during the *replacement* of *alkali feldspar* by *plagioclase*.

myrmekitic texture See *myrmekite*.

N

Nabarro-Herring creep A *creep* mechanism in which atoms diffuse within the *crystal lattice*.

nacreous lustre A *lustre* resembling that of a pearl.

nacrite ($Al_2Si_2O_5(OH)_4$) A *clay mineral* with the same composition as *kaolinite* but a different *crystal structure*.

Nafe-Drake curve An empirical relationship between *P-wave* velocity and *density*, allowing estimation of the latter from the former to an accuracy of ~0.1 Mg m^{-3}.

Nagssugtoqidian orogeny An *orogeny* affecting W Greenland during the *Precambrian* at ~2600 Ma and ~1900–1500 Ma.

nagyágite (AuTe.6Pb(S,Te)) A rare *mineral* found in *hydrothermal veins*.

Nammalian A *stage* of the *Triassic*, 243.4–241.9 Ma.

Namurian A *Carboniferous stage* covering the *Serpukhovian* and part of the *Bashkirian*.

nano- Extremely small; × 10^{-9} when attached to a unit.

nanoplankton Marine, *planktonic fossils* of ultramicroscopic size, useful in correlation over wide distances.

nanoTesla (nT) The *SI* subunit used for *magnetic anomalies*, equal to 10^{-9} *tesla* (V s m^{-2}).

naphthene series A component of *crude oil* with the general formula C_nH_{2n}.

napoleonite A *diorite* containing centimetric-scale, spheroidal structures comprising alternating shells of *hornblende* and *feldspar*.

nappe (decke) A body of rock, generally highly folded and with greater *ductile deformation* than a *thrust sheet*, which has suffered considerable horizontal *tectonic* transport in an *orogenic belt*.

Narizian An *Eocene* succession on the west coast of the USA, comprising part of the *Lutetian* and the *Bartonian*.

native element A *mineral* comprising a chemical element in an uncombined state or as an alloy with another element(s).

native metal A *mineral* comprising a metallic element in an uncombined state or as an alloy with another element(s).

natroalunite ($(Na,K)Al_3(SO_4)_2(OH)_6$) The sodium-rich equivalent of *alunite*.

natrocarbonatite A *carbonatite* composed of sodium, calcium or potassium carbonates.

natrojarosite ($NaFe_3^{3+}(SO_4)_2(OH)_6$) A hydrated sodium-iron sulphate of the *alunite* group found as a *secondary mineral* in cracks in *ferruginous* rocks and *ores*.

natrolite ($Na_2Al_2Si_3O_{10}.2H_2O$) A *zeolite*, often found in *acicular* form in *amydales* in *basaltic* rocks.

natron ($Na_2CO_3.10H_2O$) A hydrated sodium carbonate commonly found in *soda lakes*.

natron lake A lake rich in *natron*, often occurring in continental *rift valleys*.

natural arch A bridge or arch joining two rock pillars, produced by *weathering* or *erosion*.

natural gas The gaseous constituents of *petroleum*, i.e. *methane*, ethane, propane and *n*-butane.

natural remanent magnetization (NRM) The permanent magnetization of the *ferrimagnetic minerals* in certain rocks that was acquired when they formed or at some later stage in their history. Its measurement forms the basis of the study of *palaeomagnetism*.

natural strain A linear measurement of shape change during *deformation* based on the integration of instantaneous *strain* increments.

Nautiloidea/nautiloids A subclass of class *Cephalopoda*, phylum *Mollusca* characterized by coiled shells whose internal compartments are defined by smooth septae. Range *Cambrian–Recent*.

Navarro A *Cretaceous* succession on the Gulf Coast of the USA equivalent to the *Maastrichtian*.

Navier-Stokes equations A series of equations of linear momentum for a moving, viscous, compressible fluid which relate *stress* and the state of *strain*.

nebkha (coppice dune, shrub-coppice dune) An *aeolian* bedform of wind-blown *sand* collected within and behind, and stabilized by, vegetation.

Nebraskan glaciation The first of the major *glaciations* affecting North America in the *Quaternary*.

neck A volcanic *plug*.

necking A localized thinning of a *structure* during *extension*.

Nectarian An *era* of the *Precambrian*, 3950–3850 Ma.

Nectridea An order of subclass *Lepospondyli*, class *Amphibia*; small *amphibians* of sometimes bizarre morphology. Range *Permian–Carboniferous*.

needle stone (rutilated quartz) A colloquial name for clear *quartz* with *acicular* inclusions of *rutile* or, rarely, *actinolite*.

Néel temperature The temperature at which the coupling between *magnetic domains* in a *ferrimagnetic* material breaks down, so that it shows simple *paramagnetic* behaviour.

negative crystal A *pseudomorph* comprising a hollow mould from which the original crystal has been removed by *solution*.

negative grading See *inverse grading*.

Neichiashan See *Aijiashan*.

nekron mud See *gyttja*.

nektonic Free-swimming.

nematath A linear series of *oceanic islands*.

nematoblastic Descriptive of a *foliation* formed by prismatic *minerals*.

Nenetskiy A *Permian* succession in the Timan area of the former USSR equivalent to the U. *Asselian*.

neo- New.

neocatastrophism A theory concerning events of great magnitude and low frequency, applied to *mass extinctions* and *geomorphology*.

Neocomian An *epoch* of the *Cretaceous*, 145.6–131.8 Ma.

neoformation of clays The *authigenesis* of *clay minerals*, as the result of direct *precipitation* from a *pore fluid*, the *neomorphism* of one *clay mineral polymorph* to another or the *replacement* of a precursor *mineral* by a new *clay mineral*.

Neogene The younger *period* of the *Cenozoic*, 23.3–1.64 Ma.

neoglacial A small-scale *glacial* advance during the *Holocene* after the maximum *glacial* retreat of the present *interglacial*.

Neognathae A superorder of subclass *Neornithes*, class *Aves*, comprising all modern *birds* with the exception of the flightless *Palaeognathae*. Range *Cretaceous–Recent*.

neomorphism All *diagenetic* transformations between one *mineral* and itself or a *polymorph*, including differences in size and shape but excluding pore filling, in which there is simultaneous reoccupation of the space occupied by the original and the new. Occurs by the processes of *polymorphic transformation* and *recrystallization*.

Neornithes A subclass of class *Aves*; all *birds* with the exception of the very ancient forms of the *Archaeornithes*. Range *Cretaceous–Recent*.

neotectonics Late *Cenozoic deformation*.

neoteny An evolutionary reduction in the rate of morphological development, which may lead to *paedomorphosis*.

neotype See *type specimen*.

nepheline ($(Na,K)AlSiO_4$) A *feldspathoid*, found mainly in *plutonic* and *volcanic rocks* and in *pegmatites* associated with *nepheline syenites*.

nepheline monzonite An *undersaturated monzonite* containing *essential nepheline*.

nepheline syenite A *plutonic igneous rock* composed of *nepheline*, *albite* and *microcline*. High in alumina ($\sim$25%) and soda ($\sim$9%), so exploited in glass and ceramic manufacture.

nephelinite A fine-grained *igneous rock* with $>$10% *modal nepheline* and little or no *alkali feldspar*, in which the *nepheline* content exceeds that of *mafic minerals*. Occurs in association with *alkaline igneous rocks* in continental settings.

nepheloid layer A water layer, generally on the sea bed of deep oceans and a few hundred to 1500 m thick, containing abundant suspended sediment.

nephelometer (transmissometer) An instrument used for the identification of *nepheloid layers* by the scattering of light by fine particles.

nephrite (kidney stone) A tough, compact variety of *tremolite* providing much *jade*.

neptunian dyke The sediment infill of a vertical fissure, possibly recording a period of deposition not preserved in a normal stratal sequence, or injected into the *host rock*.

neptunian sill The sediment infill of a

horizontal fissure, possibly recording a period of deposition not preserved in a normal *stratal* sequence, or injected into the *host rock*.

neptunite ($KNa_2Li(Fe,Mn)_2TiO_2(Si_4O_{11})_2$) A complex, rare *inosilicate*.

neritic (sublittoral) The area between the lower limit of the *littoral* zone and the *continental shelf*.

Nerminskiy A *Permian* succession in the Timan area of the former USSR equivalent to the L. *Artinskian*.

nesosilicate (orthosilicate) A silicate in which the silicon tetrahedra share no corners with other tetrahedra.

nesting lineation A *lineation* on a *slickenside* which is perfectly matched across the *fault plane*.

net slip The distance along a *fault plane* between the original position of some marker and its new position.

Nettleton's method A method of determining the *in situ density* of an isolated hill by correcting *gravity* measurements over it for a range of *reduction densities*. The reduced *gravity* profile providing the least *correlation* with the topography then provides the best *density* estimate.

neutral fold A *fold* which closes sideways.

neutral surface Of a *fold*, the surface within a folded layer, usually parallel to its boundaries, which separates regions of *compression* and *extension*.

neutron activation analysis (INAA) An analytical method for determining a small number of *major elements* and many *trace elements* in silicate rocks. Bombardment by neutrons forms new isotopes as neutrons join the nuclei, which decay, emitting gamma rays whose energy is diagnostic of the original isotope.

neutron diffraction The scattering of thermal neutrons by atomic nuclei, the scattering being enhanced by atoms with a *magnetic moment*. The different isotopes of the same element scatter differently, allowing their distinction.

neutron log See *neutron-gamma ray log*.

neutron-gamma ray log (neutron log) A *geophysical borehole log* in which the wallrock is bombarded by neutrons, causing them to emit gamma rays, the intensity of which is a function of the *porosity* of the *wallrock*.

Nevadan orogeny An *orogeny* of late *Jurassic* to early *Cretaceous* age affecting the western USA and approximately equivalent to the *Coast Range orogeny* in Canada.

névé (firn) Compacted snow which has survived a summer.

New Red Sandstone The continental rocks of the *Permian* and *Triassic systems* in W Europe.

Newton The *SI* unit of *force*; the *force* required to accelerate a mass of 1 kg by 1 m s^{-2}.

Newton's law of gravitation The *force* of attraction between two masses is proportional to their product and inversely proportional to the inverse square of the distance between them, the constant of proportionality being the *gravitational constant*.

Newtonian flow *Flow* in which the *shear strain rate* is a linear function of *shear stress*.

Newtonian fluid A fluid whose

viscosity is constant, independent of any external *shear force* applied to it and which obeys the *Navier-Stokes equation*. cf. *non-Newtonian fluid*.

Ngaterian A *Cretaceous* succession in New Zealand covering parts of the *Albian* and *Cenomanian*.

Niagaran A *Silurian* succession in North America comprising the *Ontarian*, *Tonawandian* and *Lockportian*.

niccolite (kupfernickel, nickeline) (NiAs) A sulphide-like *mineral* with a NiAs structure; an *ore mineral* of nickel.

nickel antimony glance See *ullmannite*.

nickel arsenic glance See *gersdorffite*.

nickel arsenide group A group of *sulphide minerals* characterized by a structure in which metals and anions lie on interpenetrating simple and close-packed hexagonal sublattices respectively.

nickel bloom See *annabergite*.

nickel laterite deposit A *laterite* enriched in nickel, formed by intense tropical *weathering* of rocks with trace amounts of nickel, e.g. *peridotite*, *serpentinite*.

nickeliferous Containing nickel.

nickeline See *niccolite*.

Niedermendig lava See *Mayen lava*.

Niggli number A variant of the *norm* classification.

nigrite A member of the *asphaltite* group with a *pitch*-like appearance.

Ningguo (Ningkuo) An *Ordovician* succession in China covering part of the *Tremadoc* and *Early Arenig*.

Ningkuo See *Ningguo*.

niobite See *columbite*.

nitratine ($NaNO_3$) A nitrate *mineral* found in *caliche*.

nitre (saltpetre) (KNO_3) A nitrate *mineral* used in fertilizer manufacture.

nivation 1. *Weathering* and transport processes intensified by a late-lying snow patch. 2. Descriptive of landforms substantially modified by snow-patch related processes.

noble metal See *precious metal*.

nodal avulsion An *avulsion* occurring persistently from one point.

nodule An irregular, spherical to ellipsoidal, flattened to cylindrical body, commonly composed of *calcite*, *siderite*, *pyrite*, *gypsum* and *chert*, common in soils and *evaporite deposits*.

Noginskian The highest *stage* of the *Carboniferous*, 293.6–290.0 Ma.

noise An unwanted disturbance on a record, such as a *seismogram*. Random *seismic noise* includes *microseisms*, whereas coherent noise includes *multiple reflections* and *surface waves*.

non-associated gas *Petroleum* gas not accompanied by oil.

non-coaxial deformation Progressive *deformation* during which the lines of maximum and minimum *elongation* rotate. cf. *rotational deformation*.

non-conformity (heterolithic unconformity) An *unconformity* in which younger *strata* rest on an *erosion* surface of non-bedded *igneous rocks*.

non-contacting conductivity measurement The measurement of ground *electrical conductivity* using an *electromagnetic induction method* so that no ground contact is required.

non-cylindrical fold A *fold* which displays different *profiles* along the *hinge line*.

non-metallic lustre The *lustre* of a *mineral* that reflects light but does not shine metallically. Includes *vitreous lustre*, *silky lustre*, *resinous lustre*, etc.

non-Newtonian flow *Flow* in which the *shear strain rate* is a complex function of *shear stress*. cf. *Newtonian flow*.

non-penetrative Descriptive of a feature, such as *cleavage*, developed throughout only part of a rock at the scale of observation. cf. *penetrative*.

non-polarizing electrode An electrode designed to avoid the accumulation of charge on a metal electrode by *electrode polarization* effects, constructed of a metal in a solution of one of its salts contained in a porous casing.

non-rotational strain The *strain* experienced by a rock with a *non-coaxial deformation* history.

non-sequence (paraconformity) A *diastem*-type *unconformity* with faunal or other evidence of a time gap.

nontronite ($Fe_2(Al,Si)_4O_{10}(OH_2.Na_{0.3}(H_2O)_4)$) A *clay mineral* of the *montmorillonite* group.

norbergite ($Mg_3(SiO_4)(F,OH)_2$) A *mineral* of the *humite* group.

nordmarkite An *alkaline*, *quartz*-bearing *syenite* comprising *microperthite*, *aegirine*, sodic *ampnibole* and *accessory quartz*.

Norian A *stage* of the *Triassic*, 223.4–209.5 Ma.

norite A *gabbro* containing *orthopyroxene* and *labradorite*.

norm See *normative composition*.

normal drag *Fault drag* in which the curvature of a marker is consistent with the sense of *displacement* on the *fault*. cf. *reverse drag*.

normal fault A *dip-slip fault* with a dominant component of *dip-slip displacement* and the *hangingwall* displaced downwards relative to the *footwall*.

normal grain growth An increase in the size of a polycrystalline *aggregate* during *recrystallization*.

normal magnetization A *remanent magnetization* in the same sense as the present *geomagnetic field*. cf. *reversed magnetization*.

normal moveout *Moveout* generated by a horizontal reflector. cf. *dip moveout*.

normal polarity The orientation of a past *geomagnetic field* or a *natural remanent magnetization* in the same direction as the present *geomagnetic field*. cf. *reversed polarity*.

normal resistivity log An *electric geophysical borehole log* which provides *resistivity* information on a thick shell of the *wallrock*.

normal stress The *stress* acting at right angles to a surface.

normal twin A *twinning* phenomenon in which the *twin axis* is normal to the *composition plane*.

normal zoning *Zoning* in *plagioclase* in which the zones become more sodic towards the outside of the crystal. cf. *reversed zoning*.

normally consolidated sediment A sediment compacted by the *pressure* expected from the *overburden* thickness. cf. *overconsolidated sediment, underconsolidated sediment.*

normative composition (norm) The composition of an *igneous rock* expressed as weight proportions of idealized anhydrous *minerals* that *crystallize* from *magma*, calculated in a specified sequence from a chemical analysis of the rock.

nosean (noselite) $(Na_8(AlSiO_4)_6SO_4)$ A rare *feldspathoid* found in *phonolites* and related rocks and in *volcanic bombs.*

noselite See *nosean.*

notch A landform 1–5 m in depth at the base of a *sea cliff*, platform or *reef* flat, especially in *limestone* and on tropical coasts, characteristic of areas with a low tidal range.

nothosaurs Extinct, piscivorous *reptiles* of *Triassic* age with small heads, long necks and paddle-like limbs.

Notoungulata An order of infraclass *Eutheria*, subclass *Theria*, class *Mammalia*; an order of South American herbivores from rabbit to bear size. Range U. *Palaeocene–Pleistocene*.

novaculite A fine-grained to *cryptocrystalline* rock composed of *quartz* or other forms of *silica*, i.e. a *chert.*

NRM See *natural remanent magnetization.*

nT See *nanotesla.*

nubbin A centimetric-scale, rounded or elongate earth lump produced by *heave* associated with the growth of needle ice.

nuclear explosion seismology (forensic seismology) The detection, identification and *yield* estimation of underground nuclear explosions.

nuclear precession magnetometer See *proton magnetometer.*

nucleation The first stage in *precipitation*, before the free energy barrier resulting from the developing interface between the crystal nucleus and the solution has been overcome and *crystal growth* can take place.

nuée ardente A laterally mobile, *turbulent*, hot cloud of air, *volcanic gases* and suspended, fine-grained *tephra* generated by volcanic activity. Typical of *peléean eruptions*. Probably generated by the upward *convection* of hot gas and *tephra* from an active *pyroclastic* flow. Gives rise to thin, fine-grained deposits with low-angle *cross-bedding* overlain by air-fall *tephra.*

Nuevo Leon A *Cretaceous* succession on the Gulf Coast of the USA, covering the *Barremian* and part of the *Aptian.*

nugget A lump of *native gold*, rarely of *platinum*, generally found in *alluvial deposits.*

Nukumaruan A *Pliocene/Quaternary* succession in New Zealand covering parts of the *Piacenzian* and early *Quaternary.*

nummulitic limestone A *limestone* of *Eocene* age made up of nummulites (*Foraminifera*) quarried for some early Egyptian monuments such as the pyramids.

nunatak An isolated rock peak projecting through an *ice sheet.*

Nusselt number A hydrodynamic parameter whose value controls the possibility of *convection*.

nye channel The basal conduit of a *glacier*, incised into *bedrock*, through which meltwater discharges. cf. *Röthlisberger channel*.

Nyquist frequency (folding frequency) Half the sampling frequency of a digitized signal; the highest frequency reliably restored from that signal.

O

obduction The process whereby *oceanic lithosphere* is transported onto *continental crust.*

oblate Ellipsoidally shaped. cf. *prolate.*

oblique extinction (inclined extinction) The *extinction* of a *mineral* which takes place at an angle to its *cleavage* traces or margins.

oblique ramp A *ramp* trending at an angle between the transport direction and the normal to it in a *thrust zone.*

oblique slip *Displacement* at an angle to the *strike* of the *displacement plane* and the normal to it.

oblique slip fault A *fault* which has similar magnitudes of *strike-slip* and *dip-slip displacements.*

obrution deposit A deposit in which organisms are smothered by a rapid sediment influx which kills and buries them simultaneously, possibly giving rise to an *exceptional fossil deposit.*

obsequent stream A tributary of a *subsequent stream* which flows in a direction opposite to the regional *dip* of the land surface. cf. *anteconsequent, inconsequent, insequent, subsequent* and *resequent streams.*

observation well A *well* drilled or allocated for monitoring the extraction from a *reservoir.*

obsidian A *glassy rock* of volcanic origin of *intermediate* to *acid* composition formed by the chilling of *rhyolitic lava.* Widely used in prehistoric times for the manufacture of small tools.

occult mineral A *mineral* expected to form from a *glass* if it had crystallized completely.

Ocean Drilling Program (ODP) The current phase, since 1984, of ocean drilling using the ship *JOIDES* Resolution.

ocean island basalt (OIB) *Quartz tholeiite, alkali basalt* and *nephelinite* found on *oceanic islands.* With respect to *MORB*s, it is enriched in *LIL* elements, *LREE*s compared to *HREE*s and *incompatible elements.* Probably formed by *partial melting* of enriched *mantle.*

ocean plateau An areally extensive region of ocean floor reaching to 2–3 km of the sea surface above the surrounding seafloor, probably of volcanic origin.

ocean ridge (mid-ocean ridge) An *accretive plate margin* often, but not always, situated at the median line of an ocean basin, at which new oceanic *lithosphere* is generated by *magmatic* processes. Marked by a topographic rise 1000–4000 km in width which rises 2–3 km above the flanking ocean *basins.*

ocean trench (deep-sea trench, trench) The topographic expression of *subduction*; the large linear depression of the oceanic *lithosphere* at a *subduction zone.*

Oceanic Anoxic Event A period of about a million years in the *Mesozoic* when *black shale* was deposited and there was enhanced removal of light *carbon* isotopes from the world's oceans.

oceanic crust Thin (~7 km), young (<200 Ma) *crust* of three layers; Layer 1–the uppermost layer of sediments, Layer 2–*pillow lavas* underlain by *dykes* and Layer 3–*gabbroic* and underlying *ultrabasic* rocks.

oceanic island An intra*plate* island believed to have originated by the passage of the oceanic *lithosphere* over a nearly stationary *hotspot* in the *mantle*.

oceanic magnetic anomalies (magnetic lineations) Linear *magnetic anomalies* of alternating polarity with amplitudes of ~1000 nT, 10–20 km in width and of large lateral extent which run parallel to *oceanic ridges* and are symmetrical about the crest of the ridge. Believed to form by the process described by the *Vine-Matthews hypothesis*.

oceanite A type of *basaltic* rock with a lower proportion of alkalis and a higher proportion of *mafic minerals* than normal *basalt*, typically found as *lava flows* on *oceanic islands*

ocellar Descriptive of an *igneous rock texture* in which *aggregates* of small crystals are distributed around larger crystals to give the impression of an eye-like form.

Ochoan A *Permian* succession in the Delaware Basin, USA, equivalent to the *Longtanian* and *Changxingian*.

ochre Red, yellow and brown *iron oxides* formed by the *weathering* of iron deposits, used as pigments.

octahedrite A variety of *anatase* occurring as tetragonal bipyramids.

Octocorallia A subclass of class *Anthozoa*, phylum *Cnidaria*; *corals* represented by rare *fossils*, including the earliest *anthozoans*. Range *Ordovician–Recent*.

OD See *Ordnance Datum*.

Odontognathidae Superorder of subclass *Neornithes*, class *Aves*; toothed, flightless *birds* similar to modern divers. Range L. *Cretaceous*–U. *Cretaceous*.

odontolite See *bone turquoise*.

ODP See *Ocean Drilling Program*.

Oe See *Oersted*.

Oersted (Oe) The *cgs* unit of *magnetic induction*.

offlap A *structure* in which successive wedge-shaped *beds* do not extend to the margin of the underlying *bed*, such as would be formed in a contracting sedimentary *basin*. cf. *onlap*.

offscraping See *underplating*.

offset The horizontal *displacement* across a *fault* normal to the interrupted feature.

offset well A *well* drilled close to another further to explore or exploit a *reservoir*.

offshore bar See *submerged bar*.

Ohauan A *Jurassic* succession in New Zealand covering the upper part of the *Kimmeridgian*.

Ohm's law 'The electric current flowing between two points is equal to the ratio of the potential difference to the resistance between them.' Relevant to *electrical resistivity* surveying.

Ohre A *Permian* succession in NW Europe covering the middle part of the *Capitanian.*

OIB See *ocean island basalt.*

oikocryst A large crystal grain which encloses several randomly oriented, smaller grains of another phase or phases. Responsible for *poikilitic texture.*

oil basin A sedimentary *basin* containing commercial accumulations of oil.

oil shale A bituminous, non-marine *limestone* or *marl* (rarely marine *shale*) which contains substantial organic matter at a low level of *maturation* such that liquid oil has never been released from the *kerogen.* A very abundant *resource* of *fossil fuel*, but with great obstacles to its exploitation.

oil source rock A rock, generally *shale*, *limestone* or *coal*, in which *hydrocarbon* generation processes produce *natural gas* and *crude oil.* Contains large quantities of organic matter, a *pelagic* fauna and flora and phosphorus. *Anaerobic* conditions necessary for preservation of the organic matter indicate deposition below a *thermocline.*

oil trap A geometrical arrangement of *strata* that allows the accumulation of oil. Commonly an *anticline* but also associated with *faulting*, *stratigraphic* arrangements, *reefs* and piercement structures, e.g. *salt domes.* The *hydrocarbons* tend to move upward due to their low *density* until trapped beneath an impermeable or semipermeable *cap rock.*

oilfield A region of the *crust* containing a number of oil pools.

oilfield water The water found in *hydrocarbon*-bearing *reservoir* rocks.

Old Red Sandstone The continental *facies* of the *Devonian* in the UK.

oldhamite ($(Ca,Mn)S$) A calcium-manganese sulphide, found in *meteorites.*

Olenekian A *Triassic stage* in Siberia comprising the *Monoskaya* and *Sygynkanskaya*

oligo- Prefix indicating few or little.

Oligocene The youngest *epoch* of the *Palaeogene*, 35.4–23.3 Ma.

oligoclase A *plagioclase feldspar*, An_{10-30}.

oligomict See *monomict.*

oligomictic lake A lake which is thermally almost stable and undergoes only rare mixing. cf. *polymictic lake.*

oligotrophic lake A lake with a deficiency of nutrients and large amounts of dissolved oxygen in the bottom layers, which have only a small amount of organic matter.

olistolith A waterlain block of rock of *pebble* to *boulder* size that differs considerably in its *petrography*, composition or *texture* from the surrounding rocks. See also *olistostrome.*

olistostrome A bed or layer of *olistoliths*, commonly formed at the base of submarine *fault* scarps subject to *mass wasting.*

olivenite (Cu_2AsO_4OH) A rare, green, hydrated copper arsenate; a *secondary mineral* found in copper deposits.

olivine basalt See *alkali basalt.*

olivines A group of *orthosilicate minerals* with the general formula M_2SiO_4 where M is magnesium, iron, manganese and calcium with minor amounts of nickel. The main natural olivines derive from the *solid solution* from *forsterite* (Fo_{100})(Mg_2SiO_4) to *fayalite* (Fa_{100})(Fe_2SiO_4).

omphacite ((Ca,Na)(Mg,Fe,Al)Si_2O_6) A *pyroxene* found in *eclogite*.

oncoid (oncolith) A type of carbonate *coated grain* coated by microbial mats in which the laminae are irregular in thickness, relief and continuity.

oncolite A rock composed mainly of *oncoids*.

oncolith See *oncoid*.

onion-skin weathering See *spheroidal weathering*.

onlap (overlap) A *structure* in which successive wedge-shaped *beds* extend further than the margin of the underlying *bed*, such as would be formed in an expanding sedimentary *basin*. cf. *offlap*.

Onnian A *stage* of the *Ordovician*, 444.0–443.1 Ma.

onset time See *arrival time*.

Ontarian A *Silurian* succession in North America covering parts of the *Aeronian* and *Telychian*.

ontogeny The development of an organism through various stages. cf. *phylogeny*.

onyx (SiO_2) A layered, *microcrystalline* variety of *silica* used for decorative purposes.

onyx marble (Oriental alabaster) Banded *calcite* or *aragonite* used for decorative purposes.

ooid (oolith) A type of small (< 2 mm) carbonate or iron *coated grain* with a cortex of concentric fine laminae, lacking biogenic features, and a nucleus, often a shell fragment or *sand* grain.

oolite A rock composed mainly of *ooids*.

oolith See *ooid*.

oomicrite A *limestone* containing *ooliths* in a *micrite matrix*.

oosparite An *oolitic limestone* with a sparry cement.

ooze A *pelagic mud*.

opal ($SiO_2.nH_2O$) A hydrated variety of *silica* made up of minute (~300 nm) spheres, including the *gem* precious opal.

opal agate A variety of *opal* which is *agate*-like in structure.

opalescence The play of changing colours caused by the thin-film interference of light along planes of voids between the packed spheres forming the *mineral*.

opaque mineral A *mineral* appearing black in *thin section* in transmitted plane-polarized light.

open fold A *fold* with an *interlimb angle* between 70° and 120°.

open hole A borehole with no casing.

opencast mining (openpit mining) Mining by excavation from the surface.

openpit mining See *opencast mining*.

ophicalcite See *forsterite-marble*.

Ophiocistioidea A class of subphylum *Eleutherozoa*, phylum *Echinodermata*, seemingly intermediate between the *Echinoidea* and *Holothuroidea*. Range L. *Ordovician–Devonian*.

ophiolite An association of *ultrabasic/ultramafic–basic/mafic* rock types believed to represent a section through the *lithosphere* of a *marginal basin* emplaced onto *continental* or older *oceanic crust* by *obduction.*

ophitic texture A type of *poikilitic texture* in slow-cooling *basic igneous rocks* in which *euhedral*, randomly oriented *plagioclase* laths are wholly enclosed within *pyroxene* plates. cf. *subophitic texture.*

Ophiuroidea (brittle stars) A class of subphylum *Eleutherozoa*, phylum *Echinodermata*; the brittle stars. Range L. *Ordovician–Recent.*

Opoitian A *Pliocene* succession in New Zealand covering part of the *Zanclian.*

optic axis The direction in an *anisotropic* crystal along which there is no *birefringence.*

optical emission spectroscopy An analysis method for *major* and *trace elements* in which a powdered sample is mixed with *graphite* and vaporized in a carbon arc. The spectrum of the light emitted can be analysed to provide the concentrations of elements present. Now largely superseded by other techniques.

orbicular Descriptive of spherical to sub-spherical masses comprising concentric shells of different composition in *plutonic igneous rocks.*

orbital forcing The control of terrestrial processes by changes in the Earth's orbit. See *Milankovitch cycle.*

orbital ripple A type of steep *wave ripple.*

Ordian A *Cambrian* succession in Australia covering part of the *Atdabanian*, the *Lenian* and part of the *Solvan.*

Ordnance Datum (OD) *Sea level datum* in the UK, measured at Newlyn or Liverpool, to which all *bench marks* are related.

Ordovician A sub-*era* of the *Palaeozoic*, 510.0–439.0 Ma.

ore A metalliferous *mineral* or an *aggregate* of metalliferous *minerals* mixed with *gangue* that can be exploited at a profit.

ore microscopy The technique of examining *opaque minerals* in reflected light with a polarizing microscope.

ore mineral A *mineral* from which a useful metal can be extracted.

ore reserve (indicated reserve, inferred reserve, measured reserve) The calculable tonnage of *ore* in an *orebody*, including that believed to be present.

ore shoot A thin, ribbon-like extension of an *orebody.*

orebody A volume of rock that can be exploited commercially for its metal content.

orendite A *silica-saturated lamproite* with *sanidine* rather than *leucite*, often carrying *diamonds.*

Oretian A *Triassic* succession in New Zealand covering parts of the *Ladinian* and *Carnian.*

organic weathering The disintegration and decomposition of rock by the action of micro-organisms, plants, animals and decaying organic matter.

Oriental alabaster See *onyx marble.*

Oriental almandine A deep red *corundum* of *gem* quality.

Oriental amethyst Purple *corundum* or *sapphire*.

Oriental cat's eye See *cymophane*.

Oriental emerald *Gem*-quality *corundum* with the colour of *emerald*.

Oriental topaz A variety of *corundum* with the colour of *topaz*.

oriented core A drilled rock sample oriented with respect to the present horizontal and *true north*.

oriented lake A lake with a *preferred orientation* of its long axis.

Ornithischia An order of subclass *Archosauria*, class *Reptilia*; herbivorous *dinosaurs* with a bird-like pelvis. Range late *Jurassic*–end *Cretaceous*.

Orochenian A *Cretaceous* succession in the far east of the former USSR equivalent to the *Coniacian*, *Santonian*, *Campanian* and *Maastrichtian*.

orogen (orogenic belt) The total volume of rock deformed during an *orogeny*.

orogenesis (orogeny, tectogenesis) The process of creation of a mountain belt by *tectonic* activity, generally by the collision of continental *plates* or *microplates*. Characterized by *regional metamorphism*, igneous activity and vertical movements. cf. *epeirogeny*.

orogenic andesite association An association of *calc-alkaline* rocks found at *subduction zones*.

orogenic belt See *orogen*.

orogenic cycle A term referring to the cyclicity of *orogenic* events with time.

orogeny See *orogenesis*.

Orowan-Elsasser convection A convective model for the Earth in which the tops of *convection* cells are the oceanic *lithosphere* and *plates* are driven by the expansion of *mantle* material at *ocean ridges* and the pull of descending *lithosphere* at *subduction zones*.

orpiment (As_2S_3) A rare *ore mineral* of arsenic found in *veins* and *hot spring* deposits.

Orthida An order of class *Articulata*, phylum *Brachiopoda*; *brachiopods* with biconvex, impunctate shells, broad, straight hinge lines and no brachidium. Range L. *Cambrian*–*Permian*.

orthite See *allanite*.

ortho- Genuine, right, straight, upright.

orthoamphibole An *amphibole* with an orthorhombic structure, e.g. *anthophyllite*.

orthoclase ($KAlSi_3O_8$) A common *feldspar*.

orthoconglomerate A *clast-supported conglomerate* deposited by aqueous currents. cf. *paraconglomerate*.

orthoconic Descriptive of a shell in the form of a straight, tapering cone.

orthoenstatite An orthorhombic *polytype* of *enstatite*.

orthoferrosilite ($FeSiO_3$) An *orthopyroxene*.

orthogneiss A *gneiss* formed from an *igneous rock* parent. cf. *paragneiss*.

orthogonals See *wave rays*.

orthomagmatic (orthotectic) Descriptive of the main stage of *crystallization* from a *magma*, when up to 90% of it may crystallize.

orthomagmatic segregation deposit (magmatic segregation deposit) An *ore* deposit that has crystallized directly from a *magma*, produced by *fractional crystallization* or *liquation*.

orthophyric Descriptive of a fine- or medium-grained *syenitic* rock with closely-packed *orthoclase* crystals.

orthopyroxene A *pyroxene* crystallizing in the *orthorhombic system* ranging between *enstatite* and *ferrosilite*.

orthoquartzite (quartzarenite) An unmetamorphosed sedimentary *quartzite*. cf. *metaquartzite*.

orthorhombic system A *crystal system* whose members have three mutually perpendicular axes of different lengths.

orthosilicate See *nesosilicate*.

orthotectic See *orthomagmatic*.

Osagean A *Carboniferous* succession in the USA covering parts of the *Tournaisian* and *Viséan*.

osar See *esker*.

oscillatory flow An aqueous flow which reverses periodically.

oscillatory zoning *Zoning* within a crystal comprising alternating layers of the two end-members of a *solid solution* series.

ossicle A single *echinoderm* plate.

Osteichthyes A class of superclass *Pisces*; the bony *fish*. Range L. *Devonian–Recent*.

Osteostrachi An order of subclass *Monorhina*, class *Agnatha*, superclass *Pisces*; small, jawless *fish* with a flattened body covered by polygonal plates and a broad head covered by a bony shield. Range *Silurian–Devonian*.

Ostracoda/ostracods (ostracodes) A class of subphylum *Crustacea*, phylum *Arthropoda*; animals with a small, bi*valved*, calcareous carapace of vastly diversified morphology, useful in *biostratigraphy*. Range *Cambrian–Recent*.

Ostracodermi/ostracoderms An old or informal term for the *fossil Agnatha* (*fish*) of the *Ordovician* to *Carboniferous*.

ostracodes See *Ostracoda*.

Ostwald step rule '*Polymorphic* transformations proceed in steps from the unstable form through a series of metastable states of decreasing free energy until the stable form is achieved.'

Otaian A *Miocene* succession in New Zealand covering the *Aquitanian* and part of the *Burdigalian*.

Otamitian A *Triassic* succession in New Zealand covering the upper part of the *Carnian*.

Otapirian A *Triassic* succession in New Zealand equivalent to the *Rhaetian*.

Oteke A *Jurassic* succession in New Zealand covering part of the *Tithonian*.

otolith The ear-bone of a *fish*.

ottrelite A manganese-bearing *chloritoid* found in *schists* as the result of *contact metamorphism* of certain *argillaceous* rocks.

ouady See *wadi*.

oued See *wadi*.

out-of-phase component See *imaginary component*.

out-of-sequence thrust A *thrust fault* in an *imbricate structure* which causes deviation from a *piggyback thrust* sequence in which *shortening* takes place on the lowest, latest *thrust*.

outcrop The total area over which a particular rock unit occurs at the surface.

outcrop mapping See *exposure mapping*.

outlier An area of younger rocks surrounded by older rocks. cf. *inlier*.

overbank deposit Suspended sediment and fine *bedload* deposited by floodwater on a *floodplain* when *bankfull discharge* is exceeded.

overburden Loose, unconsolidated material resting on *bedrock*.

overconsolidated sediment A sediment which, sometime in its history, has been subjected to pressure greater than at present and so it is stronger than anticipated. cf. *normally consolidated sediment, underconsolidated sediment*.

overcoring (doorstopper technique) A type of *in situ stress measurement* in which small *strains* are produced after the release of *stress* around a cylinder at the end of a borehole.

overfold A *fold* in which the *axial surface* and both *fold limbs dip* in the same direction.

overland flow The flow of water downslope across the soil surface.

overlap See *onlap*.

overlap integral A measure of the spatial overlap between orbitals in *molecular orbital* theory.

overpressuring (geopressuring) The *pressure* in excess of *hydrostatic pressure* in subsurface sediments, caused when the flow of fluids out of sediments is impeded. Common in actively subsiding sedimentary *basins* where *compaction* is occurring.

overprinting The superimposition of a *mesoscopic* to *microscopic* younger *structure* on an older one.

oversaturated Descriptive of an *igneous rock* containing free *silica*. See also *silica saturation*.

overstep A *stratigraphic* relationship at an *unconformity* where the oldest unit of the younger sequence is in contact with more than one of the older *beds* beneath. Implies tilting or *folding* before deposition of the younger *beds*.

overstep propagation sequence A *thrust fault* sequence in an *imbricate structure* in which successively later *thrusts* form closer to the *hinterland* than existing *structures*; earlier *thrusts* and related *folds* can be truncated by later *thrusts*.

overthrust 1. *Displacement* of a *thrust sheet* over the edge of an adjacent rock. 2. See *thrust*.

overthrust fault See *thrust*.

overvoltage (electrode polarization) An *induced polarization* effect caused by the different rates of electronic and electrolytic current flow. The impedance of electrolyte flow by metallic *minerals*, in which flow is electronic, gives rise to a charge build up which gradually decays when the driving potential difference is removed, causing a decaying voltage.

overwash fan A deposit formed on the landward side of a *beach ridge* and the *backshore* zones of a *barrier island* or low *spit* when it is breached by storm waves.

oxbow lake (mortlake) A curved lake isolated from a stream when an acute meander is cut off from the channel.

Oxfordian A *stage* of the *Jurassic*, 157.1–154.7 Ma.

oxidation 1. The loss of electrons from an ion or atom. 2. The loss of hydrogen or the addition of oxygen through chemical reaction.

oxide minerals *Minerals* formed by the combination of an element or elements with oxygen, common as *accessory minerals*.

oxygen isotope analysis The determination of the ratio of the oxygen isotopes ^{18}O and ^{16}O by *mass spectrometry*, used in determining the provenance of artefacts and in *oxygen isotope stratigraphy*.

oxygen isotope stratigraphy The use of the oxygen isotope ratios of *Foraminifera* for *stratigraphic* purposes.

oxygen minimum layer A level in the ocean, generally between depths of 150 m and 1000 m, where the concentration of oxygen is at a minimum, the bounding limit being 0.2 ml l^{-1}. With a high influx of organic matter and the oxygen minimum layer at the seabed, *anoxic* sediments may accumulate.

oxy-hornblende (basaltic hornblende, lamprobolite) $(Ca_2(Na,K)_{0.5-1.0}(Mg,Fe^{2+})_{3-4}(Fe^{3+},Al)_{2-1}(Si_6Al_2O_{22})(O,OH,F)_2)$ An *amphibole* found in *volcanic rocks*.

oxyphile element See *lithophile element*.

ozocerite (mineral wax) A dark *paraffin* wax found in irregular *veins*.

P

P In *earthquake seismology*, a *P wave* that does not travel through the *core*.

P foliation A feature of a *fault gouge* comprising a *phyllosilicate preferred orientation* or layering at an acute angle to the *shear* direction.

P shear A subsidiary *fault* in a *shear zone*.

P wave A *seismic body wave* which propagates by the vibration of particles forwards and backwards in the direction of propagation, i.e. as a series of compressions and rarefactions.

Pa See *pascal*.

Pacific-type coast (concordant coast, longitudinal coast) A smooth coastline which runs parallel to the trend of topography and geological *structure*. cf. *Atlantic-type coast*.

packer test (Lugeon test) A method of measuring the *permeability* of *strata* in a borehole by isolating a section with an inflatable tube, filling it with water and timing the water loss per metre.

packstone A *grain-supported, clastic limestone* containing some lime *mud*.

paedomorphosis The retention of ancestral juvenile characters in a descendant adult.

pagodite See *agalmatolite*.

pahoehoe A thin (0.1–2m) *lava flow* with a thick, smooth, wavy surface, formed by low-*viscosity lava*.

paired metamorphic belt Juxtaposed low-temperature/high-pressure and high-temperature/low-pressure linear belts believed to be characteristic of the *ocean trench* and *island arc* environments, respectively, of a *subduction zone*. There is now evidence that paired metamorphic belts may have a more complex *tectonic* history.

paisanite A sodic *microgranite* with *riebeckite* as the main *mafic mineral*.

palaeo- (paleo-) Ancient.

palaeoautecology The study of the past ecology of individual organisms or taxonomic groups. cf. *palaeosynecology*.

palaeobiogeography The study of the distribution of ancient organisms.

palaeobiology See *palaeontology*.

palaeobotany The study of *fossil plants*.

Palaeocene The oldest *epoch* of the *Cenozoic*, 65.0–60.5 Ma.

palaeoclimatology The study of ancient climates.

palaeocurrent A current which influenced sediment deposition at some time in the past.

palaeoecology The study of the past interactions of organisms with each other and their environment. See also *palaeoautecology*, *palaeosynecology*.

palaeoflow An ancient flow, whose direction can be assessed from the geometry of the *cross-stratification* it produces.

Palaeogene The older *period* of the *Cenozoic*, 65.0–23.3 Ma.

palaeogeographical map A map showing the *outcrop* pattern at a particular time.

palaeogeography The study of ancient geography, e.g. past distributions of land and ocean.

Palaeognathae Superorder of subclass *Neornithes*, class *Aves*; large, flightless *birds*. Range late *Cretaceous–Recent*.

palaeohydrology The reconstruction of *flow regimes* for ancient river channels.

palaeomagnetic pole A location of the *geomagnetic* pole in the past calculated from *palaeomagnetic* data.

palaeomagnetism The study of the magnetic properties of rocks, particularly *natural remanent magnetizations*, which can provide information on the *palaeomagnetic pole* position and the ancient intensity of the *geomagnetic field*.

palaeontology (palaeobiology) The study of ancient organisms.

palaeopiezometry The measurement of the previous state of *stress* in a rock.

palaeosalinity An ancient salinity.

palaeosol A soil formed on a past landscape, documented for both the *Precambrian* and *Phanerozoic*.

palaeosome The apparently older part of a composite rock body.

palaeosynecology The study of the past ecology of fossil communities as a whole. cf. *palaeoautecology*.

palaeotemperature An ancient temperature, which can be determined from oxygen isotopes in sediments and estimated from some *fossils*.

Palaeozoic An *era* comprising the *Cambrian* to *Permian* systems.

palagonite A *volcanic rock* consisting of hydrated and chemically altered *basaltic hyaloclastite*, the *alteration* occurring at any temperature in the presence of seawater or *meteoric water*.

palagonitization The process of *alteration* of *basaltic hyaloclastite* to *palagonite*.

paleo- See *palaeo-*.

palimpsest structure (relict structure) A relic of an original *texture* or *structure* visible through a superimposed *texture* or *structure*.

palingenesis See *anatexis*.

palinspastic map A geological map in which *deformation* has been removed so that the rocks are displayed in their pre-*tectonic* configuration.

palinspastic reconstruction The reconstruction of the geometry of a set of rocks before *deformation*.

palladium (Pd) A very rare, platinum-group *native metal*.

pallasite A *stony meteorite* comprising metals and silicates.

palsa A *peat* mound above an ice lens in *permafrost*. See also *pingo*.

paludal Relating to a marsh.

palustrine In a *paludal* environment.

palygorskite $((MgO)_2(SiO_2)_2.4H_2O-$

$Al_2O_3(SiO_2)_5.6H_2O$) A *clay mineral* with a chain-like structure, found mainly in *hydrothermal veins* or in altered *serpentine* or *granitic* rocks.

palynology The study of carbonaceous micro-organisms, including pollen, spores, *dinoflagellates* and *acritarchs.*

palynomorph A micro-organism studied in *palynology*.

pan A closed depression, often one of a great number, in arid to semi-arid areas, caused mainly by *deflation*, but also by *solution* and animal activity.

Pan-African orogeny An *orogeny* during the *Proterozoic* from 900–600 Ma affecting the Arabian and African *shields.*

pandemic distribution See *cosmopolitan distribution.*

panfan An arid landscape of eroded hills and ridges, extensive, coalescing *pediments* and infilled *basins.*

Pangaea (Pangea) The single *supercontinent* comprising all the *continental crust* in the late *Palaeozoic*, which split into the *supercontinents Laurasia* and *Gondwanaland* at ~300 Ma.

Pangea See *Pangaea.*

panidiomorphic Descriptive of an *igneous rock* with well-developed crystals.

panplain (planplain) An area of very subdued relief, formed by the coalescence of *floodplains* and resulting from the lateral migration of streams.

pantellerite A type of *peralkaline rhyolite* with >12.5% *normative femic minerals.* Normally weakly *porphyritic* with *phenocrysts* of *alkali feldspar*, sodic *ferrohedenbergite, fayalite, aenigmatite, amphibole, quartz* and Fe-Ti oxides. Found in continental *rifts* and rarely on *oceanic islands.*

Panthalassa The ancestral Pacific Ocean which surrounded *Pangaea.*

Pantotheria An order of infraclass *Trituberculata*, subclass *Theria*, class *Mammalia*; egg-laying insectivores of shrew-like appearance. Range M.–U. *Jurassic.*

para- Parallel to, resembling.

parabolic dune A crescentic *dune* whose arms are tethered by vegetation and point upwind.

paraconformity See *non-sequence.*

paraconglomerate A *matrix-supported conglomerate*, *mixtite* or pebbly *mudstone* deposited by a *sediment gravity flow* or as *glacial till*. cf. *orthoconglomerate.*

paraffin series See *methane series.*

paragenesis A *mineral* association, generally expressed in terms of a time sequence.

paragenetic sequence The order of deposition or *crystallization* of *minerals*, commonly used in connection with *epigenetic hydrothermal deposits.*

paragneiss A *gneiss* derived from *detrital sedimentary rock.*

paragonite ($NaAl_2(AlSi_3O_{10})(OH)_2$) A *mica isostructural* with *muscovite* found mainly in *schists, phyllites, quartz veins* and fine-grained sediments.

paralic basin A sedimentary *basin* in a marginal marine environment. cf. *limnic basin.*

parallel extinction (straight extinction) The *extinction* of a *mineral* which takes place parallel to its *cleavage* traces or margins.

parallel fold A *fold* in which the layer thickness is constant everywhere perpendicular to the folded layer surface.

parallel retreat A type of slope evolution in which the form and angle of the slope remains constant as it is eroded.

parallel twin A *twinning* phenomenon in which the *twin axis* lies in the *composition plane.*

paramagnetism The magnetic property of some materials which acquire a relatively weak magnetization when placed in a *magnetic field* in the same direction as that field which is lost when the field is removed.

paramorph A *pseudomorph* formed by the conversion of one *polymorph* to another.

parasitic antiferromagnetism The magnetization arising in an *antiferromagnetic* material when the *crystal lattice* is slightly distorted and a *spontaneous magnetization* is produced. Exhibited by *haematite.*

parasitic cone (adventive cone) A small cone on the flank of a *volcano.*

parasitic fold (satellite fold) A *second order fold* with a pattern of asymmetry depending on its position with respect to *first order fold hinges.*

parataxitic texture A *eutaxitic texture* with greatly elongated *fiamme*, caused by secondary *flow.*

Paratethys A large seaway extending from north of the Alps to the east of the Aral Sea from *Oligocene* to *Pliocene* times.

paratype A specimen selected to show additional characters to a *holotype.*

parautochthon A large structural unit which has travelled a short distance from its site of origin. cf. *allochthon, autochthon.*

Pareora A *Miocene* succession in New Zealand comprising the *Otaian* and *Altonian.*

pargasite ($NaCa_2Fe_4(Al,Fe)Al_2Si_6O_{22}(OH)_2$) An *amphibole* found as a constituent of *igneous* and *metamorphic rocks.*

parna An *aeolian clay* (*loess*) deposit in the form of a *dune* or a thin, discontinuous, widespread sheet, possibly derived from the *deflation* of unvegetated, saline lake floors or other soil or alluvial surfaces.

partial area model A concept proposed to account for the low ratio of *runoff* to rainfall in *catchments* where *Hortonian overland flow* is the dominant storm *runoff* mechanism. Only small parts of the *basin* contribute to surface *runoff* while elsewhere *infiltration capacity* is sufficiently high for complete *infiltration.* cf. *variable area model.*

partial dislocation A crystal *dislocation* whose *displacement* is a fraction of the *unit cell.*

partial melting The incomplete melting of rock to produce a melt of different composition.

particle cluster A grouping of particles on a bed through mutual interference between the constituent *clasts*, characteristic of *gravel*-bed rivers.

particulate flow A *cataclastic deformation mechanism.*

parting The breaking along planes of weakness in *minerals*, e.g. *twin planes*. Similar to *cleavage*, but found only in those crystals that are *twinned* or deformed.

partition coefficient (distribution coefficient) The ratio of the concentration of an element between a *mineral* and a coexisting melt. Used to quantify *petrogenetic* models based on *trace elements*.

pascal (Pa) The *SI* unit of *pressure* or *stress* in kg $m^{-1}s^{-2}$.

pascichnia *Trace fossils* comprising grazing traces.

passive continental margin (quiescent continental margin, trailing continental margin) A *continental margin* which is not a *plate margin* and represents the site of *continental splitting* prior to the *seafloor spreading* that carried it to its present position. cf. *active continental margin*.

patch reef See *micro-atoll*.

patronite (VS_4) A *sulphide ore mineral* of vanadium.

patterned ground Symmetrical forms, such as circles, stripes and polygons, developed on soil, sediment and fractured *bedrock* by frost processes and *cryoturbation*.

pause plane The surface in a *cross-stratified bed* which represents a time of minimum flow velocity.

paved bed A layer similar to an *armoured surface* but more stable and much coarser than the substrate. Forms as a *lag* deposit of the material too coarse to be transported away.

pay streak A zone of concentrated heavy *minerals* in a *placer deposit*.

Payntonian A *Cambrian* succession in Australia covering the upper part of the *Dolgellian*.

PDR See *precision depth recorder*.

peacock coal An *iridescent bituminous coal* or *anthracite* found in upper mine levels where *groundwater* has deposited a film of *iron oxide* along *fractures*.

peacock ore A popular name for *bornite*.

peak zone See *acme zone*.

pearlspar A rhombohedral form of *dolomite* or *ankerite*.

pearly lustre An *iridescent*, pearl-like *lustre* shown on the *cleavage* surfaces of layer lattice silicates, e.g. *talc*.

peat A mass of dark brown, partly decomposed, fibrous plant debris. The precursor of *coal*, requiring substantial growth, standing water to prevent *oxidation* or bacterial destruction and an absence of introduced *detrital* sediment.

peatland See *bog*.

pebble A rounded rock fragment of 2–64 mm in diameter.

Péclet number A dimensionless number whose magnitude determines the relative importance of *diffusion* over flow in a sediment.

pectolite ($Ca_2NaH(SiO_3)_3$) A *pyroxenoid inosilicate* found commonly in cavities in *basaltic* rocks.

pediment A gently-sloping (0.5°–7°), concave-up *erosion* surface on the flank of a steep-sided hill or mountain and common at the base of *mesas* or *inselbergs*. Possibly the product of *sheetflood* *erosion* or *backwearing* by *parallel retreat*.

pediplain A landform created by coalescing *pediments*.

Pedleian A *stage* of the *Carboniferous*, 332.9–331.1 Ma.

pedogenesis The process of soil formation, depending on the interplay between parent, climate, organisms, topography and time. The main processes involved are *calcification*, *eluviation*, *ferralitization*, *gleying*, *humification*, *illuviation*, *leaching*, *pedoturbation*, *podsolization*, *rubifaction*, *salinization* and *weathering*.

pedogenetic calcrete A *calcrete* originating in a soil, mainly by biologically-induced *calcite precipitation*.

pedology The study of soils.

pedoturbation A *pedogenetic* process involving the churning of soil by physical processes.

peel method A technique used in carbonate *sedimentology* and *palaeontology* in which calcareous material is etched by weak acid to enhance relief and polyvinylacetate sheeting rolled on the surface after flooding with acetone. The resulting mould is peeled off along with a thin surface layer which can reveal structures and can be stained to show extra details.

pegmatite A very coarse-grained *igneous rock* with *phenocrysts* over 250 mm in length, usually of *granitic* composition and forming at a late stage of *crystallization*.

pegmatite deposit A *pegmatite* enriched in *minerals* of economic interest, often providing the following elements: lithium, beryllium, rubidium, cesium, niobium, tantalum and tin.

pelagic Descriptive of the deep sea environment.

pelagite A *pelagic* sediment, generally decimetres thick, formed by the slow settling of calcareous and siliceous biogenic material from suspension.

pelecypods See *Bivalvia*.

peléean eruption A *volcanic eruption* in which *ash flows* <1 km^3 in volume, accompanied by a *nuée ardente*, sweep down the flanks of a *volcano* and deposit block and *ash flows* as radiating *fans*.

Pelé's hair See *achnelith*.

Pelé's tears Cylindrical, spherical or tear-shaped drops of *pyroclastic*, *glassy basalt* 6–13 mm in length which solidified during flight.

Pelican Creek A *Permian* succession in Queensland, Australia, covering the upper part of the *Wordian*.

pelite An aluminium-rich rock formed by the *metamorphism* of *clay*-rich sediments.

pelitic With a similar chemical composition to *shale*.

pellet A small ovoid to spherical particle of *micrite* with no internal structure, 0.03–0.15 mm in diameter, possibly of faecal or *algal* origin.

Pelmatozoa A subphylum of phylum *Echinodermata*; *echinoderms* with the aboral surface expanded to form a functional stalk and the water vascular system extended into elongate appendages. Range *Cambrian–Recent*.

pelmicrite A *limestone* composed of *micrite* and small, rounded *aggregates* of sediment.

peloid A *sand*-size grain of carbonate *mud*.

Pel'skiy A *Permian* succession in the Timan area of the former USSR equivalent to the U. *Sakmarian*.

pelsparite A *limestone* composed of *pellets* in a *sparite matrix*.

Pelycosauria An order of subclass *Synapsida*, class *Reptilia*; originally small lizard-like *reptiles*, radiating in the *Permian* to large carnivorous, piscivorous and herbivorous forms, some with a dorsal 'sail'. Range U. *Carboniferous*–L. *Permian*.

Penck and Brückner model A framework for understanding the *Pleistocene* history of the Alps based on four main *glacial* phases: *Günz*, *Mindel*, *Riss* and *Würm*. Now superseded by a more complex history.

pene- Almost.

penecontemporaneous Very shortly after or before.

penecontemporaneous dolomitization The formation of *dolomite* from *limestone* by *metasomatic alteration* very soon after deposition when the *limestone* was still unconsolidated. cf. *subsequent dolomitization*.

peneplain A low-angle ground surface developing at a late stage in the *cycle of erosion*.

penetration twin A *twin* in which two crystals penetrate each other.

penetrative Descriptive of a feature, such as *cleavage* or *fabric*, developed throughout a rock at the scale of observation. cf. *non-penetrative*.

Peng Lai-Zhen A *Jurassic*/*Cretaceous* succession in Sichuan, China, covering the *Tithonian* and L. *Cretaceous*.

pennantite ($(Mn,Al)_6(Si,Al)_4O_{10}(OH)_8$) A *chlorite* group *mineral* found in manganese *ores*.

Pennsylvanian The younger sub-*period* of the *Carboniferous*, 322.8–290.0 Ma.

Pentamerida An order of class *Articulata*, phylum *Brachiopoda*; *brachiopods* with normally smooth, biconvex, impunctate shells, often with two diverging plates in the brachial *valve* and a median septum in the pedicle *valve*. Range M. *Ordovician*–U. *Devonian*.

pentlandite ($(Fe,Ni)_9S_8$) The most important *ore mineral* of nickel.

Penutian A *Palaeocene*/*Eocene* succession on the west coast of the USA, covering parts of the *Thanetian* and *Ypresian*.

peperite A *breccia* comprising *glassy lava* fragments and sedimentary material formed by *magma* flowing over or through wet sediment.

peralkaline A composition with an excess of (Na_2O + K_2O) over Al_2O_3.

peraluminous A composition with an excess of Al_2O_3 over (Na_2O + K_2O).

peramorphosis The appearance of ancestral adult characters in a descendant juvenile.

percentage contour map A subsurface map showing the percentage of one lithology in the total thickness of a *stratigraphic* unit.

percentage frequency effect (PFE) A parameter quantifying the *induced* *polarization* effect in *frequency domain* surveys, defined as $100(\rho_0 - \rho_\infty)/\rho_\infty$, where ρ_0 and ρ_∞ are *apparent resistivities* measured at <1 Hz and a few tens of Hz respectively.

perched aquifer A locally developed, water-saturated body located above the regional *water table* due to the presence of an underlying impermeable layer.

perched water table The top surface of a *perched aquifer*.

percussion mark A *lunate* scar on a *pebble* formed by a sharp blow, such as would occur in a high-velocity flow.

perennial head The highest point along the course of a river from which flow always occurs.

perennial stream A stream or river with continuous flow. cf. *ephemeral stream*.

pergelation The formation of permanently frozen ground.

peri- Near, around.

periclase (MgO) An *oxide mineral* occurring in *contact metamorphosed limestone*.

pericline A *fold* with elliptical or circular *outcrop* in which *dips* vary around the *structure*.

pericline twinning A type of lamellar *twinning*, shown by *feldspars* in which the *twin axis* is *b*[101].

peridot A clear green, *gem* variety of *olivine*.

peridotite An *ultramafic* rock with 40–90% *olivine* and *pyroxene*, including *harzburgite*, *wehrlite* and *lherzolite*.

periglacial Referring to a wide range of cold, non-*glacial* climatic and geomorphic conditions, irrespective of the proximity to a *glacier*.

period A second order geological time unit.

Perissodactyla An order of infraclass *Eutheria*, subclass *Theria*, class *Mammalia*; the ungulate, odd-toed herbivores, e.g. horses. Range *Eocene–Recent*.

peristerite A very fine-scale intergrowth of two *plagioclase feldspars* in the range An_{2-15}, invisible to the naked eye but possibly producing *iridescence* in *moonstone*.

perknite A coarse-grained, *ultramafic igneous rock* comprising *pyroxenes* and *amphiboles* but no *feldspar*.

perlite A hydrated, silicic, volcanic *glass* with curved, concentric, 'onion-skin' cracks probably arising from volume changes associated with the *hydration*. Used commercially for thermal insulation and as a rooting medium.

perlitic texture A *texture* of *glassy* and *devitrified igneous rocks* comprising curved or spherical to subspherical cracks produced by contraction during cooling.

permafrost Rock, soil and sediment in which temperatures remain below 0°C for at least two consecutive winters and the intervening summer.

permanent strain The retained *strain* resulting from *ductile deformation* before *failure*.

permanent water table The lowest level to which the *water table* falls in a given locality.

permeability The coefficient linking flow rate to the pressure gradient in a medium in the *Darcy equation*.

Permian The youngest *period* of the *Palaeozoic*, 290.0–245.0 Ma.

permineralization (petrification) The preservation of organic hard parts by *mineral*-bearing *groundwater* infiltrating porous matter after burial. The common preserving *minerals* are *silica*, *calcite* and *iron oxides*.

permitted emplacement The *emplacement* of an *igneous body* in which the *country rock* is passively displaced. cf. *forceful emplacement*.

perovskite ($CaTiO_3$) An *oxide mineral* found in *nepheline syenites* and *carbonatites*.

perthite A *macroscopic*-scale intergrowth of *potassium feldspar* and *albite*.

perthosite A sodic *syenite* mainly composed of *perthitic feldspars*.

Petalichthydia An order of class *Placodermi*, superclass *Pisces*; *fish* resembling arthrodires, but *benthonic* in habit. Range L. *Devonian*–U. *Devonian*.

petalite ($Li(AlSi_4O_{10})$) A *tectosilicate* important as an *ore mineral* of lithium found in *granite pegmatites*.

petrification See *permineralization*.

petro- Rock.

petrofabric The *fabric* elements of, usually, an *igneous rock*.

petrofabric diagram See *fabric diagram*.

petrogenesis All aspects of the formation of a rock.

petrography The systematic description of rocks in hand specimen and *thin section*.

petroleum All naturally occurring *hydrocarbons*: *asphalt*, *bitumen*, *crude oil*, *gas hydrates* and *natural gas*.

petrology The study of all aspects of rocks.

petromict conglomerate (polymict conglomerate) A *conglomerate* with *clasts* of a variety of compositions. cf. *oligomict conglomerate*.

Petrotsvet A *Cambrian* succession in Siberia covering part of the *Tommotian* and the *Atdabanian*.

petzite (Ag_3AuTe_2) A steel-grey to black *silver-gold* telluride.

PFE See *percentage frequency effect*.

phacoidal structure A *structure* in an *igneous* or *metamorphic rock* in which lens-shaped fragments are present.

phacolith A metric- to kilometric-scale, concavo-convex lens of *intrusive igneous rock* between *beds* in an *antiform*.

phaneritic Descriptive of a *texture* of an *igneous rock* in which crystals are discernible by the unaided eye.

phanerocrystalline Descriptive of an *igneous rock* in which the crystals of all *essential minerals* are discernible to the unaided eye.

Phanerozoic Post-*Precambrian* time.

pharmacolite ($CaHAsO_4.2H_2O$) Hydrated calcium arsenate, formed by the *alteration* of *arsenopyrite* and other arsenical *ores*.

pharmacosiderite ($Fe_3(AsO_4)_2(OH)_3.5H_2O$) A hydrated iron arsenate formed by the *alteration* of arsenic *ore*.

phase component system An *electromagnetic induction method* in which the relative magnitudes of the *real* and *imaginary components* of the secondary field generated by a conductor are measured. cf. *dip-angle system*.

phase diagram A graphical method of illustrating the equilibrium boundaries of different phases of a chemical system in terms of temperature, pressure and composition.

phase transformation The isochemical transformation of one *mineral* phase into another, e.g. *olivine* → *spinel* with increasing pressure.

phase velocity The velocity of a specific crest or trough of a waveform. cf. *group velocity*.

phenakite (Be_2SiO_4) A rare *nesosilicate* found in *pegmatites*.

phengite A high-*silica* variety of *muscovite*.

phenoclast A *clast* larger than 4 mm in diameter in a *sedimentary rock*.

phenocryst A large, generally *euhedral*, *mineral* grain within the fine-grained *matrix* of an *igneous rock*, probably reflecting slow cooling and growth prior to the cooling giving rise to the *matrix*. See also *porphyritic texture*.

phi (φ) unit A unit used to express *clast* sizes in *clastic rocks* according to $\varphi = -\log_2 x$, where x is the grain size in mm. See also *Udden-Wentworth scale*.

Philip equation A theoretical equation which describes the *infiltration* process.

phillipsite ($KCa(Al_3Si_5O_{16}).6H_2O$) A *zeolite* found in cavities in *basalt*, *phonolite* and related rocks, in *saline lake* deposits, in calcareous deep-sea sediments and in *hot spring* deposits.

phlogopite ($KMg_3(AlSi_3O_{10})(OH)_2$) A *mica* found mainly in metamorphosed *limestones* and *ultrabasic rocks*.

Pholidota An order of infraclass *Eutheria*, subclass *Theria*, class *Mammalia*; the pangolins. Range ?*Miocene–Recent*.

phonolite A fine-grained, commonly *porphyritic*, *felsic igneous rock* with >10% *modal feldspathoid* plus *alkali feldspar* and minor sodium-rich *amphibole* or *pyroxene*. Occurs in association with *alkali basalt-trachyte* in *alkaline* volcanic provinces and *oceanic islands*.

phosgenite (horn lead) ($Pb_2CO_3Cl_2$) A rare *carbonate mineral*.

phosphate rock An *igneous* or *sedimentary rock* with a high concentration of phosphate *minerals*, commonly the *francolite–apatite* series.

Phosphatocopida An order of class *Ostracoda*, phylum *Arthropoda*; *ostracods* with a phosphatic carapace. Range U. *Cambrian*.

phosphorite (rock phosphate) A *phosphate rock* which occurs in beds from centimetres to tens of metres thick, composed of grains of *cryptocrystalline* carbonate *fluorapatite* or *collophane* and detrital material. Forms under low latitude, marine conditions in shallow water. The major commercial source of phosphates.

photic zone (euphotic zone) The zone at the surface of the ocean, from a few metres to 150 m thick, through which light penetrates at sufficient intensity for photosynthesis. cf. *aphotic zone*, *diphotic zone*.

photo- Pertaining to light.

photogeological map A geological map constructed from *aerial photography*.

phreatic eruption The explosive ejection of *tephra* consisting of *mud* and *lithic* material, but no juvenile *magmatic* material, from a *volcanic vent*, resulting from the explosive boiling of *groundwater* in response to indirect *magmatic* heating.

phreatic gas Gas originating in the atmosphere or ocean.

phreatic zone (zone of permanent saturation) The zone beneath the *water table* in which intergranular pores and fissures are completely filled with water at *hydrostatic pressures* in excess of atmospheric.

phreatomagmatic eruption (hydrovolcanic eruption) An explosive interaction between *magma* and water, typically producing very fine-grained *tephra*.

phreatoplinian deposit A very widespread, very fine-grained air-fall *tephra*, resulting from the reaction of silicic *magma* and water.

phyletic gradualism A theory of evolutionary development of species by gradual, continuous change. cf. *punctuated equilibrium model*.

phyllarenite A *lithic arenite* whose rock fragments are dominantly of *metamorphic* origin.

phyllic alteration *Alteration* found in the *host rocks* of *porphyry copper* and molybdenum deposits.

phyllite A *regionally metamorphosed, foliated, pelitic* rock.

Phyllocarida A subclass of class *Malacostraca*, subphylum *Crustacea*, phylum *Arthropoda*, characterized by a large, bi*valved* carapace, an abdomen of seven somites and a telson with a simple paired appendage. Range *Cambrian–Recent*.

Phyllolepida An order of class *Placodermi*, superclass *Pisces*; freshwater *fish* with extensive dermal armour whose plates have a concentric pattern of ridges. Range U. *Devonian*.

phyllonite A *dynamically metamorphosed* rock rich in *phyllosilicates* in *preferred orientation*, typically formed by *retrograde metamorphism* of *schist* or *gneiss*.

phyllosilicate (layer silicate, sheet silicate) A silicate with layers formed when each silicon tetrahedron shares three corners with other tetrahedra.

phylogeny The sequence of branching events involved in the evolution of a *taxon*. cf. *ontogeny*.

phyteral Plant material in *coal* whose morphological form remains discernible.

phytogeomorphology The study of the relationships between plants and landforms.

phytokarst See *biokarst*.

phytolith A rock formed of plant material or by plant activity.

pi (π) diagram A *stereogram* on which *pi poles* are displayed. cf. *beta* (β) *diagram*.

pi (π) pole On a *stereogram*, the best-fit *great circle* through a *girdle distribution* of *poles* to a folded surface, i.e. the mean *fold axis*.

Piacenzian The higher *stage* of the *Pliocene*, 3.40–1.64 Ma.

pickeringite (magnesia alum) ($MgAl_2(SO_4)_4.22H_2O$) A hydrated aluminium-magnesium sulphate, occurring in fibrous masses and formed by the *weathering* of *pyrite*-bearing *schists*.

picotite (chrome spinel) ($(Fe,Cr)Al_2O_4$) A form of *hercynite* with appreciable chrome replacing aluminium.

picrite A dark-coloured *volcanic* or minor intrusive *rock* with abundant *olivine* (totalling >90% with other ferromagnesian *minerals*) and <10% *plagioclase*.

piedmont Descriptive of a feature at the base of a mountain.

piedmontite See *piemontite*.

piemontite (manganepidote, piedmontite) ($Ca_2MnAl_2O(SiO_4)Si_2O_7$) A *sorosilicate isostructural* with *epidote*.

piercement fold A *fold* caused by the *diapiric* intrusion of an *evaporite*.

piercement structure A *structure* arising from the piercing of *strata* by the *diapiric* intrusion of an *evaporite*.

piezometric surface See *potentiometric surface*.

piezometry The measurement of *pressure* or *stress*.

piezoremanent magnetization A *natural remanent magnetization* acquired by a rock subjected to sudden impact, such as by a *meteorite*, or prolonged *stress*. Of potential value in studying changes in seismically-induced *stress* from variations in the local *geomagnetic field*.

pigeonite ($\sim Ca_{0.25}(Mg,Fe)_{1.75}Si_2O_6$) A *pyroxene* found in rapidly cooled *igneous rocks*.

piggyback basin A *sedimentary basin* formed on a *piggyback thrust* sheet.

piggyback propagation sequence An *imbricate structure* in which new *thrust* segments develop closer to the *foreland* than existing *thrusts*, so that all the *shortening* at a particular time is taken up on the lowest, latest *thrust*.

piggyback thrust A *thrust* in a *piggyback propagation sequence*.

pillar structure A water escape structure which characteristically penetrates flat laminae lacking *dish structures*.

pillow lava A *volcanic rock* of *basaltic* composition comprising rounded, sack-like bodies, 0.2–2 m in diameter, separated from each other by fine-grained rinds. Forms when *lava flows* come into contact with water.

pilotaxitic texture A *texture* comprising a felted mass of *acicular* or lath-like crystals, which may show flow structure. See also *hyalopilitic*.

pin line A reference point on a *balanced* or *restored section*, the distance between two of which provides an estimate of the *shortening*.

pinacoid A *crystal form* whose faces are parallel to two of the axes.

pinch-and-swell structure A repetitive thinning and thickening of a body of rock, possibly formed when a *fault* crosses a series of *beds* at different inclinations so that the *fault* movement creates a series of openings.

pinger A high-frequency, shallow penetration, marine *seismic source* used in *seismic reflection* profiling.

pingo A conical hill in *permafrost* formed above a body of ice. See also *palsa*.

pinite A fine-grained, blue mixture of *muscovite* and *chlorite* formed by the *alteration* of *cordierite*, *spodumene*, *feldspar* etc.

pinnate fracture (feather fracture) A minor *fracture* which intersects a larger *fracture* at an acute angle, probably formed due to the *tensile stresses* generated by frictional sliding on the main *fracture*.

pipe 1. A subsurface channel from several millimetres to three metres in diameter, usually formed in soil or *peat* in which flow is concentrated in cracks rather than being absorbed by the *matrix*. 2. A vertical or subvertical, tubular *orebody*, often acting as a feeder to a *manto*.

pipe clay See *ball clay*.

Piripauan A *Cretaceous* succession in New Zealand covering parts of the *Santonian* and *Campanian*.

Pisces (fish) A superclass of aquatic, cold-blooded, vertebrate animals with persistent gills, first appearing in the *fossil* record in the *Ordovician*.

pisoid (pisolith) A carbonate *coated grain* over 2 mm in diameter, with an origin similar to an *ooid*.

pisolite A rock composed of *pisoids*.

pisolith See *pisoid*.

pisolitic texture A *texture* suggestive of being formed by *pisoids*.

pistacite See *epidote*.

pitch 1. **(rake)** The orientation of a line, measured as an angle from the horizontal, in a specified non-vertical plane. *Plunge* may be derived from a series of pitch measurements using a *stereogram*. 2. See *asphalt*.

pitch coal (bituminous brown coal) A *brittle*, lustrous *bituminous coal* or *lignite* with a *conchoidal fracture*.

pitch lake See *tar pit*.

pitchblende A massive, *amorphous*, *microcrystalline* variety of *uraninite*, the major *ore* of uranium.

pitchstone A hydrated, *recrystallized*, silicic, volcanic *glass*, typically with an irregular *fracture* and a dull, resinous appearance, sometimes with a *spherulitic texture*.

pixel The smallest picture element recorded by a *remote sensing* device.

place value The economic relevance of the location of a *mineral* deposit which reflects the proportion of the value of the material represented by transport costs.

placentals See *Eutheria*.

placer deposit A sedimentary deposit of economic *minerals* concentrated by natural, mechanical processes. The concentration mechanism is usually *gravity*-driven and accomplished by moving water in which the dense *minerals* sink. The placer *minerals* must be durable after being freed from the source rock, and the main ones are *cassiterite*, *chromite*, *columbite*, *copper*, *garnet*, *gold*, *ilmenite*, *magnetite*, *monazite*, *platinum*, *ruby*, *rutile*, *sapphire*, *xenotime* and *zircon*.

Placodermi A class of superclass *Pisces*; an extinct group of armoured, jawed *fish* with a mobile, bony carapace over the head and shoulders. Range early–late *Devonian*.

Placodontia An order of subclass

Euryapsida, class *Reptilia*; heavily armoured *reptiles* similar in appearance to turtles. Range *Triassic*.

plagioclase A series of *feldspars* with compositions in the range $NaAlSi_3O_8$ to $CaAl_2Si_2O_8$.

plagiogranite A *granite* relatively rich in *plagioclase*.

plagionite ($Pb_5Sb_8S_{17}$) A rare *ore mineral* of lead.

planar crystal defect A *crystal defect* in a structure comprising layer modules which are displaced irregularly in different directions relative to an adjacent layer.

planar slump A type of *mass movement* involving sliding along a planar surface.

planation surface A relatively flat plain resulting from *erosion*.

planchéite ($Cu_8(Si_4O_{11})_2(OH)_4.H_2O$) A blue, hydrated, *secondary mineral* of copper with *inosilicate* structure.

plane bed A flat sediment bed with any irregularities less than 2–3 grain diameters in height but many diameters in length, over which there is sediment transport.

plane bed lamination A thin, flat-bedded *lamination* formed by the *aggradation* of *plane beds*. cf. *grain fall lamination*.

plane strain *Strain* in three-dimensions in which the intermediate *principal strain* is 1, i.e. there is no *extension* or *contraction* in the intermediate *principal strain* direction.

planeze A wedge-shaped *lava flow* on the slopes of a dissected *volcano* protecting underlying material from *erosion*.

planktonic Floating passively in surface water.

planplain See *panplain*.

plastering See *underplating*.

plastic deformation See *plasticity*.

plastic limit An *engineering geology* term for the minimum amount of water mixed with a sediment necessary for it to deform plastically under standard conditions. cf. *liquid limit*.

plasticity (plastic deformation) *Deformation* that causes permanent, continuous *strain* without *brittle failure* or a significant change in volume arising from applied *stress* in excess of the *yield stress*.

plate A large segment of oceanic or continental *lithosphere* that is in relative motion with adjacent segments. Up to 12 major plates have been recognized and a large number of *microplates*.

plate boundary (plate margin) The lateral margin of a *plate*, represented by an *ocean ridge* (*constructive plate margin*), *subduction zone* (*destructive plate margin*) or *transform fault* (*conservative plate margin*).

plate boundary force The *force* acting on the lateral margins of a *plate*: *ridge push*, *slab pull* and *subduction suction*.

plate margin See *plate boundary*.

plate tectonics The generally fully accepted theory that the solid Earth's surface is made up of a small number of large *plates* of *lithosphere* which are in relative motion and internally largely undeformed, the majority of the Earth's *tectonic* activity taking place at the *plate margins*.

plateau Any large, relatively flat area at high altitude.

plateau lava One of a number of *lava flows* making up a great volume which covers earlier topography, creating a new, nearly flat *plateau* between 1000 km^2 and 300 000 km^2 in area.

plateau uplift A kilometric-scale *uplift* of large *plateau* regions of *continental crust*, originating in *isostatic* response to *underplating*, *tectonic* thickening, thermal expansion and *hydration* or phase changes in the lower *crust*.

platinum (Pt) A very rare *native metal ore mineral*.

playa lake (alkali flat, dry lake, inland sabkha, salina, salt flat) A continental, shallow, dried-up, *brine* lake in a *saline lake* depositional complex.

Playfair's law See *Law of accordant junctions*.

Pleistocene The older *epoch* of the *Quaternary*, 1.64–0.01 Ma.

Pleistogene See *Quaternary*.

pleochroism (dichroism) A phenomenon exhibited by some crystals in *thin section* whose colour changes on rotation in plane polarized light.

pleonaste (ceylonite) A green variety of *spinel* intermediate in composition between *spinel* and *hercynite*.

plesiosaurs Large, extinct, marine *reptiles* with powerful, paddle-like limbs and a variety of adaptations for macrophagy and microphagy. Range late *Triassic–Cretaceous*.

Pleurocanthodii An order of subclass *Elasmobranchii*, class *Chondrichthyes*, superclass *Pisces*; freshwater sharks. Range late *Devonian*–L. *Triassic*.

plicate Folded, wrinkled.

Pliensbachian A *stage* of the *Jurassic*, 194.5–187.0 Ma.

plinian eruption A continuous, high rate *volcanic eruption* of volatile-rich *magma* which is torn apart by *vesiculation* with the resulting mix of *pumice* or *scoria*, *ash* and *volcanic gas* forming a column rising rapidly above the *volcanic vent*. When mixed with air the column increases in buoyancy and can rise to up to 50 km. Generates widely-dispersed, sheet-like, air-fall deposits with good *sorting*.

plinthite A thick, well-cemented horizon which forms by the *illuvial accretion* of *ferricrete* or other hard soil crusts, often *lateritic* and protecting underlying materials from *erosion*.

Pliocene An *epoch* of the *Neogene*, 5.2–1.64 Ma.

pliosaurs Predatory *plesiosaurs* with large heads.

plis de couverture *Folds* in a cover sequence above a basal *detachment horizon* or *décollement*.

plis de fond *Folds* in the basement below a *décollement*.

plucking (glacial plucking) The removal of large particles of an irregular rock surface by a *glacier*.

plug 1. A cylindrical feeder of a *volcano* filled with solidified *magma* and/or *pyroclastic* material and subsequently exposed by *denudation*. 2. Any small, vertical to subvertical mass of *igneous* *rock* in the form of a cylinder. cf. *stock*.

plumbago An obsolete name for *graphite*.

plume 1. A curving trace on the surface of a *fracture*. *Fractures* are initiated at the convergence of plumes and propagate with the *fracture* surface approximately orthogonal to the plume. 2. See *hotspot*. 3. A streak of effluent entering water.

plumose Like a feather.

plunge The angle between a linear *structure* and a vertical plane. cf. *pitch*.

plus-minus method (Hagedoorn method) A *seismic refraction* interpretation method for *reversed profiles* over non-planar *strata* whose velocity is variable. cf. *generalized reciprocal method*.

pluton A large, thick, *igneous body* with steep lateral contacts which was emplaced and *crystallized* beneath the surface, possibly now exposed as an irregular polygonal *outcrop*.

plutonic Originating at great depth.

pluvial 1. **(lacustral)** A period of increased moisture availability as the result of increased precipitation and/or reduced evaporation. 2. Relating to rain or other precipitation.

pneumatolysis *Alteration* by hot gas, excluding water, commonly fluorine, hydrofluoric acid and boron fluorides. Occurs in the late stage cooling of an *igneous rock* and can affect the igneous material and *country rock*.

Podolskian A *stage* of the *Carboniferous*, 307.1–305.0 Ma.

podsol (podzol) An *illuvially* accumulated soil with an upper horizon from which aluminium and iron oxides and hydroxides have been *leached*.

podsolization (podzolization) A *pedogenetic* process in which iron, aluminium and organic matter move downwards to give a prominent *eluvial* horizon with residual *silica*.

podzol See *podsol*.

podzolization See *podsolization*.

poikilitic texture An *inequigranular texture* seen in *gabbroic* and *ultramafic rocks* in which an *oikocryst* encloses several randomly oriented smaller grains of another phase or phases.

poikilo- Spotted.

poikiloblastic texture A *texture* similar to *poikilitic texture*, seen in *metamorphic rocks*.

poikilotopic fabric A *fabric* of a *sedimentary rock* in which coarse cement crystals enclose smaller, detrital grains.

point bar A channel bar of *mud* to coarse *conglomerate* forming on the convex side of a channel bend due to reduced flow velocity or *flow separation*. *Epsilon cross-stratification* may develop as the channel *migrates*.

point crystal defect A *crystal defect* caused by departure of the structure from perfect regularity at a point.

point group A group of symmetry elements in a crystal.

Poisson solid A homogeneous elastic solid whose *shear modulus* and *Lamé's constant* are equal, giving a *Poisson's ratio* of 0.25.

Poisson's equation The relationship between *gravitational* and *magnetic potentials* which allows the transformation of a *magnetic field* into a *pseudogravity field* and a *gravity* field into a *pseudomagnetic field*.

Poisson's number The reciprocal of *Poisson's ratio*.

Poisson's ratio The ratio of lateral *strain* to longitudinal *strain* in an elastic body due to uniaxial longitudinal *stress*.

polar wander The true or apparent motion of the north *magnetic pole* over the Earth's surface. See *apparent polar wander, true polar wander*.

polar wobble Regular and irregular changes in the Earth's axis of rotation in space. The main regular change is annual with an amplitude of 100 marcs and caused by varying planetary gravitational attractions and changes in oceanic and atmospheric gas distributions. See also *Chandler wobble, Markowitz wobble*.

polarity epoch (polarity interval) A period of constant polarity of the *geomagnetic field*, usually $>$ 10 000 years.

polarity excursion A time interval when the *geomagnetic field* attempted unsuccessfully to reverse polarity.

polarity interval See *polarity epoch*.

polarity reversal A relatively rapid change in the polarity of the *geomagnetic field* in which the north *magnetic pole* becomes the south *magnetic pole* and vice versa.

polarity timescale A timescale based on reversals of the *geomagnetic field*.

polder A flat, low-lying area of land reclaimed from the sea or a lake by artificial drainage.

pole figure A *stereogram* showing the distribution of *crystallographic* orientations, normally as poles to *crystallographic* planes on an *equal-area plot*.

pole of a plane The normal to a plane, a more convenient method of plotting plane data from a large number of measurements on a *stereogram*.

pole of rotation See *Euler pole*.

polje A large, commonly flat-floored, closed depression in a *karst* area, of equivocal origin.

pollucite ($CsAlSi2O_6.H_2O$) A rare *feldspathoid* found in *pegmatites*.

poloidal field A field, such as the *geomagnetic field*, with both radial and tangential components. cf. *toroidal field*.

Poltava An *Oligocene/Miocene* succession on the Russian Platform covering the *Chattian* and part of the *Aquitanian*.

poly- Many.

polybasite ($Ag_{16}Sb_2S_{11}$) A rare *ore mineral* of silver found in low- to medium-temperature silver *vein* deposits.

polyclinal fold A *fold* whose *axial surface* is variably inclined along and between the folded layers.

polygonal karst A *karst* landscape of a closely packed assemblage of closed, polygonal depressions.

polygonization The formation of polygonal grains or *subgrains*. One of the processes contributing to *recovery* by reducing the surface energy of grain boundaries. Occurs by the *climb* and *cross slip* of *dislocations* towards the boundaries at high temperatures.

polyhalite ($K_2Ca_2Mg(SO_4)_4.2H_2O$) An *evaporite mineral*, a source of potassium.

polymict Descriptive of a *clastic sedimentary rock* composed of more than one *mineral* type. cf. *monomict.*

polymictic lake A lake whose waters are in continuous circulation. cf. *oligomictic lake.*

polymorph One of two or more solid state forms of a chemical compound with different *crystal structures.*

polymorphic transformation The transformation of a *mineral* into a *polymorph,* a mechanism of *neomorphism.* See also *recrystallization.*

polymorphism The ability of a chemical compound to exist as two or more *polymorphs,* e.g. *diamond* and *graphite, calcite* and *aragonite.*

polyphyletic Descriptive of a group of organisms evolved from different ancestral stock as a result of *convergent evolution.* cf. *monophyletic.*

Polyplacophora A subclass of class *Amphineura,* phylum *Mollusca*; marine *molluscs* whose dorsal surface is covered by seven or eight calcareous plates. Range *Cambrian–Recent.*

polysomatic series A range of structures found in crystals made up of layer modules which have different compositions and atomic arrangements but which can fit together, e.g. the relationship between *pyroxenes, amphiboles* and *pyriboles.*

polysynthetic twinning (laminar twinning, multiple twinning) A type of *inversion twinning* in which the two twin orientations are equally likely to be adopted and are often found as closely spaced lamellae.

polytypism A type of *polymorphism* in which a compound exists with two or more layer-like structures differing in their stacking sequences. Small differences in chemical compositions (<0.25 atoms per formula unit) are allowed.

polytypoid A compound showing extreme *polytypism* in which there are large differences in chemical composition (>0.25 atoms per formula unit).

ponor A *sinkhole* or *swallow hole* found in a *limestone* area.

pontic Deposited in deep, still water.

pool and riffle sequence Alternating deeps (pools) and shallows (riffles) along a river channel spaced at 5–7 times the channel width. The pools probably arise from *erosion* of the outer bank and deposition on the inner bank of a meander.

pop-up The relatively uplifted part of the *hanging wall* of a *back thrust.*

Porangian An *Eocene* succession in New Zealand covering the middle part of the *Lutetian.*

porcellanite An archaeological term for a rock formed by the *thermal metamorphism* of a soil horizon in *basalt.*

pore fluid (pore water) A solution occupying the pore spaces in soil or rock, whose composition reflects its origin, subsequent mixing and *diagenetic* interaction with the host sediment. Plays an important role in *diagenesis.*

pore fluid pressure The *pressure* exerted by *pore fluids,* comprising a *normal stress* with no *deviatoric* components and equal *principal stresses,* which can exceed *hydrostatic pressure* in *overpressured* rocks. Causes *failure* of rocks at lower differential *stress* than in its absence by lowering the applied *normal stress.*

pore throat A restricted opening that connects adjacent pores.

pore water See *pore fluid.*

Porifera (sponges) A phylum intermediate between the *Protozoa* and the *Anthozoa*; sac-like organisms, attached to a substrate, with a central cavity opening upwards, which comprise a jelly colloid, sometimes containing calcareous or siliceous *spicules* which may be fossilized. Range L. *Cambrian–Recent.*

porosity The ratio of the fraction of voids to the volume of rock in which they occur.

porphyrite The central part of a *porphyry copper deposit* with *porphyritic texture.*

porphyritic microgranite See *granite porphyry.*

porphyritic texture The *texture* of an *igneous rock* that contains *phenocrysts.*

porphyroblast (metacryst) A large *euhedral* to *subhedral mineral* grain within a fine-grained *matrix* formed during *metamorphic recrystallization.*

porphyry A *hypabyssal* rock containing *phenocrysts*, commonly of *feldspar.*

porphyry copper deposit A large, low-*grade stockwork* to *disseminated deposit* of copper which may also contain minor molybdenum, *gold* and *silver*, commonly in a *granitic host rock.*

porphyry gold deposit A *disseminated* and *stockwork deposit* of *gold* in an intrusive *igneous rock.*

porphyry tin deposit A *stockwork tin* deposit, generally of low *grade.*

portal A location where a meltwater stream leaves the snout or front of an ice mass.

Portland stone A yellow-white *oolitic limestone* from the Isle of Portland, S England, widely used as a *building stone.*

Portlandian A regional British name for the highest *stage* of the *Jurassic* (*Tithonian*).

portlandite ($Ca(OH)_2$) Calcium hydroxide, a rare colourless, transparent *mineral.*

Post-Idamean A *Cambrian* succession in Australia covering the lower part of the *Dolgellian.*

post-vortex ripple A term for a very flat variety of *wave ripple.*

pot earth See *potter's clay.*

potash feldspar See *potassium feldspar.*

potassic silicate alteration (potassium silicate alteration) A form of *wall rock alteration* characterized by the formation of *potash feldspar* and/or *biotite* (*biotization*) and *anhydrite.*

potassic zone A zone characterized by the *potassic silicate alteration* of the host *stock* in the *Lowell-Guilbert Model* of *porphyry copper deposits.*

potassium-argon dating A dating method based on the radioactive decay of ^{40}K to ^{40}Ar by electron capture. Potassium concentration is determined chemically and argon concentration by *mass spectrometry.* Errors can occur by the presence of ^{40}Ar in the rock on formation and the subsequent loss of the gaseous argon.

potassium feldspar (K-feldspar, potash feldspar) ($KAlSi_3O_8$) The general name for the potassium end-member of the *alkali feldspar* series.

potassium silicate alteration See *potassic silicate alteration.*

potential The energy required to bring a unit quantity from infinity to the point of measurement against the ambient *potential field.*

potential field A field which obeys *Laplace's equation,* i.e. gravity, magnetic and electrical fields.

potentiometric surface (piezometric surface) An imaginary surface indicating the static *head of water* in an *aquifer.*

pothole 1. **(swirlhole)** A deep, circular hole in a river bed or *cave* stream caused by *erosion.* 2. **(aven, swallow hole)** A vertical shaft in a *limestone* area connecting a *cave* system with the surface.

potstone A massive variety of *steatite.*

potter's clay (pot earth) Any *clay* or earth that can be used in pottery making.

Pound Quartzite An *exceptional fossil deposit* of *Vendian* age in the Flinders Ranges, Australia containing a diverse, shallow marine fauna preserved as casts and impressions.

Poundian The highest *stage* of the *Vendian,* 580–570 Ma.

powder diffractometer An instrument used in *X-ray diffraction analysis* in which a powdered sample rotates in a focused beam to determine *mineral* characteristics from the angle between the incident and diffracted beams and the intensity of the latter.

powellite ($CaMoO_4$) A molybdate *mineral* with partial *solid solution* to *scheelite.*

power spectrum The representation of a time or distance function as its power at the various frequencies or *wavenumbers* present; the square of the amplitude spectrum.

power spectrum analysis See *spectral analysis.*

pozzolan A pumiceous *ash* used mixed with lime to make cement by the Romans, nowadays used mixed with cement as a construction material.

Pragian 1. A *stage* of the *Devonian,* 396.3–390.4 Ma. 2. A *Devonian* succession in Czechoslovakia covering parts of the *Pragian* and *Emsian.*

Prandtl number A dimensionless coefficient equal to the product of specific heat and *kinematic viscosity* divided by *thermal conductivity.*

Prandtl-Reuss equations A modification of the *Levy-Mises equations* incorporating a component of elastic behaviour.

prase A dull green, *microcrystalline* variety of *silica.*

Pratt hypothesis A model which suggests that *isostatic compensation* is achieved by varying the *density* of the outer layer of the Earth in inverse proportion to its elevation, i.e. mountain ranges are supposed to be made up of relatively low *density* rocks.

pre- Before.

Pre-Payntonian A *Cambrian* succession in Australia covering the middle part of the *Dolgellian.*

Precambrian The oldest (>570 Ma)

era characterized by a paucity of organisms with hard parts capable of fossilization.

Precambrian iron formation A low-*grade*, *metamorphic* iron deposit of *Precambrian* age.

precious metal *Gold*, *silver* or *platinum*.

precipitate controlled replacement The non-specific *replacement* by highly concentrated, aggressive solutions regardless of the replaced *mineral*'s composition, structure or surface free energy. cf. *host grain controlled replacement*.

precipitation The deposition of an *authigenic mineral* from a supersaturated *pore fluid* in either solid form by *crystallization* or as a gel by *flocculation* followed by *crystallization*, giving rise to total or partial *cementation* of the *porosity* of the *host rock*. The two fundamental processes of precipitation are *nucleation* and *crystal growth*.

precision depth recorder (PDR) An *echo sounder* of high precision.

preferred orientation A concentration of linear or planar, structural or *fabric elements* in a particular attitude.

prehnite ($Ca_2Al(AlSi_3O_{10})(OH)_2$) A green *secondary mineral* found lining cavities in *basaltic* rocks.

prehnite-pumpellyite facies A *metamorphic facies* characterized by the formation of *prehnite*, *pumpellyite* and *quartz* in *basic igneous rocks* under moderate pressure and low temperature, usually during *burial metamorphism*.

prescellite See *preselite*.

preselite (prescellite) An archaeological term for the spotted *dolerite* of the Preseli Hills, SW Wales, used in the construction of Stonehenge.

pressure 1. **(hydrostatic pressure, hydrostatic stress)** A three-dimensional *stress* state in which the magnitude of *stress* is the same in all directions. 2. The *force* per unit area acting on a surface.

pressure dissolution See *pressure solution*.

pressure fringe A *structure* at the margins of the more rigid body in a *pressure shadow* formed by the growth of fibrous *minerals*.

pressure release A *weathering* mechanism which causes spalling and *exfoliation* of rocks which were once deeply buried by the release of *pressure* when they are brought to the surface.

pressure shadow The region around a relatively rigid body in a deformed rock that has undergone *extensional strain*, commonly containing new *calcite* or *quartz*, whose growth is related to *pressure solution*.

pressure solution (pressure dissolution) 1. The enhanced rate of transfer of material from a *mineral* grain into intergranular fluid caused by increasing *stress* as the external *pressure* exceeds the *pore fluid pressure*. 2. The process of *strain* by *dissolution*, *diffusion* through fluid on grain boundaries and redeposition of *minerals*. The *dissolution* is enhanced on surfaces normal to the maximum *principal stress* and deposition occurs preferentially on surfaces normal to the minimum *principal stress*.

pressure solution cleavage A *cleavage* developed by preferential solution at contact surfaces of grains or crystals when the external *pressure* exceeds the *pore fluid pressure*.

Priabonian The highest *stage* of the *Eocene*, 38.6–35.4 Ma.

priderite ($(K,Ba)(Ti,Fe^{3+})_8O_{16}$) A black-red *oxide mineral* found as an *accessory* in *leucite*-bearing rocks.

Pridoli The youngest *epoch* of the *Silurian*, 410.7–408.5 Ma.

Pridoli-Schichten A *Silurian* succession in Bohemia equivalent to the *Pridoli*.

primärumph A tectonically elevated *dome* of rock which is *eroding* at the same rate as the *uplift*.

primary creep (transient creep) The initial stage of *viscoelastic strain* characterized by a concave-downwards *strain*-time curve. See also *secondary creep*, *tertiary creep*.

primary current lineation A low (2–3 grain diameters high), extensive, linear ridge of grains parallel to the flow direction on the surface of an *upper stage plane bed*. Results from high velocity *sweep* events in the *viscous sublayer* which entrains sediment and deposits it as flow-parallel ridges.

primary migration The first phase of the upward *migration* of *hydrocarbons* within and out of the *source rock* to the *reservoir rock*. cf. *secondary migration*.

primary mineral A *mineral* formed at the same time as the rock bearing it. cf. *secondary mineral*.

primary porosity All the pore space initially present in a sediment at the time of deposition. cf. *secondary porosity*.

primary structure A *structure* formed as a rock was formed, rather than by subsequent *deformation*, e.g. *cross-bedding*, *slump folds*.

Primates An order of infraclass *Eutheria*, subclass *Theria*, class *Mammalia*; including the lemurs, monkeys, apes and man. Range *Palaeocene–Recent*.

primitive circle The *cyclographic trace* of a horizontal plane on a *stereographic projection*.

Primitive deposit A type of zinc- and copper-bearing *massive sulphide deposit* which may contain *gold* and *silver*.

principal axes of stress (stress axial cross) The normals to the *principal planes of stress*.

principal finite strain *Finite strain* normal to the axis of maximum *shortening*, along which it is assumed that *cleavage* is developed.

principal planes of stress The three mutually perpendicular planes in a *stress* system along which there are no *shear stresses*.

principal strain axes The three axes of symmetry of a *strain ellipsoid*.

principal strain planes The planes orthogonal to the three axes of a *strain ellipsoid*.

principal strains The relative sizes of the three axes of a *strain ellipsoid*, measured as *stretches* or *quadratic elongations*.

principal stresses The values of *stress* along the principal axes of *stress*, termed the maximum, intermediate and minimum principal stresses.

Priscoan See *Hadean*.

prismatic crystal A crystal with a

prism as the main *form*, i.e. one dimension significantly greater than the others.

prismatic texture A *texture* of a *metamorphic rock* characterized by *prismatic crystals*.

Proanura An order of subclass *Lissamphibia*, class *Amphibia*; extinct, frog-like *amphibians*. Range L. *Triassic*.

Proboscidea An order of infraclass *Eutheria*, subclass *Theria*, class *Mammalia*; the elephants and their extinct relatives. Range ?*Miocene–Recent*.

prochlorite See *ripidolite*.

prod-and-bounce mark A type of *sole structure* or *tool mark*. Prod marks are asymmetric, elongate, semicircular-triangular depressions in a sediment surface, the depression being broader and deeper at the downcurrent end, formed when a body in a sediment flow hits the sediment surface at an angle and momentarily stops. Bounce marks are more symmetrical depressions formed when a body impacts a bed at an acute angle and rises off it.

production well A *well* from which fluid is recovered.

proglacial lake A lake formed adjacent to the snout of a *glacier*.

progradation The forward movement of *facies* belts as a result of sediment supply.

Progymnospermopsida A class of division *Trachaeophyta*, kingdom *Plantae*; the possible ancestors of the *Gymnospermopsida*. Range *Devonian*–U. *Carboniferous*.

prokaryote The earliest type of primitive organism with no cell wall around the nucleus. cf. *eukaryote*.

prolate Rod or spindle shaped. cf. *oblate*.

propagating rift An *ocean ridge* whose axis is growing in length at the expense of an older ridge in a different direction, probably in response to a change in location of the *Euler pole* about which the adjacent *plates* are separating.

propylitic alteration *Alteration* characterized by the development of *chlorite*, *epidote*, *albite* and carbonate (*calcite*, *dolomite* or *ankerite*).

propylitic zone A zone of *propylitic alteration* around a *porphyry copper deposit* in the *Lowell-Guilbert Model*.

Proterozoic The younger *eon* of the *Precambrian*.

proto- First, foremost.

protocataclasite A cohesive *fault rock* with a random *fabric* and with 10–50% *matrix*. cf. *cataclasite*, *ultracataclasite*.

protoclastic Descriptive of the structure of an *igneous rock* whose early crystals were broken or deformed by movement of the liquid fraction.

protoenstatite An orthorhombic *polymorph* of *enstatite*.

protomylonite A *foliated fault rock* with 10–50% *matrix*. cf. *mylonite*, *ultramylonite*.

proton magnetometer (nuclear precession magnetometer) A *magnetometer* for measuring the strength of the *geomagnetic field* by measuring the precessional frequency of protons which are returning to the *geomagnetic field* direction after being deflected. The standard instrument for *magnetic surveys* and *magnetic observatories*.

protopyroxene An orthorhombic *polymorph* of *pyroxene*, not yet found naturally.

protore A *mineral* deposit which, by the action of further natural processes, may be upgraded to *ore*.

Prototheria A subclass of class *Mammalia* of which the *Monotremata* is the only order.

Protozoa Very small, primitive, unicellular animals including the *Foraminifera* and *Radiolaria*. Range *Precambrian–Recent*.

proustite (light red silver ore, light ruby silver, ruby silver ore) (Ag_3AsS_3) An *ore mineral* of silver found in *vein* deposits.

provenance The source area(s) of sedimentary material and the nature(s) of the rocks from which it is derived.

proximal Descriptive of a feature close to its source. cf. *distal*.

psammite A *metamorphic rock* rich in *quartz*.

psephite A coarse sediment or its *metamorphic* form.

pseudo- False.

pseudobreccia An irregularly or partially *dolomitized limestone* in which the growth of some coarse crystals imparts a *texture* similar to a *breccia*.

pseudobrookite ($FeTiO_5$) An *oxide mineral* found in cavities in *basalt*.

pseudogravity field The *gravity* field derived from a *magnetic field* using *Poisson's equation* for the purposes of interpretation.

pseudokarren *Karren*-type forms developed on non-carbonate rocks.

pseudokarst Landforms and landscapes resembling those of *limestone* areas but developed on non-carbonate rocks.

pseudoleucite A mixture of *nepheline*, *orthoclase* and *analcime pseudomorphous* after *leucite*.

pseudomagnetic field The *magnetic field* derived from a *gravity* field using *Poisson's equation* for the purposes of interpretation.

pseudomalachite (tagilite) ($Cu_5(PO_4)_2(OH)_4.H_2O$) Copper phosphate and hydroxide, found as a *secondary mineral*, which resembles *malachite*.

pseudomorph A *secondary mineral* which has replaced another but maintained its shape.

pseudonodule A spherical body of *sandstone* with an internal convolute *lamination* set in a *mudstone*, resulting from the sinking of *sand* from an overlying bed into soft *mud*.

pseudophite A massive variety of *chlorite* sometimes used as a substitute for *jade*.

pseudospar Recrystallized *micrite* with grains $>20\ \mu m$ in size. cf. *microspar*.

pseudosymmetry An arrangement of the atoms in a *crystal structure* which is nearly, but not exactly, symmetrical about the *twin axis* or across the *twin plane*.

pseudotachylite A very fine-grained or *glass*-like *fault rock*, formed by rapid *displacement* and melting by *shear*-generated heating.

pseudowollastonite ($CaSiO_3$) A high

temperature (>120°C) form of *wollastonite*.

psilomelane A *botryoidal* mass of manganese *oxide minerals*, of which *romanechite* is a major constituent.

pteropod ooze A *pelagite* composed primarily of the remains of *pteropods*.

pteropods Animals with thin, calcareous shells forming gelatinous zooplankton which contributes to *oozes* and *marine snow*.

Pterosauria/pterosaurs A class of subclass *Archosauria*, class *Reptilia*; flying *reptiles*. Range *Jurassic*–late *Cretaceous*.

Ptyctodontida An order of class *Placodermi*, superclass *Pisces*; small *fish* with limited armour at the back of the head, superficially resembling modern *fish*. Range M. *Devonian*–U. *Devonian*.

ptygmatic fold A rounded, generally *concentric fold* whose *amplitude* is large with respect to the thickness of the folded layer and whose *wavelength* is small, generally approaching an *elasticas fold profile*. Common in layers of high *competence* in a less *competent matrix*.

Puaroan A *Jurassic* succession in New Zealand covering part of the *Tithonian*.

pudding ball See *armoured mud ball*.

puddingstone A colloquial term for coarse *conglomerate*, particularly the silicified variety from the *Eocene* Reading Beds of Hertfordshire, England.

pulaskite A *felsic alkali syenite* mainly comprising *alkali feldspar* plus subordinate *felsic minerals* and often a small amount of *nepheline*.

pull-apart basin (rhombochasm) A small *extensional basin*, formed where two parallel *strike-slip faults* join.

pumice A light coloured, highly *vesicular*, *acid* volcanic *glass* with a low *density*, used in construction because of its insulating properties. cf. *scoria*.

pumpellyite ($Ca_2Al_2(Al,Mg,Fe)Si_2O_{10}(OH)_2(O,OH)_2$) A silicate *mineral* formed during moderate pressure and low temperature *metamorphism*.

pumping test A method of measuring the properties of an *aquifer* by pumping from one *well* and monitoring the expansion of the *cone of depression* in surrounding observation *wells*. The inclination of the cone is an indication of the *hydraulic gradient*, which depends on the pumping rate and the *transmissivity* and *storage coefficient* of the *aquifer*.

Punctuated Aggradational Cycles A model for an *allocyclic* mechanism of generating cyclic carbonate deposits on a platform.

punctuated equilibrium model A theory of evolutionary development of species involving morphological stasis and rapid evolutionary changes taking place at a rate too fast for *stratigraphic* resolution. cf. *phyletic gradualism*.

pure shear A shape change without volume change in which the *strain axes* do not rotate.

pure strain *Deformation* that can affect each point in a body differently.

purple copper ore See *bornite*.

Pusgillian A *stage* of the *Ordovician*, 443.1–440.6 Ma.

puy A volcanic hill, often a *plug*, of the Massif Central, France.

pycnocline A *density* interface in a body of water, e.g. a *thermocline.*

pyralspite An acronym for the *garnets* ***pyrope***, ***almandine*** and ***spessartite***. cf. *ugrandite.*

pyrargyrite (dark ruby silver, ruby silver ore) (Ag_3SbS_3) An *ore mineral* of silver found in *vein* deposits.

pyriboles (biopyriboles) A group of *minerals* with a *mineralogy* common to the *pyroxenes* and *amphiboles*, but also including certain non-classical chain silicates.

pyrite (iron pyrites) (FeS_2) The most common *sulphide mineral.*

pyritization The *replacement* of the hard parts of an organism by *pyrite.* cf. *haematization.*

pyro- From or by fire.

pyrochlore ($(Ca,Na)_2(Nb,Ta)_2O_6(O,OH,F)$) An *oxide mineral* of niobium and tantalum found associated with *alkaline igneous rocks.*

pyroclastic flow See *ash flow.*

pyroclastic rock A rock formed by the accumulation of material generated by the explosive fragmentation of *magma* and/or existing solid rock during a *volcanic eruption.*

pyrogenesis The intrusion and extrusion of *magma* and its products.

pyrogenetic mineral A *mineral* crystallized from an anhydrous or near-anhydrous *magma.*

pyrolite A hypothetical *peridotitic* rock proposed as representing the upper *mantle* composition from which *basaltic magma* could be derived on *partial melting.*

pyrolusite (MnO_2) The most important *ore mineral* of manganese.

pyrometamorphism Intense *thermal metamorphism* at high temperature and pressure.

pyrometasomatic deposit See *skarn.*

pyromorphite ($Pb_5(PO_4)_3Cl$) A *supergene mineral* found in the oxidized parts of lead *veins.*

pyrope (Cape ruby, false ruby) ($Mg_3Al_2Si_3O_{12}$) A deep red to black magnesium *garnet*, valued as a *gem* when clear and transparent.

pyrophanite ($MnTiO_3$) An *oxide mineral* related to *ilmenite.*

pyrophyllite ($Al_2Si_4O_{10}(OH)_2$) A rare *phyllosilicate* found in *metamorphic rocks*, with many properties and uses in common with *talc.*

pyrophyllite deposit A product of the *hydrothermal alteration* of *acid volcanic rocks.*

pyrostibnite See *kermesite.*

pyroxene-hornfels facies A *metamorphic facies* often formed adjacent to *plutons.*

pyroxenes Silicate *minerals* with an internal structure of a single chain of linked silicate tetrahedra with cations occupying sites between oxygen ions at the edges of the chains. General formula: $A_{1-x}(B,C)_{1+x}T_2O_6$, where A is commonly Na or Ca, B is Mg or Fe^{2+}, C is Al or Fe^{3+} and T is Si or Al. Stable over a wide range of temperature and pressure and so used in *geothermometry* and *geobarometry*. Common in *igneous* and *metamorphic rocks.*

pyroxenite An *ultrabasic rock* rich in *pyroxene.*

pyroxenoids A group of *silicate minerals* with a formula similar to the *pyroxenes*, possessing a Si:O ratio of 1:3, but which do not crystallize with a *pyroxene* structure, e.g. *wollastonite*, *rhodonite*, *pectolite*.

pyroxferroite ($Ca_{0.15}Fe_{0.85}SiO_3$) A *pyroxenoid* found in lunar *lavas*.

pyroxmangite ($(Mn,Fe)SiO_3$) A *pyroxenoid* found in *metamorphic* and *metasomatic rocks*.

pyrrhotite (magnetic pyrites) ($Fe_{1-x}S_x$) A common, *ferrimagnetic sulphide mineral*.

Pytyr'yuskiy A *Permian* succession in the Timan area of the former USSR, covering part of the *Capitanian*, the *Longtanian* and the *Changxinian*.

Q

Q See *quality factor.*

Q-mode analysis A technique of grouping organisms into communities on the basis of their relative similarity. cf. *R-mode analysis.*

QAP triangle A *modal* classification scheme for *granitic* rocks based on the relative proportions of *quartz* (**Q**), *alkali feldspar* (**A**) and *plagioclase feldspar* (**F**).

QAPF classification A *modal* classification scheme for *igneous rocks* with a *colour index* <90 based on the relative proportions of *quartz* to *quartz* + total *feldspars* (**Q**), *alkali feldspars* to total *feldspars* (**A**), *plagioclase* to total *feldspars* (**P**) and *feldspathoids* to *feldspathoids* + total *feldspars* (**F**).

Qionghusi A *Vendian/Cambrian* succession in China covering parts of the *Poundian* and *Tommotian.*

quadratic elongation A linear measure of change in shape based on changes in line lengths: (deformed length/undeformed length)^2; the square of *stretch.*

quadrature component See *imaginary component.*

quality factor (Q) A descriptor of *anelastic attenuation* whereby the fraction $2\pi/Q$ of wave energy is absorbed per cycle. Varies from 100 to 5000 for *P waves* in the *crust* and *mantle* respectively.

quaquaversal Dipping radially away from a centre.

quartz (SiO_2) The most common *silica mineral.*

quartz diorite A coarse-grained, *plutonic, igneous rock* comprising *essential plagioclase,* normally with a small amount of *orthoclase,* plus *biotite* and *hornblende.* The intrusive equivalent of *dacite.*

quartz index An expression of the *mineralogical* maturity of a *sandstone* in terms of the ratio of the percentage of (*quartz* + *chert*) to (*feldspar* + rock fragments + *clay matrix*).

quartz monzonite A *granitic* rock with *quartz* comprising 5–20% of the *felsic minerals* and *alkali feldspar* making up 35–65% of the total *feldspar.* The *plutonic* equivalent of *latite.*

quartz porphyry A *porphyritic microgranite, microgranodiorite* or *microtonalite.*

quartz topaz See *citrine.*

quartzarenite See *orthoquartzite.*

quartzite 1. A *metamorphic rock* usually formed by the *metamorphism* of a *sandstone.* 2 An outdated term for *sandstone.*

quartzwacke A *sandstone* comprising $>95\%$ *quartz clasts* with 15–75% *matrix.*

quasi-equilibrium An equilibrium giving the impression of *dynamic equilibrium* because of its observation over only a relatively short period of time.

quasi-planar adhesion stratification See *adhesion lamination.*

quasi-plastic mechanism A *plastic deformation* mechanism which can produce a *mesoscopically* continuous *strain*, including *crystal plasticity*, *diffusive mass transfer* and *grain boundary sliding*.

Quaternary The most recent sub*period* of the *Cenozoic* 1.64 Ma–present.

quick clay A *clay* in which the delicate packing of particles can be completely destroyed on *failure* to cause almost total loss of *strength* and, possibly, *liquefaction.*

quickflow A mixture of *overland flow* and subsurface *stormflow*. cf. *baseflow*.

quicksilver A common name for *mercury*.

quiescent continental margin See *passive continental margin.*

quiet zone See *magnetic quiet zone.*

R

R1 fault See *Riedel fault.*

R1 shear See *Riedel shear.*

R2 fault A *conjugate Riedel fault.*

R2 shear (P shear) A *conjugate Riedel shear.*

R channels See *Röthlisberger channels.*

R-mode analysis A technique in which the distributions of individual *taxa* are compared; those that co-occur are grouped and those which are mutually exclusive are placed in different communities. cf. *Q-mode analysis.*

radar See *ground-penetrating radar, remote sensing.*

radial dykes A set of *dykes* arranged radially around a *pluton* as a result of the *stress* field generated by the *pluton.*

radial joint A *joint* parallel to a *fold axis* which remains perpendicular to the folded layer on the *fold limbs.*

radiaxial Descriptive of a crystal which has grown in a cavity in a fan-like pattern approximately normal to the cavity wall.

radioactive dating See *radiometric dating.*

radioactivity log A *geophysical borehole log* making use of natural (*gamma log*) or induced (*gamma-gamma log, neutron log*) radiation.

radiocarbon dating (carbon-14 dating) A method of *radiometric dating* based on the decay of ^{14}C, which is at a known, equilibrium concentration in living organisms, to ^{12}C. On death the proportion of ^{14}C decreases at a known rate. Normally suitable for dating materials younger than ~50 000 years, although *accelerator radiocarbon dating* extends this range.

radioisotope dating See *radiometric dating.*

Radiolaria A subclass of tiny, free-floating, marine, single-celled, round *protozoans* with siliceous endoskeletons, generally of *opaline silica*, arranged in radial or tangential elements. The skeletons make an important contribution to deep-sea sediments and *cherts*. Range *Cambrian–Recent*, with the greatest diversity in the *Cretaceous.*

radiolarian ooze A *pelagite* composed primarily of the remains of *Radiolaria.*

radiolarite *Chert* containing abundant *Radiolaria.*

radiometric assay A measurement of the uranium concentration in buried bone, into which it is absorbed from percolating *groundwater*, from its radioactivity, generally for archaeological purposes. cf. *fluorine test.*

radiometric dating (radioactive dating, radioisotope dating) Techniques of determining the age of rocks or *fossils* from the relative proportions of a radioactive parent and its daughter decay product(s). Knowledge of the radioactive decay constant or *half-life* allows the proportion to be converted into an age. Methods include *argon-argon dating*, *lead-lead dating*, *potassium-argon dating*, *radiocarbon dating*, *rubidium-strontium dating*, *samarium-neodymium dating*, *thorium-lead dating* and *uranium-lead dating*.

radiometric surveying *Geophysical exploration* methods for the location of radioactive elements, in practice uranium, thorium and potassium (^{40}K), using *Geiger counters*, *scintillation counters* and *gamma ray spectrometers*, the latter two of which can be used in *airborne geophysical surveys*.

radon (Ra) The only radioactive gaseous element, produced by the decay of ^{238}U with a *half-life* of 3.8 days, and as such moves freely through pores, *joints* and *fractures* either as a gas or dissolved in *pore fluids*. Measured with a *radon emanometer*. Often indicative of buried uranium concentrations.

radon emanometer An instrument for measuring *radon* from air samples from a shallow borehole. The filtered gas is dried and passed to an ionization chamber or *scintillation counter* where ^{232}Ra is detected from its alpha particle activity.

Raibler Schichten A *Triassic* succession in the Alps equivalent to part of the *Carnian*.

rain-impact ripple A *raindrop-generated structure* produced when large raindrops are driven obliquely onto fine *sand* by a strong wind.

rain pit A *raindrop-generated structure* on a sediment surface.

rainbow quartz See *iris*.

raindrop-generated structure A sedimentary *structure* formed by rain impacting on unconsolidated sediment. Impact marks are generally circular to elliptical, <10 mm in diameter and several millimetres deep with a raised rim. See also *hailstone-generated structure*.

raindrop impact erosion (rainsplash erosion) The movement of soil particles as the result of raindrop impact by rebound, undermining and downslope movement; most effective in the absence of surface *runoff*.

rainfall erosivity factor A term in the *universal soil loss equation*.

rainsplash The mechanism of loosening soil particles by rain prior to their removal by surface *runoff*.

rainsplash erosion See *raindrop impact erosion*.

raised beach A *beach deposit* stranded at altitude by a fall in sea level.

rake See *pitch*.

Rakhmey An *Ordovician* succession in Kazakhstan covering the lower part of the *Arenig*.

rammelsbergite ($NiAs_2$) A nickel-bearing *mineral* with *loellingite* structure.

ramp The part of a *fault* that cuts across *datum* surfaces, such as *bedding*, in *thrust* and *extensional fault systems*. cf. *flat*.

Ramsaudolomit A *Triassic* succession in the Alps equivalent to the *Anisian*.

Ramsdell notation A method of describing the *unit cell* relationship in *polytypism*.

Randian An *era* of the *Precambrian*, 2800–2450 Ma.

random fabric (isotropic fabric) A *fabric* with no *preferred orientation*.

random structure (isotropic structure) A *structure* with no *preferred orientation*.

range The distance between the point of initiation of a *seismic wave* and its point of detection.

rank The degree of *coalification* of a *coal*.

rankinite ($Ca_3Si_2O_7$) Calcium disilicate, found in siliceous *limestones* subjected to high *grade metamorphism*.

rapakivi granite A *granite* exhibiting a *rapakivi texture*, characteristic of the *anorthosite-granite* associations of the *Proterozoic*.

rapakivi texture A *porphyritic texture* in which centimetric-scale, rounded *phenocrysts* are surrounded by a rim of sodium-rich *plagioclase*.

rapids A section of river channel in which flow is faster and more *turbulent* than elsewhere.

rare earth elements (REE) Elements with an atomic number between 57 and 71 plus scandium and yttrium.

ratio contour map A map showing the ratio of the total thickness of one lithological class to that of the remaining classes making up the total *stratigraphic* unit section.

Raukumara A *Cretaceous* succession in New Zealand comprising the *Arowhanan*, *Mangaotanian* and *Teratan*.

Rawtheyan A *stage* of the *Ordovician*, 440.1–439.5 Ma.

Rayleigh number (Ra) A dimensionless parameter used in fluid dynamics whose magnitude determines when *convection* is initiated in a fluid. $Ra = d^3 g \Delta T \rho \alpha / K\eta$, where d = depth of fluid, g = *gravity*, ΔT = super*adiabatic* temperature gradient, ρ = *density*, α = volume coefficient of thermal expansion, K = *thermal diffusivity* and η = *kinematic viscosity*.

Rayleigh wave A *seismic surface wave* whose particle motion is in the form of a vertical ellipse in the direction of propagation. Efficient in the transport of seismic energy and responsible for *ground roll*.

re-entrant angle The angle (>180°) between the members of a *twinned* crystal.

reaction rim A zone of *secondary minerals* surrounding a *primary mineral* as the result of late-stage *metasomatism* or reaction with fluids or solids with which it is in contact.

reaction series See *continuous reaction series* .

reactivation surface A surface which records the *erosion* of a *bedform foreset* slope resulting from a rise or fall in the level of the flow. After the *erosion* the *bedform* reforms with the same orientation and shape and continues to migrate.

real component (in-phase component) The part of the secondary field that is in phase or 180° out of phase with the primary field in *electromagnetic induction methods*.

realgar (AsS) A red *sulphide mineral* of the *ring structure group* found in *veins*, *volcanic-exhalative deposits* and *hot springs*.

reattachment point The outer margin of the *shear layer* bounding a zone of *flow separation* where the freestream again impinges on the bed.

Recent (Holocene) The youngest *epoch* of the *Quaternary*, 0.01 Ma–present.

Receptaculitids A group of green *algae*, *sponges* or a separate phylum with a globular or pear-shaped form and internal radiating rods terminating distally in plate-like facets which make up an outer wall. Range *Ordovician* to *Permian*.

recharge The precipitation reaching the *water table* and replenishing *groundwater* supply.

recharge area The area over which *recharge* is received.

reciprocal time The travel time in either direction between two points on a *reversed seismic profile*.

reclined fold (vertical fold) A *fold* whose *hinge line plunges* steeply.

reconstructive transformation The slow transformation of one *polymorph* into another in which bonds must be broken and new bonds formed. cf. *displacive transformation*.

recoverability The percentage of total metal in an *ore* that is present in the concentrate after processing.

recoverable strain See *temporary strain*.

recovery 1. The change from a strained to an unstrained state during or after *deformation*, an important phenomenon in *dislocation climb*. 2. The percentage of the scheduled tonnage actually mined.

recrystallization A process of *neomorphism* in which the overall chemical composition is unchanged.

recumbent fold A *fold* whose *hinge line* and *axial surface* are both horizontal.

red bed A sedimentary deposit whose *clasts* are coated with *haematite*, causing a red colour. The presence of such beds in the *Proterozoic* is taken to indicate a major change from an oxygen-rich hydrosphere to an oxygen-bearing atmosphere.

red bed copper A copper variety of a *sandstone-uranium-vanadium base metal deposit*, whose *host rocks* are often red.

red clay (brown clay) A red/brown *pelagite* primarily composed of *clay minerals* with minor *quartz*, *ash*, cosmic dust and *fish* teeth, the product of very slow deposition in the central parts of ocean *basins* at >4 km depth and below the *calcite compensation depth*.

red ochre A red, earthy variety of *haematite* used as a pigment.

red zinc ore See *zincite*.

redox reaction *Oxidation* coupled with *chemical reduction* in a balanced exchange of electrons.

redruthite See *chalcocite*.

reduced travel-time curve A means of displaying *seismic refraction* data by plotting the difference between the observed travel-time and the time which would have been observed if material of only one particular velocity had been present. Refractors with that velocity then plot horizontal to the distance axis.

reduction 1. See *chemical reduction*. 2. The procedure used in processing geophysical survey data to remove all non-geological sources of variation.

reduction to the pole A method of processing *magnetic anomaly* data so that the anomaly appears as it would be if located at the *magnetic pole*. Since the *geomagnetic field* is vertical at the poles, this operation facilitates the interpretation of the anomaly.

reduzate A sediment accumulated under reducing conditions, typically rich in organic *carbon* and iron sulphides.

REE See *rare earth elements*.

reedmergnerite ($NaBSi_3O_8$) A rare *feldspar* found in unmetamorphosed *dolomitic oil shales*.

reef 1. A *gold*-bearing *quartz vein*. 2. A *Precambrian gold*-bearing *conglomerate* of southern Africa. 3. See *coral-algal reef*. 4. A ridge of rocks rising to near sea level.

reef knoll A large mass of *reef limestone weathered* out to form a rounded hill.

reef trap An *oil trap* consisting of porous *reef limestone* covered by impermeable *strata*, commonly *shale* or *mudstone*.

Reeh calving A proposed mechanism for the splitting of icebergs from the front of *glaciers* and *ice sheets* in which it is suggested that *stress* is greatest and sufficient for *fracture* at a cross-section of floating *glacier* at a distance of about the ice thickness from the ice front.

reflectance 1. The strength with which the components of *coal*, or other carbonaceous materials, reflect light; a measure of *rank*. 2. The strength with which an *opaque mineral* in polished section reflects light.

reflected light optics The study of *opaque minerals* under the microscope in polished sections.

reflected seismic wave The *seismic wave* returned to the surface by reflection at a boundary where the *acoustic impedance* changes (a reflector). For a wave of the same type, the angle of reflection is equal to the angle of incidence. Forms the basis of a type of *seismic exploration*, particularly for *structures* forming potential *oil* and *gas traps*.

reflection character analysis A technique of *seismic stratigraphy* in which the changes in a *seismic reflection* wave shape from one record to another is studied in order to determine the nature of changes in *stratigraphy*, the fluid content and the properties of a *reservoir* rock.

reflection coefficient The ratio of the amplitudes of an incident *seismic wave* and the wave reflected from an *acoustic impedance* contrast. For vertical incidence the reflection coefficient is given by $(A_2-A_1)/(A_2 + A_1)$ where A_1 and A_2 are the *acoustic impedances* of the layers through which the wave is travelling and the layer across the contrast respectively.

reflective beach (swell beach) A *beach* characterized by low frequency swell waves, low *beach water tables*, high percolation and low *backwash* so that there is net onshore sediment movement.

refracted seismic wave The *seismic wave* transmitted through a boundary where the *acoustic impedance* changes (a refractor) according to the geometry predicted by *Snell's law*.

refractive index (RI) A most important optical property of a *mineral*, defined as the ratio of the velocity of light *in vacuo* to the velocity in the *mineral*.

refractometer An instrument used to measure *refractive index*.

refractory clay See *fireclay*.

refractory gold *Native gold* which occurs in *minerals* that are difficult to treat by amalgamation or cyanide.

refractory mineral A *mineral* resistant to decomposition by heat, pressure or chemical reaction.

refractory ore An *ore* from which it is difficult to extract the valuable metal.

Refugian An *Eocene* succession on the west coast of the USA equivalent to the *Priabonian*.

reg (serir) A stony desert floor of *gravels* underlain by mixed sediment sizes that have been removed from the surface by wind and water.

regelation The process whereby ice melts and refreezes as *pressure* is raised and lowered.

regimes of diagenesis A genetic classification of the various stages of *diagenesis* of carbonate and siliclastic sediments during the evolution of a sedimentary *basin*, comprising *eogenesis*, *mesogenesis* and *telogenesis*.

regional field The large-scale variation in a *gravity* or *magnetic field* over a region which reflects the effects of relatively deep *structures*. This field must be removed before interpreting any *residual anomalies*.

regional metamorphism Large-scale *metamorphism* involving heat and pressure.

regolith The superficial layer of loose, unconsolidated material which overlies *bedrock* over much of the land surface, comprising unlithified *in situ saprolite*, *ash*, *colluvium*, *alluvium* or *drift*.

regression A recession of the sea from a land area or from a shallow sea. cf. *transgression*.

Reitzi A *Triassic* succession in the Alps covering the lower part of the *Ladinian*.

rejuvenation 1. Of a river, its stimulation to increased *erosional* activity. 2. Of a *structure*, the topographic renewal of an *orogenic belt* by *uplift* after the main *orogenic* activity had ceased.

relative permeability The ratio of *effective permeability* at a stated fluid saturation to the *absolute permeability*.

relative plate motion The relative motion between *plates* determined from the *spreading rate* at an *ocean ridge* by matching *oceanic magnetic anomalies* of known age across it, by *palaeomagnetic* measurements, by matching geological features, *palaeontological* correlations, *palaeoclimatic* data etc.

relaxation time 1. The period of years or decades after a major *river flood* over which a channel form reverts to its former state. 2. The time for *remanent magnetization* to decay to e^{-1} of its original value. 3. The time taken for a disturbed system to reach equilibrium, or for a parameter to decrease to about 37% of its initial value.

releasing bend A curve in a *strike-slip fault* across which there is *extension* parallel to the straight parts of the *fault*. cf. *restraining bend*.

relict structure See *palimpsest structure*.

Relizian A *Miocene* succession on the west coast of the USA, covering parts of the *Burdigalian* and *Langhian*.

remanent magnetization The magnetization in a rock that is present in the absence of an external *magnetic field*. Principally *natural remanent magnetization*, but also includes magnetizations produced by laboratory processes.

remanié fossil (derived fossil) A *fossil* surviving the *erosion* of its enclosing rock and incorporated into a later sediment.

remote sensing The recording of images of parts of the Earth's surface using electromagnetic radiation, normally from an aircraft or satellite at sufficient height for a broad area to be covered. Includes passive *aerial photography* and imagery from *Landsat satellites* using the *multispectral scanner* and *thematic mapper*, and the *SPOT-1 satellite*, and also active techniques, such as *radar*, which shows the topographic texture of terrain in the presence of cloud cover, etc.

reniform Kidney-shaped.

Repettian A *Pliocene* succession on the west coast of the USA, covering the middle part of the *Piacenzian*.

repichnia *Trace fossils* comprising locomotion traces.

replacement In *diagenesis*, the growth of a chemically different *authigenic mineral* within the body of an existing *mineral*, either of the *host rock*, a detrital grain or an earlier *authigenic mineral*.

replacement orebody An *epigenetic orebody* formed by the *replacement* of existing rocks, particularly carbonates, e.g. *flats*, *skarn deposits*.

reprecipitation The *precipitation* of material following its *dissolution*, a mechanism for the *neomorphism* of *silica minerals*.

reptation The process of low velocity, *aeolian* grain transport by a short trajectory following a grain impact. cf. *saltation*.

Reptilia/reptiles A class of vertebrate animals with an egg typical of the *Amniota*, which are not *Mammalia* or *Aves*, and which first appeared in the *Mississippian*.

resequent stream A stream following the original direction of drainage but which developed at a later time. cf. *anteconsequent, inconsequent, insequent, obsequent* and *subsequent streams*.

reserves 1. Of an economic *mineral*, the *mineral* known to be mineable under the technical and economic conditions at the time of assessment. *In situ* reserves refer to the amount that can probably be won; marketable reserves to the amount of recovered *mineral* that can ultimately be marketed after processing. 2. Of oil, several systems of categorization are in use. One variant used in the USA distinguishes between proved (or measured) reserves (those outlined by drilling), probable (or indicated) supply (likely to be present in the extension of existing *oilfields*) and possible (or indicated) supply (from future discoveries in known productive *formations*).

reservoir A subsurface rock containing commercially exploitable quantities of oil and/or gas because of its *porosity* and *permeability*.

residual anomaly A *geophysical anomaly*, resulting from a relatively shallow source, in a *gravity* or *magnetic field*, isolated by the removal of a *regional field*.

residual breccia A deposit formed by the removal of all *carbonate minerals* from a rock by *dissolution*, causing collapse of the sedimentary *fabric*, occurring most often in the *vadose zone* where percolating rainwater is highly undersaturated with respect to *calcite*.

residual deposit An economic deposit formed by the action of *weathering* and *groundwater* on *protore*.

residual placer A *placer deposit* formed on a fairly flat surface immediately above a *bedrock* source by the chemical decay and removal of lighter materials.

residual strength The lower value of *strength* in a disturbed material.

resinite A *coal maceral* of the *exinite* group made up of resinous material.

resinous lustre A *lustre* like that of resin, as shown by *sphalerite* and *native sulphur*.

resistate mineral A *mineral* resistant to chemical *weathering*.

resistivity log See *electric log*.

resistivity method A *geophysical exploration* method for investigating the subsurface distribution of *electrical resistivity* by introducing low frequency or direct electric current into the ground via spike electrodes and measuring the resulting voltage difference with a second electrode pair. The two principal techniques are *constant separation traversing* and *vertical electrical sounding*. Widely used in *hydrogeology*, *engineering geology*, archaeology, etc.

resonance A term used in *valence bond theory* for the state in which pairs of electrons interact with other pairs to form a molecule in which the most stable structures dominate.

resorption The partial refusion or *solution* of a *phenocryst*, which may give rise to a *reaction rim*.

resource The total amount of a particular commodity (element, *mineral*, oil) estimated for the world or a nation, comprising *ore reserves*, uneconomic deposits and deposits as yet undiscovered (estimated by comparison with explored areas of similar geology).

restored section The arrangement of *beds* etc. in a section before *deformation*, essentially equivalent to a *balanced section*.

restraining bend A curve in a *strike-slip fault* across which there is *compression* parallel to the straight parts of the *fault*. cf. *releasing bend*.

retardation time The time taken by a deforming *viscoelastic* material to reach half the value of the *stress* divided by *Young's modulus*, equal to the ratio of the *viscosity* to *Young's modulus*.

reticulated With a net-like *structure*.

reticulite A very low *density* form of *basaltic pumice* formed in some *Hawaiian eruptions*.

retinite (burmite) A group of resins containing no succinic acid.

retrocharriage See *backfold*.

retrograde boiling Boiling in a system whose temperature is falling.

Occurs in an acid *magma* whose cooling promotes the *crystallization* of anhydrous *minerals* so that the *magma* becomes enriched in volatiles. If the increased vapour pressure exceeds the *confining pressure*, a rapidly boiling liquid separates, possibly with the development of *crackle brecciation*. Probably important in the development of *porphyry copper deposits*.

retrograde metamorphism (diaphthoresis) *Metamorphism* which produces a lower *grade* from a higher *grade*.

return flow 1. *Throughflow* which *exfiltrates* from the soil profile at the base of a slope. 2. Flow in the opposite direction to the mean flow direction, such as might occur in the *lee side* of a *dune*.

reverse drag *Fault drag* in which the curvature of a marker is contrary to the sense of *displacement* on the *fault*. cf. *normal drag*.

reverse fan See *divergent fan*.

reverse fault A *dip-slip fault* in which the *hangingwall* moves upwards relative to the *footwall*.

reverse grading See *inverse grading*.

reverse kink bands See *contractional kink band*.

reversed magnetization A *remanent magnetization* in the opposite sense to the present *geomagnetic field*. cf. *normal magnetization*.

reversed polarity The orientation of a past *geomagnetic field* or a *natural remanent magnetization* in the opposite direction to the present *geomagnetic field*. cf. *normal polarity*.

reversed profile A line of *seismometers* with a *shot* at either end so that *seismic waves* travelling in both directions are recorded. Employed in order to reveal true *seismic velocities* and *structure* in *seismic refraction* interpretation.

reversed zoning *Zoning* in *plagioclase* in which the zones become more sodic towards the interior of the crystal. cf. *normal zoning*.

reversing dune A *dune* with *slip-faces* on opposing sides as a result of a *bimodal* wind regime.

reworking 1. The process of altering, by *deformation* and *metamorphism*, of deep, *crustal* material formed in a previous *orogeny* by a subsequent orogenic event. 2. The *erosion*, transfer and deposition of older sediment within a sedimentary *basin*.

Reynolds number (Re) A dimensionless parameter used in fluid dynamics which expresses the influence of viscous forces within a fluid to the inertial forces acting on the flow. $Re = \rho Ul/\upsilon$, where ρ = fluid *density*, U = mean flow velocity, l = length term and υ = *kinematic viscosity*. Its magnitude controls whether flow will be *laminar* or *turbulent*.

Rhaetian (Rhetian) A *stage* of the *Triassic*, 209.5–208.0 Ma.

Rhenanida An order of class *Placodermi*, superclass *Pisces*; small, shark-like armoured *fish*. Range L.–U. *Devonian*.

rheology The study of the *flow* and *deformation* of materials.

Rhetian See *Rhaetian*.

rhexistasy The state of environmental and *geomorphic* instability which interrupts periods of *biostasy*, probably resulting from climatic change or *tectonic* activity. Vegetation cover deteriorates, soils erode and the landscape denudes. cf. *biostasy*.

rhipidistians A group of *Sarcopterygian fish* from which the *Amphibia* evolved.

rhizocretion An accumulation of *mineral* matter around the roots of a plant while living or after death.

rhizolite A rock showing evidence of having been formed largely by root activity.

rhizolith An organo-sedimentary *structure* resulting from the accumulation and/or *cementation* around or within, or the *replacement* of, plant roots by *mineral* matter.

rhodochrosite (dialogite, manganese spar) ($MnCO_3$) A rare *ore mineral* of manganese found in some *hydrothermal veins*.

rhodolite A pale red or purple *garnet*, compositionally two parts *pyrope* to one of *almandine*.

rhodolith A *nodule* of red *algae*.

rhodonite ($MnSiO_3$) A *pyroxenoid* found in manganese-rich deposits.

Rhodophyta The *calcareous* red *algae*. Range *Cambrian–Recent*.

rhodustalf See *terra rossa*.

rhomb-porphyry A medium-grained, *intermediate igneous rock* comprising numerous *phenocrysts* of *anorthoclase* in a fine-grained *groundmass*, occurring in *dykes* and minor intrusions.

rhomb-spar An obsolete name for *dolomite*.

rhombochasm See *pull-apart basin*.

rhombohedral packing One end member of the different ways perfectly sorted spheres can be arranged. A sediment with such packing would have 26% *porosity*. cf. *cubic packing*.

rhomboid ripple A surface sedimentary *bedform* found in *beach deposits* indicative of the high power of the environment.

rhourd A large, pyramidal *dune* formed by the intersection of smaller *dunes* in multidirectional winds.

Rhuddanian The lowest *stage* of the *Silurian*, 439.0–436.9 Ma.

Rhynchocephalia See *Rhynchosauria*.

Rhynchonellida An order of class *Articulata*, phylum *Brachiopoda*; *brachiopods* with biconvex, impunctate shells, usually with coarse ribs on each *valve* which meet in a zigzag fashion. Range M. *Ordovician–Recent*.

Rhynchosauria (Rhynchocephalia) An order of subclass *Archosauria*, class *Reptilia*; herbivorous *reptiles* with powerful chopping teeth. Range U. *Triassic*.

Rhynie Chert An *exceptional fossil deposit* in Aberdeenshire, Scotland of L. *Devonian* age containing *land plants* and *arthropods* preserved in *silica*.

Rhyniopsida A class of division *Tracheophyta*, kingdom *Plantae*; primitive *vascular plants*. Range M. *Silurian*–U. *Devonian*.

rhyodacite An extrusive, commonly *porphyritic, igneous rock* with *phenocrysts* of *plagioclase*, *sanidine* and *quartz* in a *glassy* to *microcrystalline groundmass*.

rhyolite (liparite) A fine-grained, *acidic volcanic rock* compositionally equivalent to *granite*. *Aphanitic* or with *phenocrysts* of *feldspar* and *quartz* in a *matrix* of *quartz* and *feldspar* with sparse *mafic minerals*. Most abundant in continental *rifts* as extensive *ignimbrites* plus rarer *domes* and *lava flows*, also occurring on *oceanic islands* and *island arcs*.

rhythmite A finely-laminated sediment in which two or three different lithologies are regularly repeated, common in *glacial* lakes.

RI See *refractive index*.

ria A submerged, 'V'-shaped, coastal valley, narrowing landwards. Originally formed subaerially and inundated by a rise in sea level.

rib structure The saw-tooth cross-section of a *plume structure*.

richness A *palaeoecological* descriptor of the number of *taxa* present in a community.

Richter earthquake scale A logarithmic, *earthquake magnitude* scale based on the amplitude of ground motion with a correction for distance to the *epicentre*.

richterite ($(Na,K)_2(Mg,Mn,Ca)_6Si_8O_{22}(OH)_2$) An alkali *amphibole* found in contact *metasomatic* deposits.

rider A block separated by *horsetail faults* in an *imbricate extensional fault system*.

ridge push The horizontal *force* applied to a *plate* at an *ocean ridge* due to the expansion of material brought to the surface from higher pressures at depth. An important *plate*-driving mechanism in *Orowan-Elsasser convection*.

ridges and runnels Alternating morphological highs and lows with an amplitude of 0.1–1.5 m and a wavelength of 50–200 m. Run parallel or subparallel to the coastline on gently-sloping, sandy *beaches* with a high *tidal* range, low to moderate *wave* energy and *fetch*-limited *waves*. Develop as the *tidal* height varies and the *swash* zone moves to new locations on the *beach* face.

riebeckite ($Na_2Fe_3^{2+}Fe_2^{3+}Si_8O_{22}(OH)_2$) A blue *amphibole*, known as *crocidolite* when in *asbestiform* habit.

Riedel fault (R1 fault) A *fault* within a *fault zone* with the same sense as the zone but inclined at a low angle ($< 15°$) to the main *fault* direction.

Riedel shear (R1 shear, synthetic shear) A *shear* within a *shear zone* with the same sense as the zone but inclined at a low angle ($< 15°$) to the main *shear* direction.

riegel A rock *bar* extending across a *glacial* trough, possibly formed where the erosive ability of the *glacier* wanes or the *bedrock* becomes harder.

riffles Topographic highs on the undulating longitudinal profile of a *gravel*-bed river, spaced at 5–7 channel widths.

rift (rift valley) A major extensional *graben* of large lateral extent, whose topographic expression may have been lost. Rifts in *oceanic crust* are found at the axes of slow-spreading *oceanic ridges*, and may be 100 km in width. Continental rifts may connect to *plate margins* and may represent sites of *continental splitting* where new oceans grow between their edges and which become *passive continental margins*. Other continental rifts are *impactogens*.

rift valley See *rift*.

rigid body rotation Rotation without changes in length or angular relationships and without *displacement* of the centre of rotation. Used in concepts such as the *strain ellipsoid*.

rigid-plastic Descriptive of a *plastic* material which undergoes no *deformation* at *stresses* below the *yield stress*. cf. *elastic-plastic*.

rigidity (modulus of rigidity, shear modulus) An *elastic constant* indicating the degree of resistance to *shear strain*, defined as the ratio of *shear stress* to *shear strain* in *elastic deformation*.

rill A centimetric-scale channel that changes direction with every *runoff* event.

rillenkarren (solution flutes) *Solution* depressions found on steep or vertical surfaces with sharp ridges between the flutes.

rim syncline An annular *syncline* surrounding a *salt dome* formed by the collapse of *strata* over the region evacuated by the rising salt *diapir*.

rimstone dam A type of *speleothem* which forms a carbonate barrier that impounds *cave* pools.

ring complex (ring intrusion) An annular, *tabular* body of cylindrical shape, exposed at the surface with a ring-shaped outcrop, formed of almost any *igneous rock* and found in *oceanic islands*, at continental margins and in certain continental volcanic provinces.

ring dyke An arcuate, *tabular* body of *igneous rock* with a vertical axis and near-vertical or outward-dipping contacts.

ring intrusion See *ring complex*.

ring silicate See *cyclosilicate*.

ring structure group A group of *sulphide minerals* characterized by a structure comprising parallel chains of atoms.

ringwoodite ($(Fe,Mg)_2SiO_4$) A *mineral* with *spinel* structure and *olivine* composition found in *meteorites*.

rinnentaler See *tunnel valley*.

rip current A strong, narrow current flowing seaward from a *beach* through the *breaker zone*.

rip-up clast (mud clast) An *intraformational clast* in a sedimentary sequence.

Riphean An *era* of the *Precambrian*, 1650–800 Ma.

ripidolite (prochlorite) A *chlorite* group *mineral* found in *metamorphic rocks* and some *ore veins*.

rippability A measure of the facility with which rock can be broken by mechanical equipment; can be estimated from the *P wave* velocity.

ripple A flow-transverse *bedform* generated at low *shear stresses* above the threshold for sediment movement in *cohesion*less *sand* of grainsize <0.6 mm. Wavelength commonly <500 mm and amplitude <30 mm.

ripple form index (ripple index) The ratio of wavelength to height of a *ripple*.

ripple index See *ripple form index*.

ripple mark The mark left by a *ripple* on a sedimentary surface.

ripple symmetry index The ratio of the horizontal components of *stoss-side* length to *lee-side* length of a *ripple*.

riser See *step*.

Riss Glaciation The third of the *glaciations* affecting the Alps in *Quaternary* times.

river capture A process in which one river undercuts the drainage area of another by more rapid incision, so enlarging its *catchment* at the expense of the other's.

river flood Water which arrives in a river channel sufficiently rapidly to produce a significant increase in *discharge* above *baseflow* levels and forms a distinct peak in *discharge*.

RMQ See *rock mass quality*.

Robin effect A mechanism whereby the removal of rock at the edge of steps above a *glacial* cavity is reinforced by a heat pump process resulting from the variable *stress* distribution in the ice and at the *glacier* bed as the *glacier* moves over an irregular surface.

rocdrumlin (tadpole rock) A *drumlin*-like feature of *bedrock* moulded into a streamlined shape.

roche moutonée A hump of rock with one side moulded by ice and the other steepened; a product of *glacial erosion*.

rock crystal A colourless, crystalline variety of *quartz*.

rock doughnut A doughnut-shaped rock formed when a veneer of *case-hardened* material is breached and exposes more easily eroded rock.

rock fall A rapid form of *mass movement* by falling under *gravity*.

rock flour Rock debris in a *glacier* that has been ground down to a fine mixture of *silt* and *clay*.

rock head The surface between *bedrock* and overlying unconsolidated material.

rock magnetism The study of the magnetic properties of naturally-occurring magnetic *minerals*, including the processes by which they became magnetized. Important terrestrial magnetic *minerals* include iron oxides, hydroxides and sulphides, while iron and nickel are found in *meteorites* and on the Moon.

rock mass quality (RMQ) An engineering classification based on the number of major discontinuities (e.g. *joints*) present, their orientation and spacing.

rock phosphate See *phosphorite*.

rock quality designation (RQD) A quantitative assessment of the intactness of material recovered from a borehole in *site investigation*.

rock rose See *desert rose*.

rock slide A type of *mass movement* by sliding down a surface.

rockburst A rapid release of *lithostatic pressure* in a deep mine, which can generate an *earthquake* up to a *magnitude* of 4 or more.

rocksalt See *halite*.

rodding See *rods*.

roddon A sinuous ridge of *silty* material above the general level of the *peat fens* of East Anglia, England, which represents the remains of ancient river systems flowing between *levées* above the surrounding land level, or were formed by *peat* wastage caused by drainage activities.

Rodentia An order of infraclass *Euth-*

eria, subclass *Theria*, class *Mammalia*; the rodents. Range U. *Palaeocene–Recent*.

rodingite A rock comprising *grossular* and *prehnite* ± *wollastonite*, *diopside* and *hydrogrossular*, formed by calcium *metasomatism* of *gabbro* or *dolerite*.

rods (rodding) Cylindrical *structures* of a single *mineral* in a deformed rock, similar in form to *boudins* and *mullions*, but distinguished by their monomineralic composition. Commonly of *quartz*, but rods of *calcite* and *pyrite* have been described. Probably formed by the *deformation* of *mineral* segregations, the *detachment* and isolation of *fold hinges*, where they define a *lineation*, or by the *elongation* of *conglomerate clasts*.

rogen moraine A *moraine* field, with ridges 10–30 m high and spaced at 100–300 m, orthogonal to the direction of former ice advance, which probably formed beneath the *glacier*.

rollover anticline An *accommodation structure* accompanying the movement on a *listric fault* in an *extensional fault system*.

romanèchite ($BaMn^{2+}Mn_8^{4+}O_{16}(OH)_4$) An *ore mineral* of manganese.

roméite ($(Ca,Fe,Na)_2(Sb,Ti)_2(O,OH)_7$) Hydrated calcium antimonite, sometimes with manganese and iron, found in manganese deposits and *placer deposits*.

roof fault The *fault* at the top of the extensional *duplex* formed when the *sole fault* migrates into the *footwall* of an *extensional fault system*.

roof pendant A large mass of *country rock* in the roof of an igneous intrusion.

roof rock See *seal*.

roof thrust The *fault* into which *thrust faults* merge at the top of a *duplex* in an *imbricate structure*. cf. *sole thrust*.

room and pillar mining (bord and pillar mining, stoop and room mining) A mining method in flat-lying deposits in which *minerals* are extracted from rooms separated by a series of pillars which support the roof.

root cast See *root mould*.

root-mean-square velocity The *seismic velocity* down to a reflector expressed as the square root of the sum of the products of layer thickness and squared velocity divided by the sum of the thicknesses. Usually estimated from *normal moveout* measurements.

root mould (root cast) The cast of a root which has subsequently decayed.

root petrification The formation of a *rhizolith* by the impregnation or *replacement* of the organic matter of a root.

root tubule A *rhizolith* comprising a cemented cylinder around a *root mould*.

root zone Of a *nappe*, the area from which the *nappe* originated and from which it is now usually separated due to *erosion* and/or *deformation*.

roscoelite ($KV_2(AlSi_3O_{10})(OH)_2$) An *ore mineral* of vanadium with a *mica* structure.

rose diagram A type of circular histogram used to display directional data, e.g. the trends of *glacial* striations.

rose opal A red variety of *opal*.

Rossi-Forel Scale A ten-point scale of *earthquake intensity*.

Rostroconchia A class of phylum *Mollusca*, comprising a pseudobi*valve*d shell whose two parts are connected dorsally by a continuous calcareous strip. Range *Cambrian–Permian*.

rotational deformation *Deformation* in which the initial and final positions of the *strain axes* appear to have rotated with respect to an external reference frame. cf. *non-coaxial deformation*.

rotational fault A *fault* with a curved *fault plane* causing rotation of displaced material. See also *listric fault*.

rotational slip (rotational slump) A type of *mass movement* in *cohesive* sediments along a *failure* plane with the shape of an arc of a circle, common where river valleys incise weak *clay* deposits.

rotational slump See *rotational slip*.

rotational strain *Strain* resulting from a non-*coaxial strain* path, described by *vorticity*.

Röthlisberger channels (R channels) A series of conduits at the bed of a *glacier* incised upwards into the ice, probably by the heat generated by the flow of *turbulent* water.

Rotliegendes 1. The older *epoch* of the *Permian*, 290.0–256.1 Ma. 2. A *Permian* succession in NW Europe equivalent to the *Asselian*, *Sakmarian* and *Artinskian*.

roundness (angularity) A measure of the sharpness or otherwise of the corners of a *clast*, used in the description of a *clastic rock*.

royalty The percentage of the revenue from mining or quarrying or *hydrocarbon* recovery paid to the owner of the *mineral* rights.

RQD See **r***ock* **q***uality* **d***esignation*.

rubefied Descriptive of a reddening of a soil or rock by the *oxidation* of iron during *weathering*.

rubellite A red to pink variety of *tourmaline*, sometimes used as a semi-precious *gem*.

rubicelle A yellow or orange variety of *spinel*.

rubidium-strontium dating A *radiometric dating* technique based on the decay of ^{87}Rb to ^{87}Sr with a *half-life* of ~50 Ga.

ruby A red *gem* form of *corundum*.

ruby copper A variety of *cuprite* forming ruby-red *transparent* crystals.

ruby silver ore See *proustite*, *pyrargyrite*.

ruby spinel (almandine spinel) ($MgAl_2O_4$) A red, *gem* quality *spinel*.

rudaceous With the appearance of a *rudite*.

rudistid A cone-shaped *bivalve* resembling a solitary *coral*.

rudite A *clastic rock* with >30% *clasts* of *gravel* grade (>2 mm) with or without a *matrix* of *sand* and/or *mud* grade.

Rugosa (Tetracorallia) A subclass of class *Anthozoa*; solitary or colonial *corals* with bilaterally symmetrical, calcareous *corallites* in which septae are inserted in four loci. Range *Cambrian–Permian*.

Runangan An *Eocene* succession in New Zealand equivalent to the *Priabonian*.

rundkarren Subsoil *karren* made up of rounded *solution* runnels.

runoff The flow of water on hillslopes, comprising *baseflow* and *quickflow*.

runzelmarken (wrinkle marks) Parallel or reticulate ridges, <1 mm high and spaced at a few millimetres, on *cohesive* muddy sediment surfaces. Form in intertidal environments when strong wind *stress* affects the surface when covered by a very thin film of water, or by the action of subaqueous currents, or by sediment loading.

Rupelian The lower *stage* of the *Oligocene*, 35.4–29.3 Ma.

Rustler A *Permian* succession in the Delaware Basin, USA, equivalent to the upper part of the *Longtanian*.

rusty gold *Native gold* with a stained surface.

rutilated quartz See *needle stone*.

rutile (TiO_2) An *ore mineral* of titanium and a common *accessory mineral*.

ruware A low, dome-shaped *bedrock exposure* protruding from a cover of *alluvium* or *weathered bedrock*; a relict or incipient *inselberg*.

S

S 1. In *earthquake seismology*, an *S wave* that does not travel through the *core*. 2. See *siemen*.

s fold An *asymmetrical fold* with one *limb* shorter than the other whose profile defines an 'S' shape. cf. *z fold*.

S surface The planar, curved or irregular surface to which a *foliation* is parallel.

S tectonite A rock with a tectonic *foliation*. cf. *L-S tectonite*.

S wave (secondary wave, shear wave, transverse wave) A *seismic body wave* whose associated particle motion is at right angles to the direction of propagation. Its *seismic velocity* is generally ~0.58 of the *P wave* velocity in the same medium.

S wave splitting The phenomenon whereby cracks of common alignment within a rock cause *S waves* to be split into a fast wave polarized parallel to the cracks and a slow wave polarized at right angles to them. A means of observing *dilatancy* and important in *earthquake prediction*.

Saale glaciation The second of the *glaciations* of N Europe in the *Quaternary*.

sabkha (sebkha) A broad plain or *salt flat* in an arid or semi-arid region containing *evaporites* at a level dependent on the local *water table*.

saccharoidal With the appearance of sugar.

saddle dolomite A coarsely-crystalline *dolomite* with pronounced lattice curvature, believed to form at temperatures between 50–100° C.

saddle reef A saddle- to triangle-shaped accumulation of *minerals* or *coal* in the *hinge* zone of a *fold*, forming in the gap produced when *competent* beds are interbedded, particularly when thin horizons of *shale* are present.

safflorite ($CoAs_2$) A *mineral* with a *loellingite* structure found in *mesothermal vein* deposits.

sagenite *Quartz* containing oriented needles of *rutile* in an attractive pattern.

St. David's An *epoch* of the *Cambrian*, 536.0–517.2 Ma.

Saint Venant equations *Flood routing* equations which describe varying, unsteady flow in open channels.

Sakaraulian A *Miocene* succession on the Russian Platform covering parts of the *Aquitanian*, *Burdigalian* and *Langhian*.

Sakmarian A *stage* of the *Permian*, 281.5–268.8 Ma.

Salado A *Permian* succession in the Delaware Basin, USA, equivalent to the middle part of the *Longtanian*.

salar A *basin* of inland drainage in an arid or semi-arid region, dry for the majority of the time.

salcrete A surface or near-surface crust, principally of sodium chloride, which cements a *sand* surface or other permeable soil, formed through the evaporation of moisture.

salic mineral A *normative mineral* composed of *silica* and alumina.

salina See *solar pond.*

salinity The total quantity of dissolved solids in sea water, usually expressed in parts per thousand when carbonate has been converted to oxide, bromide and iodide to chloride and all organic matter completely oxidized.

salinization A *pedogenetic* process in which salts accumulate in the soil.

salite See *sahlite.*

salps A gelatinous zooplankton contributing to *marine snow*.

salt diapir A *structure* formed by the buoyant ascent of relatively low *density evaporite* through a sediment *overburden* by *halokinesis.* Often constitutes an effective *hydrocarbon trap* in combination with the deformed sediments.

salt dome A *salt diapir* with a dome-like shape.

salt flat A flat stretch of salt-encrusted ground.

salt heave A source of damage to man-made structures arising from the presence of soluble salts which increase in volume on *hydration.*

salt marsh A vegetated, intertidal, *mud* flat on a temperate, low energy coastline.

salt pan A deposit of salt developed in a *salt marsh.*

salt pillow The protuberance forming on a salt layer when *halokinesis* is initiated.

salt pond See *solar pond.*

salt weathering The *weathering* resulting from physical changes associated with salt *crystallization*, *hydration* or thermal expansion, all of which can generate substantial *pressures.*

saltation The process of *bedload transport* comprising a series of ballistic jumps in which grains ascend steeply ($>45°$) from the bed and return at a low angle ($<10°$). Height and impact effects are dependent on the fluid characteristics, being greater in air than in water.

saltpetre See *nitre.*

samarium-neodymium dating A *radiometric dating* technique based on the decay of ^{147}Sm to ^{143}Nd with a *half-life* of $\sim 2.5 \times 10^5$ Ma, which is resistant to the effects of *alteration* and *metamorphism.*

Samfrau fold belt A L. *Palaeozoic orogenic belt* formed along the southern edge of *Gondwanaland* and now found in Australia, Tasmania, Antarctica, S Africa and Brazil.

sand *Mineral* or rock grains of 0.625–2 mm diameter, often composed of *quartz.*

sand flow cross-strata *Cross-stratification* of the deposits on the *slip-face* of a *dune* produced by grain flows of dry *sand*, so that they show *inverse grading.*

sand rose A *rose diagram* showing the volume of *sand* moved by the wind from different directions, usually over a one-year period.

sand sea See *erg*.

sand sheet A rippled to unrippled, flat to irregular area of *aeolian sand* where *dunes* with *slip-faces* are generally absent. Forms in *ergs* where conditions are unsuitable for *dune* formation, i.e. a high *water table*, periodic floods, surface *cementation*, the presence of vegetation or a significant component of coarse-grained particles.

sand volcano A mound of *sand/mud*, <30 cm high and about 25–50 mm in diameter with a conical depression at the apex from which fine-grained sediment in suspension with water is emitted. Develops by the extrusion of interstitial water as the sediments compact. cf. *mud volcano*.

sandbed channel A channel of largely sandy material which is transported at a wide range of *discharges*, generally wider and shallower than channels in more *cohesive* sediment.

Sander's Symmetry Principle 'The symmetry of a *fabric* is the same or less than the kinetic symmetry producing the *fabric*.' Used to infer the kinematics of *deformation* from observed *fabrics*.

sandstone A *clastic sedimentary rock* with $>25\%$ by volume of *clasts* of *sand* grade (0.625–2 mm diameter).

sandstone-uranium-type deposit A uranium-rich *sandstone-uranium-vanadium base metal deposit*.

sandstone-uranium-vanadium base metal deposit A terrestrial sedimentary deposit of probable *fluviatile* origin laid down under arid conditions so that *host rocks* are often red. Generally contains one or two metals in economic quantities.

sandstorm The elevation of *sand*-grade particles by a strong wind so that visibility is reduced to <1000 m. Particles rarely rise >15 m above the surface. cf. *dust storm*.

Sandugan A *Silurian* succession in the Mirnyy Creek area of NE Siberia equivalent to the *Telychian* and *Wenlock*.

sandur A large outwash plain created by the meltwater from an ice mass.

sandwave 1. Any large-scale periodic *bedform*. 2. A large-scale, linear-crested, two-dimensional periodic *structure* with a large wavelength : height ratio. 3. A large-scale *bedform* growing to equilibrium over longer time-scales than individual steady or quasi-steady flows. Formed by reversing flows of *tidal* frequency.

sanidine ($KAlSi_3O_8$) A high-temperature form of *potash feldspar*.

sanitary landfill (landfill) A land site where municipal solid waste is buried in a fashion such as to cause minimum disturbance or pollution to the environment. The cheapest method of disposal, but subject to problems of *leaching* and *seepage* of waste substances into *groundwater*.

sannaite An *alkaline lamprophyre*, similar to *camptonite* but with *alkali feldspar* more abundant than *plagioclase*.

Santonian A *stage* of the *Cretaceous*, 86.6–83.0 Ma.

sanukite A type of *hypersthene andesite*, black and *glassy* when fresh, commonly used for stone tools in prehistoric Japan.

saponite (bowlingite) ($(Mg,Fe)_3(Al,$

$Si)_4O_{10}(OH)_2(Ca,Na)_{0\cdot3}(H_2O)_4)$ A swelling *clay mineral* of the *montmorillonite* group.

sapphir d'eau (water sapphire) An intense blue variety of *cordierite*.

sapphire A blue *gem* variety of *corundum*.

sapphirine $(Mg_{3\cdot5}Al_{9\cdot0}Si_{1\cdot5}O_{20})$ A bluish, magnesium-aluminium silicate with some iron found in *silica*-poor *metamorphic rocks*.

sapping The undercutting of the base of a cliff and subsequent *failure* of the cliff face caused by wave action, lateral stream *erosion*, *groundwater* outflow etc.

saprolite A fine-grained *clay* material formed by the *in situ deep weathering* of *bedrock*, particularly *crystalline igneous* and *metamorphic rocks*, under humid tropical and sub-tropical conditions.

sapropel An organic-rich deposit of *Neogene* age found in the Mediterranean and the Black Sea.

sapropelic coal A fine-grained, faintly stratified to homogeneous, massive *coal* deposit. Generally dark coloured, tough and exhibiting *conchoidal fracture*. The two main types are *cannel coal* and *boghead coal*.

Saratovan A *Palaeocene* succession in the former USSR, covering parts of the *Danian* and *Thanetian*.

Sarcopterygii A subclass of class *Osteichthyes*, superclass *Pisces*; lobe-fin *fish*. Range *Devonian–Recent*.

sard A brown variety of *chalcedony*.

sardonyx A variety of *microcrystalline silica* composed of interlayered *onyx* and *sard*.

Sarka An *Ordovician* succession in Bohemia equivalent to the *Llanvirn*.

Sarmatian See *Meotic*.

sarsen (gray wethers, grey wethers) Blocks of *silica*-cemented *sandstone* (*silcrete*), *breccia* or *conglomerate* believed to have formed on the surface or in the near-surface of S England during warm phases of the *Tertiary* and broken down by frost action and mass movement. Used in the construction of Stonehenge, England and earlier structures. See also *puddingstone*.

sassolite (H_3BO_3) The rare *mineral* boric acid.

sastrugi A small ($<$ 50 mm high), irregular ridge parallel to the wind direction, commonly with a concave upwind face, produced by wind *erosion* of a moist or salt-cemented sandy surface.

satellite fold See *parasitic fold*.

satellite geodesy The use of satellites for geodetic observations, such as in *satellite laser ranging* and *satellite radiopositioning*.

satellite imagery *Remote sensing* from an artificial satellite.

satellite laser ranging (SLR) A technique involving the simultaneous measurement of the distance to two or more terrestrial stations from an artificial satellite by the travel time of a laser beam so as to determine the distance between the stations to an accuracy just sufficient to monitor *continental drift*.

Satellite Probatoire pour l'Observation de la Terre (SPOT) A satellite used in *remote sensing* which produces stereo pairs of images.

satellite radiopositioning A technique of three-dimensional positioning by radio interferometry using the *Global Positioning System*, accurate enough to measure rates of *continental drift* and *tectonic* disturbance.

satin spar A fibrous variety of *gypsum*.

saturated Descriptive of an *igneous rock* with neither an excess nor deficiency of *silica*. See also *silica saturation*.

saturated wedge A zone of soil saturation by *throughflow* formed when *infiltration* rates exceed the rate of percolation into lower horizons, with a wedge shape due to its extension upslope.

saturated zone The zone below the *water table* saturated with *groundwater*.

saturation magnetization The maximum intensity of *isothermal remanent magnetization* caused by a strong (>0.1 *Tesla*), external *magnetic field*, measured after removal of the field.

saturation overland flow *Runoff* from an area of soil on a shallow slope saturated by *throughflow* which accumulates at the base of the slope.

Saucesian A *Miocene* succession on the west coast of the USA, covering parts of the *Aquitanian* and *Burdigalian*.

sauconite ($Na_{0.33}Zn_3(Si,Al)_4O_{10}(OH)_2.4H_2O$) A swelling *clay mineral* of the *smectite* group.

Saurischia A class of order *Rhynchocephalia*, subclass *Archosauria*, class *Reptilia*; *dinosaurs* with a lizard-like pelvis. Range *Triassic*–end *Cretaceous*.

Sauropoda Herbivorous members of the *Saurischia*.

Sauropterygia An order of subclass *Euryapsida*, class *Reptilia*; the *plesiosaurs* and *nothosaurs*. Range L. *Triassic*–U. *Cretaceous*.

saussurite A fine-grained mixture of *zoisite* and other *minerals* formed by the *alteration* of *feldspar*.

Saxonian metallogenic epoch An *epoch* of *epigenetic* mineralization from M. *Triassic* to *Jurassic* in the *Variscan orogenic belt* of NW Europe.

saxonite A coarse-grained, *ultrabasic rock* comprising *olivine* and *orthopyroxene* (commonly *hypersthene*), i.e. a *hypersthene peridotite*.

scabland A landscape of bare rock surfaces, thin soils and sparse vegetation underlain by flat *basalt* flows and dissected by dry channels caused by *glacial* floodwaters.

scanning electron microscopy (SEM) A technique of *electron microscopy* used in the investigation of surface structure and composition by irradiating a bulk sample with a scanning, focused beam of electrons and imaging backscattered and secondary electrons on a cathode ray tube scanned synchronously with the specimen.

scanning transmission electron microscopy (STEM) A form of *scanning electron microscopy* using transmitted electrons.

Scaphopoda A class of phylum *Mollusca*; *molluscs* with a *univalved*, tusk-shaped shell open at both ends, a differentiated head region and radula but rudimentary gills. Range *Ordovician*–*Recent*.

scapolite series ($3NaAlSi_3O_8.NaCl$–$3CaAl_2Si_2O_8.CaCO_3$) A group of *metamorphic tectosilicates* forming a *solid solution* between *marialite* and *meionite*.

scar A steep, rocky cliff in massively-bedded *limestone*.

scar fold A *fold* formed by the *flow* of material around a *boudin*.

scarpslope See *escarpment*.

scheelite ($CaWO_4$) A major *ore mineral* of tungsten found in *pneumatolytic* and *hydrothermal veins* and *metamorphic aureoles*.

schiller The play of light on a crystal face at a certain angle of illumination.

schillerspar See *bastite*.

schist A rock exhibiting *schistosity*, generally of high *metamorphic grade*.

schistosity A *foliation* produced by *deformation* in which *tabular minerals*, coarse enough to be visible to the unaided eye, have a *preferred orientation*.

schlieren Pencil-shaped, discoidal or blade-like inclusions made up of *aggregates* with a greater concentration of *mafic minerals* than the *host rock*.

Schlumberger configuration An electrode arrangement in the *resistivity method* in which the potential electrodes are much closer together than the current electrodes.

Schmid factor See *critical resolved shear stress*.

Schmidt net A type of *stereographic net* using *equal area projection*, in which areas are portrayed undistorted. Mainly used for the statistical analysis of orientation data. cf. *Wulff net*.

schollen Large blocks of *country rock* found in *till* sheets, resulting from *glacial erosion*.

schorl The black, iron-rich variety of *tourmaline*.

schorlomite A black variety of *andradite* with Ti > Fe^{3+}, which is even richer in Ti than *melanite*.

Schottky crystal defect See *vacancy*.

Schumann resonance frequencies The frequencies that are preferentially enhanced as *sferics* propagate in the Earth–ionosphere cavity.

schuppen structure An *imbricate structure* in an *orogenic belt*.

scintillation counter (scintillation meter, scintillometer) An instrument used in *radiometric surveying* for measuring gamma rays by counting the flashes they produce on a screen coated with e.g. zinc sulphide.

scintillation meter See *scintillation counter*.

scintillometer See *scintillation counter*.

scissor fault A *fault* with a reversal of *displacement* direction along the *fault plane*, so that there is pivoting about an axis of zero movement analogous to the opening of a pair of scissors.

Scleractinia (Hexacorallia) An order of subclass *Zoantharia*, class *Anthozoa*; *corals* secreting calcareous, commonly *aragonitic* exoskeletons. Range M. *Triassic–Recent*.

Sclerospongia A class of phylum *Porifera*; *sponges* with an *aragonitic* skeleton. Range *Ordovician–Recent*.

sclerotinite A *coal maceral* of the *inertinite* group made up of the coalified remains of *fungal* material.

scolecite ($CaAl_2Si_3O_{10}.3H_2O$) A *zeolite* found in cavities in *basaltic* rocks.

scoria A rusty red to black, highly *vesicular*, *mafic* volcanic *glass*. cf. *pumice*.

scoriaceous Descriptive of a *lava* or *pyroclastic rock* containing empty cavities.

scorodite ($FeAsO_4.2H_2O$) A hydrated iron-aluminium arsenate found as an *accessory mineral* formed by the *alteration* of arsenic-bearing *minerals*.

Scorpionida An order of class *Arachnida*, subphylum *Chelicerata*, phylum *Arthropoda*; the scorpions. Range *Silurian–Recent*.

scorzalite ($(Fe,Mg)Al_2(PO_4)_2(OH)_2$) A rare phosphate *mineral* occurring in *metamorphic rocks* and *pegmatites*.

Scottish topaz A yellow, transparent variety of *quartz*.

Scottville A *Permian* succession in Queensland, Australia, covering the lower part of the *Wordian*.

scour mark A depression caused by *erosion* at the base of a *turbidity current* or on other erosional surfaces.

Scourian An *orogenic* episode affecting NW Scotland at 2900–2300 Ma.

scrap mica See *flake mica*.

scratch See *tool track*.

scree (talus) A sloping accumulation of loose *clasts* of *granule* grade or larger, generally in the form of a wedge, metres to hundreds of metres in height, at the base of a steep rock face from which the *clasts* fall as a result of *weathering* and *erosion*.

screw dislocation A *crystal defect* in which the atoms are shifted relative to each other in a direction parallel to the *dislocation line*.

scroll bar A sediment ridge deposited parallel to the contours of a *point bar* in an alluvial channel.

Scyphozoa *Cnidaria* with a polypoid or medusal form with few hard parts. Range U. *Precambrian–Recent*.

Scythian An *epoch* of the *Triassic*, 245.0–241.1 Ma.

sea cliff A steep coastal slope caused by the interaction of marine and subaerial processes.

sea level datum The height of the mean sea level surface to which all elevations are related. Frequently used as the reference level in geophysical surveys.

seafloor spreading The hypothesis that *continental drift* takes place through the growth of ocean basins between the continents. *Oceanic lithosphere* is created at *oceanic ridges*, generating the source of *oceanic magnetic anomalies* according to the *Vine-Matthews hypothesis*, and recycled into the *mantle* at *subduction zones*.

seal (roof rock) The impervious capping which prevents the upward *migration* of *hydrocarbons* from a *reservoir*, often comprising *clays*, *shales* or *evaporites*.

seamount An approximately conical, submarine peak which does not rise above sea level. Often found in chains formed as the *lithosphere* passes over a *hotspot*.

seat earth A thin horizon beneath a *coal seam* containing *fossil* rootlets which represents the soil in which the vegetation grew. Often rich in *kaolinite* formed by the *leaching* of cations by acidic soil waters and the *eluviation* of *clays*.

sebkha See *sabkha.*

second order fault (splay fault) A *minor fault* at an acute angle to a *major fault* with the same sense of slip.

second order fold A *fold* which is smaller than a *first order fold* and whose *enveloping surface* is folded by a *first order fold.*

secondary creep The second stage of *viscoelastic strain* characterized by steady-state *viscous flow* and a constant *strain*-time slope. See also *primary creep, tertiary creep.*

secondary enrichment See *supergene enrichment.*

secondary migration The *migration* of *hydrocarbons*, usually lateral, through or out of the *reservoir* rock. cf. *primary migration.*

secondary mineral A *mineral* formed after the formation of its enclosing rock, usually by the *alteration* of a *primary mineral.*

secondary porosity The pore space generated in *sedimentary rocks* after deposition and during *diagenesis* by the *dissolution* of detrital framework grains, *matrix* or earlier *authigenic* cements, by the *replacement* of other carbonates by *dolomite* in carbonate rocks or by fracturing or shrinking associated with the dehydration of hydrous *minerals.* cf. *primary porosity.*

secondary quartz *Quartz* forming as a cement in *diagenesis.*

secondary wave See *shear wave.*

sectile mineral A *tenacity* descriptor indicating a *mineral* that can be cut by a knife but powders under a hammer.

secular variation The long-term, progressive and predictable change in the *geomagnetic field*, probably arising from the changing circulation patterns of charged particles in the fluid outer *core.*

sedarenite An *arenite* whose *clasts* are dominantly of sedimentary origin.

sedentary rudaceous deposit A *rudite* formed *in situ* by *weathering.*

sedifluction The subaerial or subaqueous movement of unconsolidated sediment.

sedigraph A device which records the relative *settling velocities* of particles in water so as to study the *grain size distribution* of a sediment.

sediment delivery ratio The ratio of gross *erosion* to *sediment yield* for an entire drainage *basin.*

sediment drift An elongate, lobate mound of fine-grained *sands* and *muds* constructed by a *contour current*, i.e. a *contourite.*

sediment gravity flow An *aggregate* of grains moving under the influence of *gravity* mostly independently of the overlying medium, including *grain flows, debris flows, liquefied flows* and *turbidity currents*, which overcome the friction between particles in different fashions.

sediment transport equation One of a number of equations designed to predict the *sediment yield* of a given flow from its *shear velocity* or *shear stress*, the *grain size distribution*, the *hydraulic radius* of the channel and the characteristics of the fluid.

sediment yield The total quantity of

sediment exported from a drainage *basin* over a given period of time.

sedimentary breccia A *breccia* formed of *sedimentary rocks*.

sedimentary exhalative deposit See *volcanic-exhalative deposit*.

sedimentary facies A *sedimentary rock* body with specific and distinctive characteristics, such as defined by its biota, chemistry or physics. Also used in a genetic sense for sediment deposited by a particular process or in an environmental sense for sediment deposited in similar settings.

sedimentary rock A rock formed by the consolidation of sediment.

sedimentology The study of sediments and their deposition and accumulation.

seepage 1. An escape of gas or oil. 2. A diffuse flow of water from an *aquifer*, not sufficiently localized to be termed a *spring*.

seepage velocity The apparent velocity of *groundwater* calculated from the *Darcy equation*, which is higher than the true velocity because of the effects of *porosity* and the *tortuosity* of the flow path.

seiche A *standing wave* generated on the surface of an enclosed or partly enclosed body of water by winds, atmospheric gradients, *tides*, *earthquakes* or sediment *slides*.

seif A linear *dune*, up to 100 m in height, whose orientation is controlled by the prevailing winds, commonly occurring in a series of parallel ridges.

seismic array 1. A group of *seismometers*, whose output can be combined to improve the *signal to noise ratio*. In *seismic prospecting* arrays are generally arranged in linear patterns along survey lines. In *earthquake seismology* and *forensic seismology* arrays may be distributed over hundreds of square kilometres, and the direction of the causative *epicentre* of the arrivals can be located from the different *arrival times* at individual sensors. 2. Less commonly, a group of *seismic sources*, whose output can be combined to improve the *signal to noise ratio*.

seismic belt A linear grouping of *earthquake epicentres*, commonly along a *plate boundary* but also on a local scale.

seismic body wave See *body wave*.

seismic creep A form of non-instantaneous *deformation* in which a residual *strain* remains after removal of *stress* by an *earthquake*, which decreases or increases over a period of time dependent on the initial *stress*, *rigidity* and a time-dependent *creep* function.

seismic discontinuity An abrupt boundary between two media over which the *acoustic impedance* changes significantly, e.g. *Mohorovicic discontinuity*, *Gutenberg discontinuity*.

seismic event A sudden disturbance on or within the Earth that generates *seismic waves*, such as an *earthquake*, chemical or nuclear explosion, *rockburst* or mine *failure*, but excluding sources of *microseisms*.

seismic exploration (seismic prospecting) The use of seismic methods to locate and delineate geological *structures* or bodies in the subsurface. The two principal techniques use *seismic reflection* and *seismic refraction*.

seismic facies analysis A technique of

seismic stratigraphy which determines the characteristics of *seismic reflections* (continuity, amplitude, frequency, *interval velocity*) which distinguish them from adjacent reflections and uses them to infer the depositional environment and thus the presence of possible *reservoirs*. See also *reflection character analysis, seismic sequence analysis*.

seismic gap (Mogi Doughnut) A space in a *seismic belt* where no *earthquake* has occurred over a stated period of time, commonly 10 years. Sometimes the anticipated site of a large *earthquake* and so useful in *earthquake prediction*.

seismic head wave See *head wave*.

seismic migration (migration) The process of displaying seismic reflectors on a *seismic reflection seismogram* in their true position in space, rather than immediately below the *shot*-detector location. A necessary operation when *strata* are dipping.

seismic moment The scalar size of the set of couples that represent the *forces* causing *failure* in an *earthquake* or explosion; a measure of *earthquake* 'size', equal to the product of the *rigidity*, slip length along the *fault plane* and the slip area on the *fault plane*.

seismic multiple See *multiple reflection*.

seismic noise The *noise* appearing on a *seismic record*.

seismic prospecting See *seismic exploration*.

seismic record See *seismogram*.

seismic reflection The *seismic wave* or energy returned to a receiver from a *seismic source* after reflection according to *Snell's Law* at a discontinuity where there is a contrast in *acoustic impedance*. The basis of an important form of *seismic exploration*.

seismic refraction The *seismic wave* or energy that has been refracted according to *Snell's Law* at a discontinuity where there is a contrast in *acoustic impedance*. The *head wave* returning to the surface from a *seismic source* after *critical refraction* is the basis of an important form of *seismic exploration*.

seismic sequence analysis A technique of *seismic stratigraphy* in which a *seismogram* is divided into units of common reflection and depositional characteristics by the identification of *unconformities* or changes in the seismic patterns. Each sequence thus represents a three-dimensional set of contemporaneous sediments in the same system of depositional processes and environment. See also *reflection character analysis, seismic facies analysis*.

seismic shooting The detonation of a *shot* or other *seismic source* so as to collect information for *seismic exploration*.

seismic slip Slip on a *fault plane* accompanied by *seismicity*. cf. *aseismic slip*.

seismic source Any mechanism which generates *seismic waves*, e.g. *earthquakes*, chemical and nuclear explosions.

seismic stratigraphy (seismostratigraphy) The study of the *stratigraphy* and distinct, genetically-related, depositional units from *seismic reflection* data by the analysis of non-structural information on *seismograms*. The techniques utilized are *seismic sequence analysis, seismic facies analysis, reflection character* *analysis* and the recognition of direct *hydrocarbon* indicators, such as *bright spots*.

seismic surface wave See *surface wave*.

seismic tomography (tomography) The technique of mapping *seismic velocity* variations in a subsurface volume by the comparison of *arrival times* at a network of *seismometers* of *seismic waves* traversing the volume in different directions.

seismic velocity The velocity with which a medium transmits *seismic waves*, controlled by its *elastic moduli* and *density*.

seismic wave An *elastic wave* travelling within the Earth, falling into one of the two general categories of *body* and *surface waves*.

seismicity The distribution of *earthquake*-induced seismic activity in time, location, *magnitude* and depth, of great significance to *earthquake prediction* and studies of the dynamic behaviour of the Earth.

seismogram A recording of the variation of ground displacement, velocity or acceleration with time, measured with a *seismometer*.

seismograph See *seismometer*.

seismology The study of *seismic waves* within the Earth, including their generation by *earthquakes* or explosions, how they propagate and the information they provide about the structure and nature of the Earth's interior.

seismometer (seismograph) The general name for an instrument that monitors *seismic waves*, e.g. *geophone*, *hydrophone*.

seismostratigraphy See *seismic stratigraphy*.

seismoturbidite See *megaturbidite*.

Selachii An order of subclass *Elasmobranchii*, class *Chondrichthyes*, superclass *Pisces*; the modern sharks and their ancestors. Range *Jurassic–Recent*.

selenite A variety of *gypsum* yielding broad, colourless, transparent, *cleavage* folia.

self potential (spontaneous polarization) The phenomenon whereby natural electrical fields are generated in the subsurface by electrochemical reactions between electrically conductive *minerals* and *groundwater*.

self potential log A *geophysical borehole log* which measures the *self potential* effects generated when *pore fluids* of different ionic concentration come into contact; particularly sensitive to *shale* horizons.

self-potential method See *spontaneous polarization method*.

self reversal A rare mechanism by which a *ferromagnetic mineral* can acquire a *natural remanent magnetization* of the opposite polarity to the ambient *geomagnetic field* by interaction with the *magnetic field* of an existing *mineral* with *normal polarity*.

SEM See *scanning electron microscopy*.

semi-metal The *native elements arsenic*, *antimony* and *bismuth*, which are *brittle minerals* with lower *electrical* and *thermal conductivities* than metals.

semifusinite A *coal maceral* of the *inerti-*

nite group, similar to *fusinite* but with a lower *reflectance.*

semseyite ($Pb_9Sb_8S_{21}$) A rare *ore mineral* of lead.

Senecan A *Devonian* succession in E North America covering part of the *Gavetian* and the *Frasnian.*

Senonian A division of the *Cretaceous,* 88.5–65.0 Ma.

sepiolite (meerschaum) ($Mg_4(OH)_2Si_6O_{15}H_2O + 4H_2O$) A *clay*-like, hydrous, *secondary mineral* of magnesium found in association with *serpentine.*

septarian nodule (septarium) A *nodule* with irregular internal cracks or *veins.*

septarium See *septarian nodule.*

Serevovinskiy A *Permian* succession on the eastern Russian Platform covering parts of the *Longtanian* and *Changxinian.*

seriate texture A *texture* in an *igneous rock* in which there is a continuous range in crystal sizes.

sericite A fine-grained variety of *muscovite.*

sericitization A type of *wall rock alteration* whose products are dominantly *sericite* and *quartz.*

series A third order *chronostratigraphic* unit.

serir See *reg.*

serpentine asbestos See *chrysotile.*

serpentine group ($Mg_3Si_2O_5(OH)_4$) A group of common, hydrated *phyllosilicates* with three common *polymorphs*: *antigorite, lizardite* and *chrysotile.* Often found as an *alteration* product of magnesium silicates, particularly *olivine, pyroxene* and *amphibole.*

serpentinite A rock composed mainly of *serpentine minerals* with minor *brucite, talc,* Fe-Ti oxide and Ca-Mg carbonates. Forms by *serpentinization* with either a massive *fabric* with relict *textures* of the parent rock or a sheared *fabric* in which original *fabrics* have been destroyed by intense *penetrative deformation.*

serpentinization The *alteration* of *ultramafic rocks* such as *dunite, peridotite* and *pyroxenite* into *serpentinite.*

Serpukhovian An *epoch* of the *Carboniferous,* 332.9–322.8 Ma.

Serravallian A *stage* of the *Miocene,* 14.2–10.4 Ma.

service reservoir See *distribution reservoir.*

sessile Descriptive of a non-mobile organism.

settling velocity The terminal velocity at which a grain settles through a static fluid, dependent on the *density,* size, shape and bulk concentration of the particles and the *density, viscosity* and *rheological* properties of the fluid. May be measured to determine sediment grain size.

sferic A natural electromagnetic field in the audiofrequency range of $\sim 1–10^3$ Hz originating from thunderstorm activity and the source field of the *magnetotelluric* and *AFMAG* techniques of *geophysical exploration.*

Sha-Xi-Miao A *Jurassic* succession in Sichuan, China, equivalent to the *Bajocian, Bathonian* and *Callovian.*

shadow zone A distance *range* from a *seismic source* in which some given type of *seismic wave* does not emerge due to a subsurface refraction.

shakehole See *doline*.

shale An *argillaceous* rock with closely-spaced, well-defined laminae.

shallow-focus earthquake An *earthquake* with a depth of *focus* shallower than 70 km. cf. *deep-focus earthquake*, *intermediate-focus earthquake*.

shallowing-upward carbonate cycle A phenomenon of sedimentation on *carbonate platforms* in which carbonate production exceeds the rate of subsidence.

Shannon-Weaver dominance diversity equation An equation used to calculate the *dominance diversity* (H) in a *palaeoecological* study from $H = -\sum_{i=1}^{s} P_i \log P_i$, where P_i is the proportion of the ith species in the sample and s the total number of species.

Shaodong A *Carboniferous* succession in China equivalent to the *Famennian*.

shape fabric A planar or linear structure defined by *preferred orientation* of the shapes of components of a material, requiring an *anisotropic* shape of the components and the alignment of the *anisotropy*.

shard An abraded fragment of *pumice*.

shatter cone A centimetric- to metric-scale conical *structure* believed to form during *meteorite* impact.

shattuckite ($Cu_5(SiO_3)_4(OH)_2$) An *inosilicate* found in the oxidized zones of copper deposits.

shear *Deformation* in which the angular relationship between material lines in a body change, i.e. a rotational *stress* or *strain*.

shear band A narrow band of localized *strain* developed in deforming, *anisotropic*, *foliated* rocks, e.g. *mylonite*, *phyllonite*, in which the *deformation* folds the existing *metamorphic foliation*.

shear direction The direction in which *displacement* takes place during *shear*.

shear fault A *fault* on which *displacement* has been produced by *shear*.

shear folding See *flow folding*.

shear joint A *joint* forming by *shear* at an acute angle to the maximum *principal stress*.

shear layer The zone of velocity gradient formed by the juxtaposition of two fluids of different velocity or *density*.

shear modulus See *rigidity*.

shear plane The surface to which points undergoing *simple shear* move parallel, i.e. the plane of a *shear zone*.

shear resistance See *shear strength*.

shear strain *Deformation* involving a change in the intrinsic angular relationship between lines, defined as the tangent of the change in angle.

shear strength (shear resistance) The *strength* of a material subjected to *shear stress*, a function of *normal stress*.

shear stress The *force* per unit area exerted tangentially to a given surface.

shear velocity (U^*) The velocity which expresses the velocity gradient of the lowest regions of a boundary layer (the lowest 15% of the flow depth). $U^* = \sqrt{(\tau/\rho)}$, where τ = *shear stress*, ρ = fluid *density*.

shear wave See *S wave*.

shear zone A zone of *ductile deformation* between two undeformed blocks that have suffered relative *shear displacement*; the *ductile* analogue of a *fault*.

sheath fold A highly *non-cylindrical fold* with a strongly curved *hinge line* within the *fold axial* surface, formed during high *strain*, *ductile deformation*.

sheet erosion (sheetflood erosion) The removal of fine-grained, superficial material by sheets of flowing water rather than channelized streams, occurring, for example, in deserts where sporadic rainfall causes *runoff* by *sheetflow* on *fans* and *pediments* lacking incision.

sheet joint A *fracture* parallel to the ground found in *granitic* intrusions, possibly originating during cooling.

sheet silicate See *phyllosilicate*.

sheeted dyke complex A series of originally horizontal *basaltic dykes* in an *ophiolite* which fed *pillow lavas*. Equated with layer 2 of the *oceanic crust*.

sheetflood The unconfined flow of water over the land surface resulting from channel overtopping or *levée* breaching, causing deposits of *mud*, *silt*, *sand* or *gravel* with a sharp base. cf. *crevasse splay*.

sheetflood erosion See *sheet erosion*.

sheetflow (sheetwash) *Overland flow* over a smooth soil surface.

sheeting The process by which shells of rock split off along *joints* running approximately parallel to the rock surface, resulting from *pressure relief* on *denudation*.

sheetwash See *sheetflood*.

Sheinwoodian A *stage* of the *Silurian*, 430.4–426.1 Ma.

shell pavement An accumulation of shells left as a *lag* deposit after the *winnowing* of fine-grained material.

Shemshinskiy A *Permian* succession on the eastern Russian Platform covering the upper part of the *Ufimian*.

sherbakovite ($Na(K,Ba)_2(Ti,Nb)_2(Si_2O_7)_2$) A dark brown, *vitreous mineral* found in *pegmatite*.

sheridanite ($(Mg,Al)_6(Si,Al)_4O_{10}(OH)_8$) A *chlorite* group *mineral* poor in both iron and silica, found in *schist* and other *metamorphic rocks*.

Shidert A *Cambrian* succession in Siberia covering part of the *Maentwrogian* and the *Dolgellian*.

shield An extensive area of exposed *bedrock* with long-term *tectonic* stability, generally of *Precambrian* age and forming the central core of a continent.

shield volcano A type of *volcano*, 10–100 km in diameter, comprising many usually *basaltic lava flows* and subordinate *pyroclastic rocks*, with a circular to elliptical shape and flanks dipping at 2°–8° away from a *crater* or line of *craters* and/or a *caldera*. Often found in overlapping groups.

Shields beta (β) A dimensionless measure of *shear stress* representing the ratio of the fluid *forces* promoting sediment movement to the submerged weight of a single layer of grains, or the *gravity* forces opposing motion. $\beta = \tau/((\rho_s - \rho_f)gd)$, where τ = *shear stress* per unit area of bed, ρ_s and ρ_f = *densities* of sediment and fluid respectively, g = *gravity* and d = grain size.

Shields diagram A graphical representation of the threshold of sediment entrainment by a plot of *Shields beta* against particle *Reynold's number.* Used to interpret the *forces* influencing the entrainment.

shipborne geophysical survey *Geophysical exploration* undertaken from a ship utilizing *gravity, magnetic, seismic* or *side-scan sonar* techniques.

shock metamorphism *Metamorphism* at very high, transitory temperature and pressure generated by *meteorite* impact.

shock remanent magnetization A *natural remanent magnetization* resulting from rapidly applied *stress,* such as in a *meteorite* impact.

shoestring sand A *stratigraphic oil trap* comprising a long, narrow body of *sandstone* enclosed by *shale,* probably originating as an *offshore bar* or by a meandering river.

shonkinite A coarse-grained, *feldspar*-rich *syenite* comprising *pyroxenes* and some *olivine.*

shore platform (marine abrasion platform, wave-cut bench, wavecut platform) A low gradient, intertidal rock surface caused by the recession of a *sea cliff.*

shortening (shortening strain) A linear measure of *strain* defined as $(L_u - L_d)/L_u$, where L_u = undeformed length, L_d = deformed length.

shortening strain See *shortening.*

shoshonite A *potassic, silica*-poor *trachyandesite.*

shot An artificial *seismic source.*

Shreve's magnitude A measure of *stream order* which adds the order values at each confluence to provide the magnitude of the downstream link.

shrinkage crack A small fissure resulting from contraction of *cohesive* sediment.

shrub-coppice dune See *nebkha.*

shrub-coppice mound See *nebkha.*

SI Système **I**nternational; the modern system of units, based on m, kg, s, A, V.

sial An outmoded term for the *upper continental crust* derived from its **si**lica and **al**uminium-based composition. cf. *sima.*

sichelwannen Metric-scale, crescentic marks cut into a flat or gently sloping rock surface by a *glacier,* the horns pointing away from the snout.

side-looking airborne radar (SLAR) An airborne *radar* system which scans sideways and detects *radar* backscattered from the ground surface in a similar way to *side-scan sonar.*

side-scan sonar A marine, sideways-scanning, acoustic method of *geophysical exploration* in which the seabed to either side of a ship's track is insonified by high-frequency (30–110 kHz) sound beams. Seafloor features either reflect energy back to the ship or away from it, producing a characteristic *sonograph.*

siderite 1. **(spathic iron)** ($FeCO_3$) A *carbonate mineral* of iron, sometimes of importance as an *ore.* 2. See *iron meteorite.*

siderolite See *stony iron meteorite.*

sideromelane The brown *glass* component of *basaltic hyaloclastite*.

siderophile Descriptive of an element with an affinity with oxygen or sulphur and soluble in molten iron.

siderophyllite An iron- and aluminium-rich *biotite*.

sief A linear or longitudinal *dune*, up to 20 km long, commonly in a group oriented in the direction of the prevailing wind.

siemen (S) The *SI* unit of electrical conductance, the inverse of resistance.

sieve deposit 1. A deposit originating from the deposition of fine sediment into existing coarse sediment because of the rapid *discharge* reduction of a surface flow due to percolation into the coarse sediment. 2. A coarse sediment lobe on a semi-arid, *alluvial fan* originating from rapid *discharge* reduction due to the *infiltration* of water.

signal to noise ratio (SNR) The ratio of the amplitude or energy of a particular signal to that of the unwanted background signal.

silcrete A nodular, highly indurated or massive, surface or near-surface layer composed of *silica*, resulting from the *cementation* and replacive introduction of *silica* into soils, rock or sediment.

Silesian A *Carboniferous* succession in NW Europe equivalent to the *Serpukhovian*, *Bashkirian*, *Moscovian*, *Kasimovian* and *Gzelian*.

silica minerals A group of *minerals* with the formula SiO_2. *Quartz* is the most common, but other *polymorphs* include *agate*, *chalcedony*, *chert*, *coesite*, *cristobalite*, *flint*, *jasper*, *onyx*, *opal*, *stishovite* and *tridymite*.

silica saturation The concentration of *silica* in an *igneous rock* relative to the other constituents which combine with *silica* to form silicate *minerals*. See *oversaturated*, *saturated*, *undersaturated*.

silication The process by which a rock is converted into, or replaced by, silicates, in particular the conversion of carbonate rock to *skarns*.

siliceous ooze A *pelagic*, biogenic sediment comprising the skeletal remains of siliceous micro*fossils*.

siliciclastic (terrigenous) A *clastic rock* descriptor indicating a rock whose *clasts* are predominantly of *silicate minerals*.

silicification 1. The *alteration* by *cementation* or *replacement* of sediment by *amorphous*, *cryptocrystalline*, *microcrystalline* or *macrocrystalline silica*. 2. A type of *wall rock alteration* involving an increase in *quartz* or *cryptocrystalline silica* in the altered rock.

silky lustre A *lustre* with a silk-like appearance, due to the reflection of light from a fine *aggregate* of parallel fibres.

sill A concordant, *tabular* or sheet-like *igneous body* from a few centimetres to hundreds of metres in thickness.

sillar A weakly-consolidated, fine-grained *ash-flow tuff*.

sillénite (BiO_3) Cubic bismuth trioxide, occurring as a *secondary mineral*. cf. *bismite*.

sillimanite (fibrolite) (Al_2SiO_5) A *nesosilicate* found in *metamorphic rocks* rich in alumina.

sillimanite minerals A group of anhydrous, aluminium silicates comprising *andalusite*, *kyanite* and *sillimanite*. Used to make a *mullite-silica* mixture, which is highly *refractory* and has a small coefficient of thermal expansion.

silt A sediment with particles in the size range 4–62.5 μm.

siltstone A lithified *silt*.

Silurian A *period* of the *Palaeozoic*, 439.0–408.5 Ma.

silver (Ag) A rare *native metal*.

silver glance See *argentite*.

silver lead ore *Galena* containing *silver*.

sima An outmoded term for the lower *crust*, which is rich in **si**lica and **ma**gnesium. cf. *sial*.

Simferopolian An *Eocene* succession in the former USSR, covering parts of the *Ypresian* and *Lutetian*.

similar fold A *fold* whose inner and outer arcs have the same curvature, so that all *dip isogons* are parallel to the *axial plane* and the orthogonal thicknesses of the *fold limbs* are less than at the *hinge zone*.

simple shear A shape change in which all particles of the deforming body move in parallel lines, the amount of movement depending on the distance of each particle from a given plane in which there are no *displacements*.

Sinemurian A *stage* of the *Jurassic*, 203.5–194.5 Ma.

singing sand (booming dune) *Dune sands* which emit audible sounds when in motion, described as booming, roaring, squeaking, singing or musical. Probably arises from some unique combination of grain properties, such as high *sorting*, uniform size and high *roundness*.

sinhalite ($Mg(Al,Fe)BO_4$) A rare *gem*.

Sinian The higher *era* of the *Precambrian*, 800–570 Ma.

sinistral The sense of movement across a boundary, such as a *fault*, in which the side opposite the observer moves to the left. cf. *dextral*.

sinkhole An approximately circular depression in *limestone* terrain into which water drains and collects.

sinter See *geyserite*.

sinuosity A measure of the degree to which a curvilinear feature winds about the shortest route between two points, defined as the ratio of feature length to shortest distance.

sinusoidal fold A *fold* with a *profile* approximating a sine curve.

Sirenia An order of infraclass *Eutheria*, subclass *Theria*, class *Mammalia*; the dugongs and manatees. Range early *Eocene–Recent*.

site investigation A procedure which provides the information about a proposed site upon which decisions can be based regarding its suitability for engineering works; the primary means by which ground conditions and properties are considered during civil engineering design and construction work using the techniques of *engineering geology*.

skarn (pyrometasomatic deposit) An irregularly-shaped, *replacement ore* deposit, usually formed at high temperatures in metamorphosed, carbonate-rich sediments at the contacts with medium to large *igneous bodies*. The *ore-*

bodies are characterized by the development of calc-silicate *minerals* such as *actinolite*, *andradite*, *diopside* and *wollastonite*, and provide iron, copper, *graphite*, lead, molybdenum, tin, tungsten, uranium and zinc.

skewness A statistical measure of the shape of a distribution which has more members at one end than the other. Used in the description of sedimentary *grain size distribution*.

skin depth In *electromagnetic induction methods*, the *depth of penetration*, which is dependent on the product of the inverse square root of the frequency used, the *electrical conductivity* and the *permeability* of the medium.

skip mark A *tool mark* formed by the low-angle impact of a particle with a muddy sediment; a trail may be formed as the particle skips along the bed of a flow.

skutterudite (smaltite) ($(Co,Ni)As_3$) An *ore mineral* of cobalt and nickel.

slab failure A form of *weathering* common on hard rock cliffs, occurring when *erosion* releases the lateral *confining pressure* and tension *joints* open until the *tensile strength* of the rock at the edges of a slab is exceeded and it falls.

slab pull The mechanism whereby the gravitational pull of a downgoing slab in a *subduction zone* exerts a lateral *force* on the *plate* attached to the slab. An important *plate* driving mechanism in *Orowan-Elsasser convection*.

slack A flooded *interdune*.

slacking The process whereby *lignites* and *sub-bituminous coals* crack and fall apart when they dry out after excavation.

slaking The disintegration of loosely-consolidated material when water is introduced or it is exposed to the atmosphere.

SLAR See *side-looking airborne radar*.

slate A fine-grained, low-*grade* or *regionally-metamorphosed mudrock* with a well-developed *penetrative cleavage*. The *cleavage* is a *foliation* in which sub*microscopic phyllosilicate minerals* are in well-developed, parallel alignment so that the rock splits into platy sheets.

slate belt A linear zone within an *orogenic belt* in which *pelitic* rocks are dominantly in the form of *slate*, characterized by the lowest *grades* of *regional metamorphism* and a steeply-dipping *cleavage* subparallel to the axis of the belt.

slaty cleavage The *cleavage* developed in a *slate*.

Slendoo A *Permian* succession in Queensland, Australia, covering the middle part of the *Artinskian*.

slickencryst A fibrous *mineral* growth on the *bedding planes* within a *fold limb*, indicative of *interlayer slip*.

slickenline (slickenside lineation, slickenside striation) A millimetric to metric scale *lineation* on a *slickenside* parallel to the *displacement* direction of the *fault*.

slickenside A smooth or polished *fault* surface.

slickenside lineation See *slickenline*.

slickenside step A short, narrow offset at one end of the termination of a *slickenline*, usually approximately orthogonal to the *slickenline*.

slickenside striation See *slickenline*.

slickolite formation The process occurring during *slickenside* formation in which solution of rock oblique to the *fault plane* forms oblique, incongruous *stylolites* in the direction of compression.

slide 1. Any downslope *mass movement* of sediment under the influence of *gravity deformation*. 2. A *fault* developed during major *recumbent folding*.

Slingram A system used in the *compensator electromagnetic method*.

slip A process of *displacement* within individual crystals during *plastic deformation*.

slip clay A *kaolin*-rich *clay* used for glazes.

slip-face The face on the *lee slope* of a *dune* where *sand* accumulates at its *angle of repose* (30°–34°).

slip line A curve indicating the direction of maximum *shear stress* at any point during *plastic deformation* and, in a perfectly *plastic* material, the direction of maximum *strain rate*. Forms an angle of 45° with the *stress trajectory*.

slip-off slope A low gradient slope on the inside of a meander bend.

slip sheet A *gravity collapse structure* in the form of a detached sheet.

slip system The *shear plane* and *shear direction* within the plane of *simple shear*, including all symmetrically equivalent combinations.

slip vector (displacement vector) The vector connecting a point to its original location before *displacement*.

slope apron The region between the margin and floor of a *basin*.

slope gradient factor A term in the *universal soil loss equation*.

slough channel A channel in a *braided stream* with relatively little flow and deposition is predominantly by the settling of fine material from *suspension*.

SLR See *satellite laser ranging*.

slump A type of sediment *slide* in which material moves downward as a unit or series of units, often with backward rotation about a horizontal axis parallel to the causative slope. Generally highly deformed internally.

slump fold A *fold* developed in a *slump*.

small circle Any circle on a sphere that does not correspond to a circumference. cf. *great circle*.

Smallfjord A *stage* of the *Vendian*, 610–600 Ma.

smaltite See *skutterudite*.

smaragdite A fibrous green variety of *actinolite* forming *pseudomorphs* after *pyroxene* in *eclogitic* rocks.

smectite clays (bentonite clays) A group of *clays* comprising *montmorillonite* and *attapulgite*, exploited in *drilling mud* for their swelling properties and for the increase in *viscosity* they cause in *drilling fluids* and as bonding agents.

smithsonite (calamine) ($ZnCO_3$) A zinc *ore mineral* of *supergene* origin.

smokey quartz A brown to black variety of *quartz*.

SMOW **S**tandard **M**ean **O**cean **W**ater, the international standard for ^{18}O:^{16}O ratios.

Snell's Law A law of optics that can

be applied to *seismic waves* impinging on an *acoustic impedance* contrast: $(\sin i)/V$ is constant, where i = angle of incidence, refraction or reflection and V = the *seismic velocity* of the medium in which i is measured.

SNR See *signal to noise ratio.*

soapstone (steatite) A compact, massive variety of *talc.*

soda lake A lake whose water has a high content of sodium, found in *rift valleys.*

sodalite ($Na_8(AlSiO_4)_6Cl_2$) A rare *feldspathoid* found mainly in *nepheline syenites* and related rock types, in *metasomatized* calcareous rocks and in cavities in *ejecta.*

soft sediment deformation See *wet sediment deformation.*

soil creep A slow type of *mass movement.*

soil crust See *clay pan.*

soil erodibility factor A term in the *universal soil loss equation.*

soil erosion The removal of material from the soil zone by natural processes, generally by wind, water and *soil creep.*

soil loss A term in the *universal soil loss equation.*

soil mechanics The techniques of *engineering geology* applied to soils.

soil profile (soil structure) The sequence through the soil zone from the surface to *bedrock,* comprising prominent horizons distinguished by different colours, structures and compositions.

soil structure See *soil profile.*

Sokol'yergorskiy A *Permian* succession on the eastern Russian Platform equivalent to the L. *Asselian.*

solar luminosity The intensity of electromagnetic radiation received from the Sun. Variation in this factor may influence the origin of *ice ages.*

solar pond (salina, salt pond) An artificial pond in which evaporation of saline water by the Sun takes place, used as a guide to the processes of *evaporate basin* evolution.

sole fault The *detachment fault* in an *extensional fault system.*

sole mark (sole structure) A cast of a sedimentary structure on the base of a *bed,* often developed where relatively coarse-grained rocks overlie finer-grained, *cohesive* sediment.

sole structure See *sole mark.*

sole thrust The *detachment fault* in a *thrust belt.*

Solenoporaceae A heterogeneous group of red *algae.* Range *Cambrian–Tertiary.*

solfatara A *fumarole* in an active volcanic area whose escaping *volcanic gases* contain much *sulphur,* which commonly forms a bright yellow deposit.

solid Earth The *crust* of the Earth and its interior, i.e. the Earth excluding the hydrosphere, atmosphere and biosphere.

solid solution The substitution of one ion for another in a *crystal lattice* with no change in *structure,* i.e. the composition of *minerals* in the series varies continuously between the two end-members.

solidus The temperature below which

a *mineral assemblage* is entirely crystalline. cf. *liquidus*.

solifluction The slow, gravitationally-driven, downslope movement of water-saturated, seasonally thawed materials.

Solikamsky A *Permian* succession on the eastern Russian Platform covering the lower part of the *Ufimian*.

Solnhoffen Limestone An *exceptional fossil deposit* of U. *Jurassic* age in Bavaria, Germany, containing a range of fauna, including articulated vertebrates, shrimps, limulines and *insects*.

solution 1. An important process of chemical *weathering* in which a *mineral* in contact with a solvent is dissociated into its component ions. 2. A fluid containing ions, such as a *pore fluid*.

solution breccia See *collapse breccia*.

solution cleavage A *spaced cleavage* found in *quartzite* and *limestone* which contains relatively soluble *minerals*, indicating the action of *pressure solution*.

solution flutes See *rillenkarren*.

solution gas drive See *depletion drive*.

solution mining A mining method for soluble *minerals* in which water is introduced into the deposit and the dissolved *minerals* pumped to the surface.

solution transfer A *deformation* mechanism by *diffusion* in the dissolved state.

solutional erosion The *erosion* by *solution* of the component *minerals* of a rock.

Solva A *Cambrian* succession in S Wales equivalent to the *Solvan*.

Solvan A *stage* of the *Cambrian*, 536.0–530.2 Ma.

solvus The line in a *phase diagram* separating a stable *solid solution* from two or more phases that form from the solution by unmixing.

somaite An *alkaline igneous rock* resembling *essexite* but with *leucite* replacing *nepheline*.

sonde A device lowered down a borehole for measuring or testing purposes.

sonic log (continuous velocity log) A *geophysical borehole log*. A type of *acoustic log* in which the *seismic velocities* of formations are determined from the travel time of an acoustic pulse from a *seismic source* at one end of a *sonde* to a detector at the other.

sonobuoy A buoy used in marine *seismic exploration* containing instruments for detecting seismic information and transmitting it to a recording ship.

sonograph A display of *side-scan sonar* data.

sorosilicate A *crystal structure* classification in which the *coordination polyhedra* are Si tetrahedra which link by pairs sharing one corner.

sorting (grading) The standard deviation of the *grain size distribution* of a sediment, which is controlled partly by transport and depositional processes and is thus of diagnostic significance.

Soudleyan A *stage* of the *Ordovician*, 457.5–449.7 Ma.

sour crude oil *Crude oil* with detectable hydrogen sulphide. cf. *sweet crude oil*.

source rock The rock in which *hydrocarbons* were generated.

South African jade See *Transvaal jade.*

Southland A *Miocene* succession in New Zealand comprising the *Clifdenian, Lillburnian* and *Waiauan.*

South West US and East AnTarctica hypothesis (SWEAT hypothesis) A controversial reconstruction of the *Precambrian supercontinent* in which it is proposed that the southwestern coast of North America and the east coast of Antarctica were juxtaposed in Late *Proterozoic* times. The rifting of this continent would have resulted in its turning inside out (or '*extraversion*') and the drifting of its components into their more modern configuration.

sövite A coarse-grained, light-coloured, calcium carbonate *carbonatite.*

SP method See *spontaneous polarization method.*

spaced cleavage A *cleavage* in which the *cleavage planes* are separated by uncleaved slices of rock.

Spanish topaz An orange-brown variety of *quartz.*

spar A mining term for any white or light-coloured *mineral* with a well-developed *cleavage* and a *vitreous lustre.*

sparagmite An *arkose* of *Precambrian* age.

sparite Sparry *calcite*, occurring as the cement in some *limestones* and formed by nucleated *precipitation* and growth in a pore space.

sparker A marine *seismic source* in which a capacitor is discharged in the water, creating a plasma bubble which oscillates in the same fashion as an explosion.

sparry limestone A *limestone* with a *sparite* cement.

spastolith A grain composed of soft material which was deformed by mechanical *compaction* during burial.

Spathian A *stage* of the *Triassic*, 241.9–241.1 Ma.

spathic iron See *siderite.*

spatter (driblet) Agglutinized masses of primary *ejecta*, larger than *lapilli*, which were erupted as fluids.

spatter cone A low mound (< 15 m high) formed by fountains of *basaltic lava.* Mainly composed of *spatter* but also including *bombs* and *cinders.*

specific form A crystal *form* whose faces bear a specific relationship to the symmetry operators of the crystal. They may be parallel or perpendicular to a rotation axis or a mirror plane. cf. *general form.*

specific retention The water remaining in an area of soil at *field capacity* after draining under the influence of *gravity.* cf. *specific yield.*

specific yield The water that has drained from an area of soil at *field capacity* under the influence of *gravity.* cf. *specific retention.*

spectral analysis (power spectrum analysis) The determination of the relationship between frequency or *wavenumber* and energy for a waveform. The slope of linear segments of the power spectrum of a *potential field* can be used to determine the *limiting depth* of its source.

spectral band A defined band of frequencies or *wavenumbers* of a waveform.

spectrographic analysis The chemical analysis of a material by the matching of its spectral lines with those of elements.

spectrometer An instrument for measuring the spectrum of a waveform.

specular lustre See *splendent lustre*.

specularite A platy, metallic variety of *haematite*.

speleology The scientific study of *caves*.

speleothem The generic name for chemical precipitates found in *caves*, including, *cave coral*, *cave drapery*, *cave onyx*, *cave pearl*, *curtains*, *flowstone*, *helictites*, *moonmilk*, *rimstone dams*, *stalactites* and *stalagmites*. Dominantly formed of *calcite*, but also, more rarely, *gypsum* and *halite*.

sperrylite ($PtAs_2$) A rare, white *ore mineral* of platinum with a *metallic lustre*.

spessartine ($Mn_3Al_2Si_3O_{12}$) A *garnet* found mainly in *granite pegmatites*, *gneiss*, *quartzite* and *schist* and as *lithophysae* in *rhyolites*, less commonly in *skarns*.

spessartite 1. A *calc-alkaline lamprophyre* containing *hornblende*, *clinopyroxene* ± *olivine* with *plagioclase* in excess of *alkali feldspar*.

sphalerite (zincblende) (ZnS) The major *ore mineral* of zinc.

sphalerite group A group of *sulphide minerals* characterized by a unit cube structure with metals at the corners and face centres and the four sulphur atoms coordinated so that each lies at the centre of a regular tetrahedron of metals and each metal is at the centre of a regular sulphur tetrahedron.

sphene See *titanite*.

sphenochasm A triangular gap of *oceanic crust* between two *cratons* whose faulted margins converge to a point, originating as one *craton* rotated with respect to the other. cf. *sphenopiezm*.

Sphenodontia An order of subclass *Lepidosauria*, class *Reptilia*; small *reptiles* with diverse dentition resembling, and related to, lizards. Range *Triassic–Recent*.

sphenopiezm A triangular region between two *cratons* which is being squeezed as one *craton* rotates with respect to the other. cf. *sphenochasm*.

Sphenopsida A class of division *Tracheophyta*, kingdom *Plantae*; the horsetails and their relatives. Range M. *Devonian–Recent*.

spherical harmonic analysis The equivalent of *Fourier analysis* in spherical coordinates.

sphericity A measure of how closely a grain resembles a sphere, defined as $\sqrt[3]{(S^2/LI)}$, where S, L, I are the short, long and intermediate grain diameters respectively.

spheroid The ellipsoid whose flattening and polar and equatorial radii provide the best approximation to the *geoid*.

spheroidal structure A *structure* shown by some *igneous rocks* comprising large, rounded masses surrounding concentric shells of the same material. Probably a large-scale form of *perlitic texture*.

spheroidal weathering The process of mechanical spalling and surficial chemical *weathering* of a *boulder*. Often associated with the *deep weathering* of *basalt*, *dolerite* and *granite*, in which water promotes chemical *weathering* within *cooling joints*.

spherule A small, spherical particle.

spherulite A subspherical *structure* in a silicic *volcanic rock*, 1–20 mm in diameter, comprising wholly or partly radially-disposed, *acicular* crystals, normally of *alkali feldspar*. Probably the product of the *devitrification* of hydrated silicic *glasses* in *rhyolitic*, *rhyodacitic* and *trachytic glassy rocks*.

spherulitic texture The *texture* of a rock containing many *spherulites*.

spicular chert *Chert* made up of *sponge spicules*.

spicule A small spine or needle.

spiculite A sediment or *sedimentary rock* composed of *sponge spicules*.

spider diagram A plot used to show variations in a variety of elements between two *igneous rocks*.

spike 1. An incongruous *stylolite* growing in the direction of compression during *slickolite formation*. 2. A mathematical waveform of infinitely small duration containing an infinite number of frequencies, used in the manipulation of waveforms, especially in *seismology*.

spilite A *mafic volcanic rock*, usually *pillow lava*, consisting of *albite* and *chlorite*. Probably represents a strongly altered *basalt*.

spilite suite The association of *spilite*, *keratophyre* and *quartz keratophyre*, believed to represent altered *mafic*, *intermediate* and silicic *volcanic rocks* respectively.

spillway A *glacial* drainage channel.

spinel ($MgAl_2O_4$) The manganese end-member of the *spinels*.

spinels A group of isostructural *oxide minerals* with the general formula M_3O_4.

spinifex texture A *texture* characterized by large, randomly oriented or subparallel, skeletal, plate or lattice *olivine* grains, or *acicular pyroxene* grains. Formed by the rapid *crystallization* of *ultramafic* liquids and common in *komatiites*.

spinner magnetometer A *magnetometer* for the determination of *remanent magnetizations* in which a sample is rotated within a magnetic detecting system. The generated signal's amplitude is proportional to the *intensity of magnetization* and the phase directly related to its direction.

spinoidal decomposition *Exsolution* without a *nucleation* step, which can occur in a *solid solution* with an unstable composition.

Spiriferida An order of class *Articulata*, phylum *Brachiopoda*; *brachiopods* with biconvex, usually impunctate shells with broad hinge lines. Range U. *Ordovician–Jurassic*.

spit A long, narrow accumulation of *beach deposits* with one end attached to the shore and the other projecting into a large body of water, forming as a result of *longshore drift*'s carrying sediment beyond an abrupt change in coastline orientation.

splay fault See *second order fault*.

splendent lustre (specular lustre) *Lustre* of the highest intensity caused by an intense reflection of light.

splintery fracture A *fracture* with a splinter-like appearance.

spodumene ($LiAlSi_2O_6$) A *pyroxene* found in lithium-rich *pegmatites*.

sponges See *Porifera*.

spontaneous magnetization A natural magnetization exhibited by *ferromagnetic* substances in the absence of an external *magnetic field*.

spontaneous polarization See *self-potential*.

spontaneous polarization method (self-potential method, SP method) A *geophysical exploration* method making used of the naturally produced potential differences at the surface generated by electrochemical interactions between an *orebody* and *groundwater* at depths down to ~30 m.

sporinite A *coal maceral* of the *exinite* group comprising coalified spores and pollen.

SPOT see *Satellite Probatoire pour l'Observation de la Terre*.

spread A pattern of *seismographs* which record a single *shot*.

spreading rate The rate of opening of an ocean by *seafloor spreading*. cf. *half spreading rate*.

spreiten *Trace fossil* structures comprising repeated, 'U'-shaped, blade-like, sinuous or spiral sedimentary *laminae* caused by intense sediment reworking by animal locomotion, excavation or excretion.

spring A localized flow of water from an *aquifer* where the *water table* intersects the surface.

spring sapping *Sapping* by *groundwater* outflow at a *spring*.

spur-and-groove topography Undulating morphology on the front of a *coral-algal reef*.

Squamata An order of subclass *Lepidosauria*, class *Reptilia*; the lizards and snakes. Range U. *Triassic–Recent*.

Squamariaceae A group of red *algae*. Range U. *Carboniferous–Recent*.

squid magnetometer See *cryogenic magnetometer*.

Srbsko A succession in Czechoslovakia equivalent to the upper part of the *Devonian*.

stable sliding *Displacement* on a *fault* at a constant or slowly varying *strain rate*. cf. *stick-slip*.

stable tectonic zone A region of the *continental crust* characterized by a lack of *tectonic* activity

stack 1. A *seismic record* produced by adding together (*stacking*) a number of different *seismograms*. 2. A coastal pillar of rock above the high tide level. cf. *stump*.

stacking The process of adding together a number of different *seismograms* to produce a *stack*.

stacking fault A *planar crystal defect* in which there is an error in the regular stacking sequence of layer modules which lies parallel to the plane of the layer.

stacking velocity The *seismic velocity* obtained from *normal moveout* measurements using *common depth point seismograms*.

stadial A short cold period with a smaller volume of ice than in a *glacial*.

stage 1. The water depth or elevation of the water surface at a location on a river system. 2. A fourth order *chronostratigraphic* unit.

stage hydrograph A graph of river *stage*.

stagnation deposit A deposit formed under conditions of stagnation, which may inhibit the decay of organic matter and give rise to an *exceptional fossil deposit*.

staining A simple chemical technique used to distinguish *minerals* in *thin section* by painting it with an organic or inorganic reagent. Particularly useful in distinguishing the different *carbonate minerals*, whose optical properties are similar.

staircase trajectory See *stairstep trajectory*.

stairstep trajectory (staircase trajectory) A *fault* surface comprising *flats* and *ramps*.

stalactite A *speleothem* composed of calcium carbonate which grows downwards from the roof of a *cave*.

stalagmite (dripstone) A *speleothem* composed of calcium carbonate which grows upwards from the floor of a *cave*.

Standard Mean Ocean Water See *SMOW*.

standing wave A stationary surface *wave*.

stanniferous Containing *tin*.

stannite (bell-metal ore, tin pyrites) (Cu_2FeSnS_4) An *ore mineral* of tin found mainly in *hydrothermal vein* deposits and more rarely in *pegmatites*.

star dune A pyramidal *dune* with three arms radiating from a high central dome.

starlite A *gem* variety of *zircon* produced by heating in a reducing environment to provide a blue colour.

stasis A period of no evolutionary change.

Stassfurt Evaporites A *Permian* succession in NW Europe covering the middle part of the *Wordian*.

stassfurtite A massive variety of *boracite*.

static correction A constant time correction applied to seismic data to compensate for the different elevations of *shots* and detectors, and for the varying thickness and velocity of the *weathered layer*.

staurolite ($(Fe,Mg)_4Al_{17}(Si,Al)_8O_{44}(OH)_4$) A *nesosilicate* found in aluminium-rich rocks subjected to *regional metamorphism*.

steady-state creep *Creep* which takes place without any change in the physical state of the material.

steatite See *soapstone*.

steinkern The sediment infilling the space left by a shell after its removal by *dissolution*, which provides an internal and external cast of the shell.

Steinmann trinity The association of *spilite*, *serpentine* and *radiolarian chert* found in deep-sea sediments.

STEM See *scanning transmission electron microscopy*.

step A *fault* surface feature approximately orthogonal to its *lineations* and pointing either towards or opposite to the sense of motion of the block sliding over them forming *wear grooves* and *growth fibres* respectively.

step faults A set of parallel *normal faults* which *dip* and *downthrow* in the same direction.

Stephanian A *Carboniferous stage* in NW Europe equivalent to the *Kasimovian* and *Gzelian.*

stephanite (Ag_5SbS_4) A rare silver *ore mineral* commonly found in *veins* associated with other silver *minerals.*

stepout See *moveout.*

steptoe An isolated hill surrounded by *lava.*

stereogram See *stereographic projection.*

stereographic net A type of protractor used in *stereographic projection* which gives the *cyclographic traces* of a complete set of *great circles* about a common axis and the *cyclographic traces* of a family of *small circles* about the same axis, which are used to graduate angles along the *great circles.* See also *Schmidt net, Wulff net.*

stereographic projection (stereogram) A graphical method of showing three-dimensional geometrical data in two dimensions and of solving three-dimensional geometrical problems, particularly those involving the angular relationships between lines and planes.

stereoscope An instrument for viewing *aerial photographs* which overlap by ~60% that allows a viewer with binocular vision to obtain an apparently three-dimensional image of the topography.

Sterlitamakskiy A *Permian* succession on the eastern Russian Platform equivalent to the U. *Sakmarian.*

stibnite (antimonite, antimony glance) (Sb_2S_3) The major *ore mineral* of *antimony* found mainly in low-temperature *hydrothermal vein* or *replacement deposits.*

stichtite ($Mg_6Cr_2CO_3(OH)_{16}.4H_2O$) A lilac or pink, hydrated magnesium-chromium carbonate found in *serpentinite* and commonly associated with *chromite.*

stick-slip *Fault* behaviour comprising alternations of slow *strain rate* and rapid *slip* when the *stress* overcomes the static *coefficient of friction* on the *fault plane.* cf. *stable sliding.* See also *elastic rebound theory.*

stiffness See *Young's modulus.*

stilbite (desmine) ($CaAl_2Si_7O_{18}.7H_2O$) A *zeolite* found mainly in *amygdales* and cavities in *basalt, andesite* and related *volcanic rocks,* but also as a *hydrothermal mineral* in crevices in *metamorphic rocks,* in cavities in *granite pegmatites* and in some *hot spring* deposits.

stilpnomelane(~$K_{0.6}(Mg,Fe^{2+},Fe^{3+})_6Si_8Al(O,OH)_{27}.2\text{–}4H_2O$) A *phyllosilicate* found widespread in *schists* in association with *chlorite, epidote, lawsonite* and other *minerals* and in iron *ore* deposits.

Stinkschiefer See *Hauptdolomit.*

stishovite (SiO_2) A high pressure (>10 GPa) form of *silica* found in areas such as *meteorite* impacts.

stock An *igneous body* smaller than a *batholith* with a subcircular section.

stockwork A closely-spaced, interlocking network of veinlets, commonly in and around intermediate *plutonic* igneous intrusions. An important control of various forms of disseminated mineralization.

Stokes law A law controlling the settling of a particle in a fluid: $v = (2r^2 g \Delta\rho)/(9\eta)$, where v = settling velocity, r = radius, g = *gravity*, $\Delta\rho$ = *density contrast* with fluid, η = *viscosity* of fluid.

Stokes surface A subhorizontal *aeolian bounding surface* produced by *deflation* over all or part of an *erg*, which causes scouring of the *cohesive* or cemented *sand* at the *water table*.

stolzite ($PbWO_4$) A tungstate *mineral* found as a *secondary mineral* in tungsten-bearing *ore* deposits.

stone lattice See *alveoles*.

stone-line A horizon of angular *gravel* and *cobbles* within the soil zone, often running parallel to the surface. Probably a *lag* deposit originally formed at the surface by the removal of fine particles by wind and wash *erosion* and subsequently buried.

stone pavement A flat desert area covered with a surface layer of rounded *pebbles* or *gravel*, often a *lag* deposit caused by the removal of fine particles by wind and wash *erosion*.

stonefield See *blockfield*.

stony iron meteorite (siderolite) A *meteorite* comprising metal and silicate.

stony meteorite (aerolite) A *meteorite* formed solely of rock-forming silicates.

stoop and room mining See *room and pillar mining*.

stope An excavation within a mine made to extract *ore*.

Stopes-Heerlen system A system for the classification of *coal macerals* on the basis of physical appearance, chemistry and biological affinity into three groups: *vitrinite*, *exinite* and *inertinite*.

stoping The mechanical incorporation of blocks of *wall rock* into a *magma* during its ascent, an important mechanism in the *emplacement* of *plutons*.

storage coefficient (storativity) The volume of water given up per unit horizontal area of an *aquifer* and per unit fall of the *water table*. Equal to *specific yield* for an *unconfined aquifer*, but dependent on the elastic compression for a *confined aquifer*.

storativity See *storage coefficient*.

stored deformation energy The potential energy within a deformed material, stored by the *elastic* distortion of a *crystal lattice* or by imperfections in the lattice. The driving force in the process of *recovery* and primary *recrystallization*.

storm beach See *dissipative beach*.

storm deposit (tempestite) A sediment body deposited as a result of one storm event.

storm setup The build-up of water against the coast during a storm.

storm surge A localized elevation of sea level by extreme meteorological events, caused by *storm setup* and extremely low atmospheric pressure.

storm surge ebb The return of waters after a *storm surge*.

stoss side The upstream side of a body where the fluid velocity and *shear* *stresses* commonly increase as the flow moves over the body. cf. *lee side*.

stoss slope The upstream slope of a body where the fluid velocity and *shear stresses* commonly increase as the flow moves over the body. cf. *lee slope*.

Strahler system A classification of *stream order* whereby a stream of order $n + 1$ is initiated at the confluence of two streams of order *n*, so that entry of a stream of lower order does not increase the order of the main stream.

straight extinction See *parallel extinction*.

strain A change in shape or internal arrangement of rock caused by *tectonic* activity.

strain axes The orientations of the *principal strains*.

strain band A planar zone of high *strain* in a set of asymmetrically folded layers corresponding to the superimposed short *limbs* of the *folds* and bounded by the *axial surfaces* of the *folds*.

strain ellipse The shape assumed by a unit circle after *deformation* which facilitates visualization of the shape change associated with the *strain*.

strain ellipsoid The shape assumed by a unit sphere after *deformation* which facilitates visualization of the shape change associated with the *strain*.

strain hardening 1. A phenomenon occurring during *dislocation climb* in which *dislocations* and *crystal defects* impede the movement of a *dislocation* in its own lattice plane so that *deformation* becomes increasingly more difficult. cf. *strain softening*. 2. The hardening of a *fault plane* caused by *minerals* precipitating on it.

strain increment Each small change in the shape of a body within a more extensive *deformation* sequence.

strain marker An object of known initial shape, geometry or shape distribution present in a deformed rock which allows *strain* to be measured, e.g. *fossils*, worm burrows, *ooliths*.

strain parallelepiped The shape taken by a unit cube after a vanishingly small *strain increment*; a geometrical construction used in the analysis of *infinitesimal strain*.

strain path A line showing the successive states of *strain* only during *deformation*. cf. *deformation path*.

strain rate The rate of change of *strain* with time, in units of time^{-1}.

strain-slip cleavage See *crenulation cleavage*.

strain softening (work softening) A phenomenon whereby *deformation* occurs at an increased rate as *strain* increases, or a decrease in the applied *stress* is needed for constant rate *deformation* as *strain* increases. cf. *strain hardening*.

strain superposition The application of successive *strain increments* to a rock body to produce the observed *finite strain*.

strandflat A partly submerged, undulating, rocky lowland close to sea level, probably formed either by frost action on a shoreline or by marine and *glacial erosion*.

strandline A shoreline, commonly an ancient one.

strandplain A *beach* of multiple *beach ridges* formed when the rate of coastal sedimentation kept pace with a slow rate of sea level rise.

strata-bound mineral deposit A deposit restricted to a small *stratigraphic* range in a group of *strata*.

stratification map A map showing the distribution of *dunes*, *interdunes* and *aeolian sand sheets* in a section through an *aeolian* deposit, used in combination with *dip* measurements to infer the nature of the *bedform*(s) which deposited the sediment.

stratiform mineral deposit A deposit with a large lateral extent parallel to the principal planar *structure* of the *host rock*, such as the *bedding* or an igneous *lamination*, but of limited thickness.

stratigraphic trap An *oil or gas trap* resulting from lateral and vertical variations in the *stratigraphy* of the *reservoir rocks* in terms of thickness, *texture*, lithology and *porosity*. cf. *structural trap*.

stratigraphy The study of layered *sedimentary* or *metamorphic rocks*, especially their relative ages and correlation between different areas.

stratotype The type representative of a particular *stratigraphic* unit or boundary.

stratovolcano (composite volcano) A *volcano* comprising layers of *lava* and *tephra* of dominantly *andesitic* composition increasing in thickness towards a central vent.

stratum See *bed*.

stratum contour See *structure contour*.

streak The colour of a *mineral* in finely powdered form, used in *mineral* identification.

streak plate A piece of unglazed white porcelain on which a *mineral* is rubbed so as to determine its *streak*.

stream flow (discharge) The volume of water passing a specified point in a specified period of time, measured by estimates of flow velocity and stream cross-sectional area or by use of a permanent structure which provides a set relation between *stage* and *discharge*.

stream order A topological classification of the links within a stream network. See also *Strahler system*, *Shrieve's magnitude*.

stream tin *Cassiterite* occurring as detrital grains in a *placer deposit*.

streamer An array of *hydrophones* towed behind a ship.

streamlines Lines depicting the motion of a fluid, drawn tangential to the flow direction at any point.

strength (yield strength) The maximum *principal stress* that a material can bear before *failure*.

stress The *force* per unit area acting on the surface of a solid plus the equal and opposite reaction of the material.

stress components The nine quantities which describe the state of *stress* at a point, the *normal* and *shear stresses* on the faces of an infinitesimally small cube around the point.

stress deviator A component of *deviatoric stress*.

stress ellipsoid A construction showing the complete variation in *stress* with direction, comprising an imaginary ellipsoid with axes parallel and proportional to the directions and magnitudes of the three *principal stresses*, the length of any radius of which gives the magnitude of the *normal stress* on the *conjugate* plane to the radius.

stress invariants The three quantities I_{1-3}, defined $I_1 = \sigma_1 + \sigma_2 + \sigma_3$; $I_2 = -(\sigma_{12} + \sigma_{31} + \sigma_{23})$; $I_3 = \sigma_1\sigma_2\sigma_3$, where $\sigma_1, \sigma_2, \sigma_3$ are the maximum, intermediate and minimum *principal stresses* respectively.

stress trajectories Lines giving the orientations of the *principal stresses*.

stretch A linear measure of *strain*; the ratio of deformed length to undeformed length, and equal to one plus the *elongation*. Stretch is >1 for *extension*, <1 for *contraction*.

stretching direction The direction of maximum *extension* in a given plane, typically a *cleavage* plane.

stretching lineation A *lineation* in the *stretching direction* produced by the *elongation* of minute grains, grain *aggregates* or *fossils* on a *cleavage* surface, which can be used to determine the *principal strain axes*.

strewnfield An area in which a specific group of *tektites* and micro*tektites* are found which are distinct in their chemistry and probably represent a particular impact event.

striation A marking on the surface of a *pebble* or *bedrock* produced by ice movement.

strike The direction of a horizontal straight line constructed on an inclined planar surface, at a direction of 90° from the *true dip* direction.

strike fault A *fault* parallel to the *strike* of the *bedrock*, *bedding* or *cleavage*.

strike line A straight line on a geological map parallel to the *strike* of a rock unit or *structure*. cf. *structure contour*.

strike separation The offset of a planar *structure* in a horizontal direction parallel to a *fault*.

strike-slip The *displacement* parallel to the *strike* of the *displacement plane*.

strike-slip duplex A part of a *strike-slip fault system* where there are several, subparallel bend segments of *faults*.

strike-slip fault A *fault* with a dominant component of *strike-slip*.

strike-slip fault system (wrench fault system) A set of related *faults* on which the individual *displacements* produce a net *strike-slip displacement* on the system as a whole.

strike-slip tectonic regime A *strike-slip fault system* associated with a *conservative plate margin* or within a continental *plate*.

strip mining A type of *open pit mining* in which strips of land are worked, with the excavation being backfilled as mining proceeds.

stripping ratio The ratio of waste to *ore* removed in *open pit mining*.

stromatolite A laminated, calcareous, microbial structure, formed principally by *cyanobacteria* and *algae*. One of the oldest known *fossils*, first occurring in the *Archaean*.

stromatoporoids Problematic, extinct organisms with a mesh-like skeleton forming sheet-like domes or discoidal or dendroid masses, often forming *reefs* with *corals*. Especially common in *Ordo-*

vician to *Devonian* carbonate sediments. Range *Cambrian–Cretaceous.*

strombolian eruption A *volcanic eruption* characterized by continuous small explosions.

stromeyerite ($(Ag,Cu)_2S$) A rare *ore mineral* of silver and copper.

stromotactis A type of cavity structure of uncertain origin with a smooth sediment floor and irregular roof, the infilling cement usually being of fibrous and/or *drusy calcite.*

strontianite ($SrCO_3$) A *carbonate mineral* exploited as a source of strontium.

strontium isotope analysis A method of analysis by *mass spectrometer* in which strontium isotopes are used to distinguish the provenance of sources of *obsidian, alabaster* and *marble* used in antiquity.

Strophomenida An order of class *Articulata*, phylum *Brachiopoda*; *brachiopods* with convex pedicle *valves* and, usually, a planar or concave brachial *valve.* The hinge is broad and straight and the pedicle foramen usually closed. Range L. *Ordovician*–L. *Jurassic.*

structural geology The branch of the Earth sciences dealing with rock *structures* formed by *deformation.*

structural terrace A local flattening in an area of generally more inclined *strata.*

structural trap An *oil or gas trap* resulting from the *structure* of the *reservoir* and *seal* rocks, commonly *anticlines, faults* and *piercement structures* such as *salt domes.* cf. *stratigraphic trap.*

structure Any geological feature that can be defined geometrically.

structure contour (stratum contour) A curved line on a geological map which follows a constant height on a geological surface, i.e. parallel to *strike* at each point on the line. cf. *strike line.*

structure contour plan A map showing the *structure contours* of an *orebody* in horizontal or vertical plane projection for steeply dipping deposits, constructed for mining purposes.

structure grumeleuse A *limestone texture* characterized by *micrite* clots surrounded by coarser, *granular calcite* or *microspar.* Probably the result of selective *recrystallization* in which larger grains grow at the expense of smaller ones.

struvite ($NH_4MgPO_4.6H_2O$) Magnesium-aluminium phosphate, found in *guano.*

Stump A *Jurassic* succession in Utah/Idaho, USA, covering parts of the *Callovian* and *Oxfordian.*

stump A coastal pillar of rock below the high tide level. cf. *stack.*

Sturtian The older *period* of the *Sinian*, 800–610 Ma.

stylolite A surface within a rock along which *dissolution* by *pressure solution* has occurred, common in carbonate rocks and often marked by seams of insoluble *clay minerals* which remained after dissolved *calcite* had moved away.

sub- Under.

sub-bituminous coal A *brown coal*, with a *calorific value* of < 19.3 MJ kg^{-1} and a *fixed carbon* content of 46–60%.

subalkaline Containing no alkali *minerals* other than *feldspars.*

subarkose A *sandstone* with insufficient *feldspar* to be termed an *arkose.*

subautomorphic See *subhedral.*

subcretion See *underplating.*

subcritical translatent stratification A *wind ripple lamination* produced in fine to medium *sand* by *ripples* climbing at an angle below the angle necessary for the preservation of *stoss side* deposits. Each *ripple* deposits a millimetric-scale, planar, *tabular, inverse graded* lamina above an *erosion* surface so that the *lamination* is very well defined. cf. *supercritical translatent stratification.*

subcrop 1. A subsurface *outcrop*, e.g. where a formation intersects a subsurface plane such as an *unconformity*. 2. In mining, any near-surface development of a rock or *orebody*, usually beneath superficial material.

subduction The process of underthrusting an oceanic *plate* into the *mantle* at a *destructive plate margin.*

subduction complex See *accretionary prism.*

subduction orogeny An *orogeny* at a *subduction zone* characterized by the formation of a *paired metamorphic belt* and *calc-alkaline* volcanic activity.

subduction suction (trench suction) The mechanism whereby the overriding *plate* in a *subduction zone* is thrown into tension, caused by the downgoing slab's descending at an increasing angle with depth near the surface.

subduction zone A region where oceanic *lithosphere* descends into the *mantle*, ideally characterized by the following set of features: *lithospheric* bulge, *trench* with *accretionary complex*, *forearc basin*, sedimentary arc, volcanic *island arc*, *back-arc basin* and *back-arc ridge.*

suberinite A *coal maceral* of the *exinite* group made up of the corky cells of *plants.*

subgrain A region of a *crystal lattice* which differs in orientation from the surrounding *mineral* and bounded by *crystal defects*, but without internal distortion.

subhedral (hypidiomorphic, subautomorphic) Exhibiting some traces of crystal form.

subjacent Bottomless.

sublittoral See *neritic.*

submarine canyon A steep-sided, canyon-like trench cut into the continental shelf and sometimes cutting across it. Probably related to a past or present river valley.

submarine fan (deep-sea cone, deep-sea fan, deep-water fan) A broad, convex-upwards, deep water, *siliciclastic* or carbonate system with positive relief above the adjacent *basin* floor, developing from a localized source, such as a *submarine canyon.* Generally of radial geometry from 10–3000 km wide.

submerged bar See *offshore bar.*

submerged forest An area of forest vegetation, usually a layer of *peat* with tree stumps in their growth positions, which has been inundated by the sea. Indicative of a sea level rise since growth.

submetallic lustre A *lustre* of a *mineral* intermediate between *metallic* and *nonmetallic.*

submicroscopic Of a size smaller than *microscopic*, imaged by techniques such as *electron microscopy*.

subophitic texture An *ophitic texture* in which the enclosure of *plagioclase* crystals by *pyroxene* is only partial.

subplinian eruption A less violent type of *volcanic eruption* than a *plinian eruption*, producing less widely-dispersed *tephra*.

subsequent dolomitization *Dolomitization* subsequent to the lithification of a *limestone* by the entry of magnesian fluids along *joints* and *faults*. cf. *penecontemporaneous dolomitization*.

subsequent stream A stream which follows a course controlled by the structure of the local bedrock. cf. *anteconsequent, inconsequent, insequent, obsequent* and *resequent streams*.

subsolidus A chemical system below its melting point and in which reactions are solid state.

subsolvus granite A *granite* which crystallized below the *solvus* temperature and thus contains two types of *alkali feldspar*, K-rich and Na-rich. cf. *hypersolvus granite*.

substage A fifth order *chronostratigraphic* unit.

subsurface flow See *throughflow*.

subsurface fluid migration The motion of *pore fluids* within a sediment along permeable *carrier beds* from regions of high pressure to regions of low pressure at a rate dependent on the sediment's *permeability*, the type of fluid, the relative balance of opposing flows, the sedimentary geometry, the *hydraulic conductivity*, the *aquifer* gradient and the overall hydrodynamic regime.

subsurface mapping The production of subsurface geological maps based almost totally on borehole data, including *geophysical borehole logging*.

subsurface stormflow See *interflow*.

subtidal Descriptive of the part of the *tidal-flat* environment below the normal level of mean low water of spring *tides*. cf. *supratidal*

Suchanian A *Cretaceous* succession in the far east of the former USSR equivalent to the *Hauterivian, Barremian, Aptian* and *Albian*.

sudoite ($Mg_2(Al,Fe^{3+})_3Si_3AlO_{10}(OH)_8$) A rare *phyllosilicate* of the *chlorite* group found disseminated in some *haematite* deposits and as a *hydrothermal mineral*.

suevite A *breccia* formed by *shock metamorphism*, whose angular fragments are set in a *glass matrix*.

suffosion The *erosion* of unconsolidated surface sediment by slumping into an underground cavity produced by bedrock *dissolution*.

Sui Ning A *Jurassic* succession in Sichuan, China, equivalent to the *Oxfordian* and *Kimmeridgian*.

sulphate reduction An important, bacterially-controlled process in fine-grained marine sediments during early *diagenesis*, a major control on the fate of detrital iron, since iron sulphide is highly insoluble.

sulphide minerals Metal sulphide compounds which make up the single most important group of *ore minerals*. Classified by *crystal structure* into ten major groups: *disulphide, galena,* *sphalerite, wurtzite, nickel arsenide, thiospinel, layer sulphides, metal excess, ring structure, chain structure.*

sulphur (S) A *mineral* occurring as a *native element*, common in volcanic regions.

sunstone A *translucent* variety of *aventurine oligoclase* with a red glow formed by minute, platy inclusions of *haematite* in parallel orientation.

super- Over, above.

supercontinent A large grouping of continents, e.g. *Pangaea.*

supercritical flow A water flow whose velocity is greater than the velocity of propagation of a long surface wave in still water.

supercritical translatent stratification A rare *wind ripple lamination* occurring when rates of deposition are sufficiently high to preserve *stoss side* deposits. Lamina contacts are gradational. cf. *subcritical translatent stratification.*

supergene Descriptive of processes involving water percolating down from the surface.

supergene enrichment (secondary enrichment) An increase in the *grade* of mineralization by processes subsequent to the primary processes of formation, especially applicable to the enrichment of sulphide *ores*, and also oxide and carbonate *ores*. Metals are usually carried down into the *ore*, leaving an upper *gossan*, where they precipitate and increase the metal content, or *gangue minerals* are mobilized and carried away.

supergroup The largest *lithostratigraphic* subdivision, comprising a series of *groups*.

superimposed drainage (epigenetic drainage) A drainage pattern unrelated to the existing *bedrock* due to its initial development on a rock cover subsequently removed by *denudation.*

superimposition of structures The *deformation* of existing *structures*, with the exception of *faults*.

Superior-type iron formation A type of *banded iron formation* comprising thinly banded rocks of the oxide, carbonate and silicate facies and free of clastic material. The banding is of iron-rich and iron-poor *chert* layers of centimetric- to metric-scale and can be followed in a layer up to several hundred metres thick over wide areas, associated with *quartzite* and *black shale*.

superparamagnetism The phenomenon in which *ferromagnetic* particles <0.5 mm in size can retain a remanence for only a few minutes.

superplasticity A *deformation mechanism* involving sliding along grain boundaries, accommodated by *diffusion*, or by *solution transfer*.

superposition of strata The basic *stratigraphic* principle that each *bed* in a stratified sequence is younger than the underlying *bed* and older than the overlying *bed*.

supra- Above, beyond.

suprafan The most rapidly *aggrading* part of a *sand*-rich *submarine fan*, which appears as a convex-up mound above the surrounding *fan*.

suprastructure The part of an *orogenic belt* deformed at a relatively shallow level. cf. *infrastructure*.

supratidal Descriptive of the part of the *tidal-flat* environment above the normal level of mean high water of spring *tides*. cf. *subtidal*.

surf The mass of broken, foaming water that forms as a wave breaks on the shore.

surf zone The area between the *breaker zone* and the *swash zone*, characterized by bores of shoreward-moving water from swirlling or plunging breakers.

surface wave (seismic surface wave) A *seismic wave* which travels along the top of a half-space, including *Rayleigh waves* and *Love waves*.

surge An inflated, dilute, *turbulent flow* of gas and *pyroclasts*.

surtseyan eruption A *volcanic eruption* of *mafic magma* into shallow water (1–2 km) causing explosive *magma*-water interactions in an almost continuous series of blasts, producing characteristic clouds in the shape of a cock's tail. The *tephra* produced is fine-grained and deposited as finely-bedded *base surge* and air-fall *beds*, occasionally interbedded with massive *beds* similar to *mudflows*, and a *tuff cone* or *ring* is formed.

survivorship curve A graph of the number of survivors of a single age group against age obtained in a *palaeoecological* study from the size-frequency distribution of individuals.

suspect terrane A *terrane* for which there are grounds for considering it to be a *displaced terrane*.

suspended load The solid material transported in a fluid which is not in contact with the channel bed and supported by the vertical component of velocity or *turbulent* fluid eddies.

suspension The transport of sediment within a fluid body, possible when upward fluid velocities exceed the *settling velocity* of the particles.

suture A surface within the *lithosphere* representing the contact between two *plates* or *microplates* at a *continent–continent collision*.

Svecofennian orogeny An *orogeny* affecting the Baltic *shield* in the *Proterozoic* at ~1700 Ma.

Sveconorwegian orogeny See *Dalslandian orogeny*.

swale 1. An area of low-lying, often marshy, land. 2. A shallow trough between storm ridges on a *beach*.

swaley cross-stratification A variety of *hummocky cross-stratification* in which hummocks are rare or absent and *swales* are preferentially preserved.

swallet (swallow hole) A vertical to steeply-sloping shaft in a *limestone* area down which surface water disappears underground.

swallow hole See *swallet*.

swallowtail twinning *Twinning* in which a crystal divides along a *twin plane* producing a *twin* with a 'V' shape.

swamp A waterlogged area with characteristic vegetation in both coastal and inland environments.

swash The shoreward-moving uprush of *turbulent* water formed as a wave breaks on a *beach*, with a velocity generally greater than the *gravity*-driven *backwash*, which moves coarse sediment onshore.

swash bar (longshore bar) A *sand* ridge on a *beach*, probably formed by the steepening of a low gradient foreshore within the *swash zone* by constructional waves.

swash zone The portion of a *beach* alternately covered by *swash* and exposed by *backwash*.

Swazian An *era* of the *Precambrian*, 3500–2800 Ma.

SWEAT hypothesis See *South West US and East Antarctica hypothesis*.

sweep The inrush of water that penetrates downwards into the *viscous sublayer* of the *turbulent boundary layer* with a velocity higher than the time mean average and a vertical component towards the *bed*.

sweet crude oil *Crude oil* low in *sulphur*. cf. *sour crude oil*.

swell beach See *reflective beach*.

swirlhole See *pothole*.

syenite A *phaneritic alkaline igneous rock* comprising *alkali feldspar* exceeding 67% of the total *feldspar* and *accessory quartz* or *nepheline* which makes up 5–20% of the rock. Compositionally equivalent to *trachyte*. Found in *alkaline* intrusive complexes.

syenodiorite See *monzonite*.

syenogabbro A coarse-grained *igneous rock* comprising *essential alkali feldspar* and Ca-rich *plagioclase* in equal proportions, *augite*, *biotite* and *accessory apatite*.

syenoid A *syenitic* rock in which *feldspathoids* take the place of *alkali feldspar*.

Sygynkanskaya A *Triassic* succession in Siberia equivalent to the *Spathian*.

sylvanite (yellow tellurium) ($(Au,Ag)Te_2$) A rare *ore mineral* of gold and silver found in low- to high-temperature *vein* deposits.

sylvine See *sylvite*.

sylvinite A mixture of *sylvite* and *halite*.

sylvite (sylvine) (KCl) An *evaporite mineral*.

Symmetrodonta An order of infraclass *Trituberculata*, subclass *Theria*, class *Mammalia*; small, probably predatory *mammals* which may be ancestral to the marsupial and placental *mammals*. Range *Jurassic*–early *Cretaceous*.

sympatric speciation The evolutionary divergence of populations of the same area into distinct species.

symplectic texture A *texture* produced by the intergrowth of two *minerals*. e.g. *graphic texture*, *ophitic texture*, *poikilitic texture*.

symplectite A textural feature of some *gabbroic* rocks in which bulbous, *myrmekite*-like extensions of *plagioclase* crystals contain *vermicular* inclusions of *orthopyroxene*. Formed during the late stages of *crystallization* or *recrystallization*.

syn- Together, with, resembling.

synaeresis crack See *syneresis crack*.

Synapsida A subclass of class *Reptilia*; *mammal*-like *reptiles*. Range *Carboniferous*–*Jurassic*.

syncline A *fold* with the younger rocks in its core.

synclinorium A large, composite *syncline* made up of smaller *folds*. cf. *antilinorium*.

syndiagenetic Descriptive of a *mineral* deposit formed during the early or late stages of *diagenesis* of a sediment.

syneresis crack (synaeresis crack) A subaqueous shrinkage crack formed in response to *dewatering* of a *mud* layer as it compacts under its own weight, as flocs break down or as some *clay minerals* shrink when overlying waters change *salinity*.

synform A *fold* with a downwards *closure*.

syngenetic deposit A deposit forming at the same time as its *host rocks*.

synkinematic See *synorogenic*.

synneusis The moving together and attachment of crystals in a *magma*. The mechanism behind the formation of *glomeroporphyritic texture*.

synoptic profile The degree of original relief of an organism such as a *stromatolite*.

synorogenic (synkinematic, syntectonic) Descriptive of a process or event taking place at the same time as *orogenic deformation*.

synsedimentary fault See *growth fault*.

syntaxial growth See *syntaxial vein*.

syntaxial vein (syntaxial growth) A *vein* filling which grows from the walls towards the centre, characterized by host and veinfill having different compositions, a lack of crystallographic continuity across the *vein*, an oblique contact of veinfill and *vein* wall, and single *crystallographic* veinfills and inclusions of *wall rock* fragments lie along the median line and in subparallel bands. cf. *antitaxial growth*.

syntaxy The cement on a crystal growing in *crystallographic* continuity with its substrate, forming a single crystal.

syntectite A rock formed by *syntexis*.

syntectonic See *synorogenic*.

syntectonic sediment A sediment deposited at the same time as *tectonism* affected the depositional area, e.g. the *erosion* of a *thrust sheet* forming a sedimentary deposit in a *foreland basin*. The sediment need not show contemporaneous *deformation*.

syntexis *Magma* generation by the melting of at least two rock types. cf. *anatexis*.

synthetic Descriptive of a *structure* or *fabric* with the prevailing orientation or *vergence*. cf. *antithetic*.

synthetic seismogram A computer-generated *seismogram* obtained by using mathematical expressions which describe the propagation of *seismic waves* through a particular *structure*, used as an aid in the interpretation of seismic data.

synthetic shear See *Riedel shear*.

Syrian garnet A variety of *almandine* of *gem* quality.

system A second order *chronostratigraphic* unit.

Syzranian A *Palaeocene* succession in the former USSR, covering the lower part of the *Danian*.

T

T See *tesla.*

T wave A *P wave* trapped in the ocean.

T-ΔT method A statistical technique for determining the *root-mean-square velocity* from *seismic reflection* data.

table mountain A large *mesa.*

table reef A large-scale *coral-algal reef.*

tabular Descriptive of a crystal or feature with a broad, flat, commonly rectangular, form.

Tabulata A subclass of class *Anthozoa*, phylum *Cnidaria*; colonial *corals* with calcareous tubular *corallites*, prominent tabulae and inconspicuous or absent septae. Massive, foliacious, *dendroid*, phaceloid or creeping in habit. Range L. *Ordovician*–U. *Permian.*

tachylite A *basaltic glass*, a component of rapidly quenched *basaltic magma* in the chilled margins of *sills* and on the rims of *pillow lavas.*

Taconian orogeny See *Taconic orogeny.*

Taconic orogeny (Taconian orogeny) An *orogeny* affecting the northern parts of the *Appalachian fold belt* in the *Ordovician-Silurian*, corresponding to the *Caledonian orogeny* of Europe. See also *Acadian orogeny.*

taconite *Banded iron formation* suitable for the concentration of *magnetite* and *haematite* by fine grinding and *benefication.*

tadpole plot A method of plotting data from a *dipmeter log* with a dot on a depth-*dip* graph and a tail pointing in the *strike* direction.

tadpole rock See *rocdrumlin.*

Taeniodonta An order of infraclass *Eutheria*, subclass *Theria*, class *Mammalia*; a small group of North American herbivores of up to pig size of uncertain relationships. Range L. *Palaeocene*–U. *Eocene.*

taenite (FeNi) A natural alloy of iron and nickel found in *meteorites.*

tafoni Pits and hollows on rock surfaces, particularly *granites* in desert environments. Possibly form when salt *weathering* and wind scour small depressions on the surfaces of exposed rock, which enlarge when moisture accumulates in them. Some may form when a *case-hardened* surface is breached, exposing the softer rock beneath to *erosion.*

Taitai A *Cretaceous* succession in New Zealand covering part of the early *Cretaceous.*

talc ($Mg_3Si_4O_{10}(OH)_2$) A *phyllosilicate* formed by the *hydrothermal alteration* of *ultrabasic rocks* or the *thermal metamorphism* of siliceous *dolomites.*

talc schist A rock consisting mainly of *talc*, formed by the *regional metamorphism* of an *ultrabasic igneous rock.*

talnakhite ($Cu_9Fe_8S_{16}$) A *sulphide mineral* of the *sphalerite group*.

talus See *scree*.

taluvium A hillslope deposit comprising coarse rubble and finer material.

tamarugite ($NaAl(SO_4)_2.6H_2O$) A hydrated sodium-aluminium sulphate usually found as a *secondary mineral*.

tangent-arc method (Busk method) A method of constructing a vertical *cross-section* through a set of *folds*, which assumes that they are *parallel*, by drawing normals to the *fold* surface at several points where *dips* are available and regarded as tangents to an arc of the *fold* surface. The normals intersect at the centre of curvature for that *fold* segment.

tangent modulus A variable parameter relating *stress* to *strain* for a perfectly elastic material where the *stress-strain* relationship is not necessarily linear.

tangential longitudinal strain A *fold mechanism* recognized by a pattern of layer-parallel extensional *strain* on the outer part of the folded layer and layer-parallel contractional *strain* on the inner arc, so that the maximum *principal strain* direction is tangential to the *fold* surface.

tangential velocity The velocity of a point on a rotating sphere, which varies from zero at the *pole of rotation* to a maximum at 90° to the pole. cf. *angular velocity*.

tagilite See *pseudomalachite*.

tantalite ($(Fe,Mn)Ta_2O_6$) An *ore mineral* of tantalum, found in *pegmatites*, *granitic* rocks and *alluvial deposits*.

tanzanite A blue, *gem* variety of *zoisite*.

taphonomy The study of the *post-mortem* history of an organism.

taphrogenesis Vertical movements of the *crust* causing the formation of major *faults* and *rift valleys*.

tapiolite ($(Fe,Mn)(Ta,Nb)_2O_6$) A tantalate of iron and manganese found in *albitized granite pegmatites* or in *placer deposits* derived from them.

tar See *asphalt*.

tar mat A layer of plastic, *malleable asphalt* at or near the surface or at former surfaces now represented by *unconformities*.

tar pit (pitch lake) A pool of *bitumen* which frequently contains the remains of animals trapped in it.

tar sand A bituminous *sandstone* or *limestone* impregnated with oil too viscous and heavy to be extracted by conventional drilling.

Taranaki A *Neogene* succession in New Zealand comprising the *Tongaporutian* and *Kapitean*.

tarbuttite ($Zn_2PO_4(OH)$) A hydrated zinc phosphate found as a *secondary mineral* in the oxidized zone of zinc *ores*.

Tarkhanian A *Miocene* succession on the Russian Platform covering parts of the *Langhian* and *Serravallian*.

tarn A small lake in a *corrie*.

Tastubskiy A *Permian* succession on the eastern Russian Platform equivalent to the L. *Sakmarian*.

Tatar A *Jurassic* succession in W Siberia covering parts of the *Bathonian* and *Callovian*.

taxa Plural of *taxon.*

taxon A group of organisms of any rank, e.g. family, genus, species.

taxonomic uniformitarian analysis A *palaeoenvironmental analytical* technique used when direct comparison with modern analogues is not possible. The tolerance for various environmental parameters of each *taxon* is estimated from that of a comparable living form and the value of each environmental parameter is estimated from the overlap of values from all *taxa* or by comparing the *fossil* community with the most comparable modern analogue.

Taylor A *Cretaceous* succession on the Gulf Coast of the USA equivalent to the *Campanian.*

Taylor number A dimensionless hydrodynamic parameter depending on the scale of a *convection* cell, its *kinematic viscosity* and the rate of rotation. If less than unity, rotation of the convecting system does not affect the pattern of the *convection.*

tear fault See *wrench fault.*

tectite A *glassy meteorite.*

tectogene A narrow, linear belt of downfolded rocks possibly related to mountain building.

tectogenesis See *orogenesis.*

tectonic 1. Descriptive of a *structure* produced by *deformation.* 2. Relating to a major Earth *structure* and its formation.

tectonic attenuation The thinning of *strata* by *extensional deformation.*

tectonic earthquake An *earthquake* resulting from *tectonic* activity. cf. *volcanic earthquake.*

tectonic erosion The process in which *erosion* results in the *detachment* of a sheet which then slides under *gravity.*

tectonic province A region of the *continental crust* distinguished by a characteristic *tectonic* history.

tectonic release The mechanism whereby the outward *pressure* from an explosion allows the regional *stress* to cause a small *earthquake* beneath, and simultaneous with, the explosion.

tectonic style The aspects of *fold profile* geometry used to characterize *folds* of different *type* and origin.

tectonic transport The direction of relative movement of a rock mass as the result of *tectonic* activity.

tectonics The processes responsible for *tectonic* activity.

tectonite See *S-tectonite, L-S tectonite.*

tectosilicate A silicate in which each silicon tetrahedron shares all four corners with others to form a three-dimensional framework.

tectosphere The stable, subcontinental zone in which *tectonics* originate.

tektite A *glassy* object probably formed by the melting of terrestrial material by the impact of an extra-terrestrial body, such as a *meteorite.*

Teleostei An order of subclass *Actinopterygii,* class *Osteichthyes,* superclass *Pisces*; the most recently-evolved bony *fish* with loosened bony articulations round the face and jaw bones so that the mouth can be used to suck prey from thick sediment. Range *Jurassic–Recent.*

telescoping A phenomenon observed in *hypothermal deposits* where steep temperature gradients cause the superimposition of zones that would have been distinct at deeper levels.

teleseism A *seismic event* recorded at an *epicentral angle* range of 30°–100°.

telethermal deposit A deposit formed at a very low temperature at a great distance from the source of its causative *hydrothermal solutions*.

telinite A *coal maceral* of the *vitrinite* group in which *microscopic* cell structure is visible.

telluric bismuth (tellurobismuth, wehrlite) (Bi_2Te_3) A natural alloy of bismuth and tellurium.

telluric current An electric current induced in the Earth by *geomagnetic field* variations, which gives rise to a variable horizontal potential gradient of ~-10 mV km^{-1}. The field can be used in *geophysical exploration* by seeking structures of anomalous *electrical conductivity* which distort this gradient and is effective to depths of some 7 km.

tellurite (TeO_2) A tellurium *oxide mineral* found in *hydrothermal vein* deposits.

tellurium (Te) A rare *native metal*.

tellurobismuth See *telluric bismuth*.

telogenesis *Diagenesis* which occurs when existing *eogenetic* or *mesogenetic pore fluids* are flushed from the host sediment by the ingress of *meteoric waters*.

Telychian A *stage* of the *Silurian*, 432.6–430.4 Ma.

TEM See *transmission electron microscopy*.

Temaikan A *Jurassic* succession in New Zealand covering the *Bajocian*, *Bathonian* and part of the *Callovian*.

Temnospondyli An order of subclass *Labyrinthodontia*, class *Amphibia*; *amphibians* with *fish*-like vertebrae seemingly designed to aid walking. Range L. *Carboniferous*–U. *Triassic*.

temperature log A *geophysical borehole log* in which downhole temperature gradients are measured for *geothermal* exploration purposes.

tempestite See *storm deposit*.

Templetonian A *Cambrian* succession in Australia covering the lower part of the *Solvan*.

temporary strain (recoverable strain) A transient shape change followed by reversion to the initial shape.

tenacity The resistance offered by a *mineral* to mechanical *deformation* or disintegration, depending on the nature of chemical bonding, the *crystal structure* and the *microstructure* of the *mineral*. See also *brittle mineral*, *ductile mineral*, *elastic mineral*, *flexible mineral*, *malleable mineral*, *sectile mineral*.

tennantite ($Cu_{12}As_4S_{13}$) An *ore mineral* of copper, *isomorphous* with *tetrahedrite* and found mainly in low- to high-temperature *hydrothermal ore veins*.

tenor See *grade*.

tenorite (CuO) A black, *supergene* copper *mineral* found mainly in the oxidation zone of copper deposits.

tensile Descriptive of a *force*, *stress* or *strain* which acts in a direction away from a reference point.

tensile fracture (extension fracture, tension fracture) A *fracture* developed under *tensile* conditions.

tensile strength The *strength* under *tensile stress.*

tensile stress The *stress* developed under *tensile* conditions.

tension The *stress* state produced by two equal *forces* acting in opposite directions from a reference point.

tension fracture See *tensile fracture.*

tension gash A linear opening in a rock developed under *tensile* conditions.

tension joint See *extension joint.*

tepee An overthrust sheet of *limestone* with the form of an inverted 'V' in section, found in tidal areas, in *calcrete* and around *playas* as a result of *deformation* or desiccation related to fluctuating water levels or changes in the nature of chemical *precipitation.*

tephra A *pyroclastic* material which was explosively fragmented during a *volcanic eruption*, comprising *ash*-sized *glass* shards, *pumice*, *scoria* and *lithic clasts* and *phenocryst*-derived crystals. Formed by the explosive decompression of *magma* on reaching the surface as dissolved *volcanic gases* rapidly *exsolve.*

tephrite An *olivine*-free, *alkaline volcanic rock* comprising *plagioclase*, *clinopyroxene* and *feldspathoids*. The presence of *olivine* causes gradation into a *basanite*, and of *alkali feldspar* into a *phonolite*. Occurs in *oceanic islands* and continental settings.

tephrochronology The study of widespread *tephra* deposits in order to establish age relations.

tephroite (Mn_2SiO_4) The manganese end-member of the *olivines* found mainly in iron-manganese *ore* deposits and their associated *skarns.*

Teratan A *Cretaceous* succession in New Zealand covering part of the *Turonian*, the *Coniacian* and part of the *Santonian.*

Terebratulida An order of class *Articulata*, phylum *Brachiopoda*; *brachiopods* with biconvex, punctate shells with a rounded hinge and pedicle foramen. Range L. *Devonian–Recent.*

terminal fan A type of *alluvial fan* in which the quantity of surface *runoff* decreases down the *fan* due to *infiltration* and evaporation so that normally little or no water leaves the *fan* as surface flow.

termite activity The contribution of termites to the *geomorphological* development of a region.

terra rossa A red *clay* soil developed on *limestone* associated with *karst* features in regions of strong seasonal variation in precipitation. Forms when *limestone dissolution* occurs in the wet season and *clay* and iron compounds are released. The red colour originates from the conversion of hydrated ferric oxides to *haematite* during the dry season.

terrace A nearly flat landform with a steep edge formed by a variety of processes.

terracette A small ridge or *terrace* which extends across a slope, usually at right angles to its direction. Possibly the expression of animal trackways or a consequence of soil instability on a steep slope.

terrain See *terrane.*

terrain correction See *topographic correction.*

terrane (terrain) A region of *crust* with well-defined margins, which differs significantly in *tectonic* evolution from neighbouring regions.

terrigenous See *siliciclastic.*

Tertiary A *period* of the *Cenozoic* comprising the *Palaeocene* to *Pliocene*, 65.0–1.64 Ma.

tertiary creep The third and final stage of *viscoelastic strain*, characterized by accelerating *strain* which leads to *failure.* See also *primary creep, secondary creep.*

teschenite An *andesine*-bearing *gabbro* or *dolerite* comprising calcium-rich *plagioclase, clinopyroxene* and *analcite*, with *accessory* Fe-Ti oxide, *amphibole, biotite* ± *olivine.* Occurs as minor intrusions, coarse *pegmatitic schlieren* or fine *veins* in *differentiated* minor intrusions.

tesla (T) The *SI* unit of *magnetic field* strength, with dimensions V s m^{-2}.

test A protective shell covering the soft parts of certain invertebrates.

Tethys An ocean which lay between *Gondwanaland* and *Laurasia* in pre-Neogene times.

Tetracorallia See *Rugosa.*

tetradymite (Bi_2Te_2S) An *ore mineral* of tellurium, commonly found in *gold-quartz veins.*

tetragonal system A *crystal system* whose members have three mutually perpendicular axes, the vertical of which is shorter or longer than the equal length horizontal axes.

tetrahedrite (fahl ore, fahlerz, grey copper ore) ($Cu_{12}Sb_4S_{13}$) An *ore mineral* of copper found mainly in low- to medium-temperature *hydrothermal ore veins.*

Tetrapoda Vertebrates with four limbs.

Teuriamn A *Palaeocene* succession in New Zealand covering the *Danian* and part of the *Thanetian.*

textural maturity The state of a *clastic rock* with high *roundness* and *sorting*, inferred to have these properties because of a long history of transport, *reworking* and thus *abrasion.*

texture The general character of a rock, shown by its component particles in terms of grain size and shape, degree of crystallinity and arrangement.

thalassostatic Descriptive of phenomena related to a period of static sea level.

thalweg A line connecting the points of deepest flow in successive downstream channel cross-sections, i.e. the planform pattern of maximum channel depth.

thanatocenosis (death assemblage, thanatocoenosis) An assemblage of *fossils* brought together only after death. cf. *biocenosis.*

thanatocoenosis See *thanatocenosis.*

Thanetian The higher *stage* of the *Palaeocene*, 60.5–56.5 Ma.

thaw lake An *oriented lake* in a *permafrost* region.

Thecodontia An order of subclass *Archosauria*, class *Reptilia*; very primitive *archosaurs.* Range U. *Permian*–U. *Triassic.*

Thematic Mapper A powerful imaging system used for *remote sensing* from the second generation of *Landsat satellites*, which records six spectral bands in the 450–2350 nm wavelength region with *pixels* ~30 m across and a thermal infrared channel with 120 m *pixels*.

thenardite (Na_2SO_4) An *evaporite mineral* whose *hydration* to *mirabilite* is accompanied by a significant increase in volume which assists in the *weathering* of a rock.

theralite (kylite) The *phaneritic* equivalent of *basanite*.

Therapsida An order of subclass *Synapsida*, class *Reptilia*; *reptiles* with an improved gait and a *mammalian* pattern of jaw-closing muscles. Range late *Carboniferous*–early *Permian*.

Theria A subclass of class *Mammalia* comprising the marsupials and placentals. Range end *Cretaceous–Recent*.

thermal conductivity A measure of the ability of a material to conduct heat; the flow of heat across unit area in unit time under zero temperature gradient normal to the area. Unit: W m^{-1} K^{-1}.

thermal diffusivity A parameter which controls the rate of heat propagation through a material; the ratio of *thermal conductivity* to the product of *density* and specific heat.

thermal maturation The process by which organic matter is transformed into *hydrocarbons* by heating.

thermal metamorphism *Metamorphism* involving heat only.

thermal resistivity The inverse of *thermal conductivity*; unit m K W^{-1}.

thermobaric flow A form of *mass solute transfer* out of a compacting *basin* under the influence of heat and *pressure*.

thermoclastis See *insolation weathering*.

thermocline The depth within a body of water at which the rate of decrease of temperature with depth is at a maximum.

thermogravimetric thermal analysis A method of studying *clay minerals* in which the weight changes with increasing temperature are monitored.

thermohaline current A current driven by *density* differences caused by seasonal variation in temperature and/or *salinity* and responsible for deep oceanic circulation.

thermokarst Topographic depressions arising from the thawing of ground ice.

thermoluminescence dating A dating technique for material containing small amounts of radioactive substances. The radioactivity causes ionization and some of the electrons produced become trapped in *crystal defects*. Heating to ~500°C releases these electrons in the form of visible light and the amount of this thermoluminescence per rad of ionizing radiation provides an estimate of the latest time the sample was heated to 500°. Used to date *speleothems* and ancient pottery and to determine if *flint* was heated in antiquity.

thermoremanent magnetization (TRM) A magnetization acquired when a *ferromagnetic* substance heated above its *Curie temperature* cools in a *magnetic field*. Present in many *igneous* and *metamorphic rocks*.

Theropoda Carnivorous forms of the *Saurischia*.

thickness-grade plan A plan derived by *underground mapping* showing the product of *grade* and thickness of a deposit.

thin section A slice of *mineral* or rock ~30 μm thick mounted on a microscope slide which is transparent or translucent to light and can be used to study the optical properties of the material.

thiospinel group A group of *sulphide minerals* characterized by a *spinel* structure based on a cubic close-packed anion sublattice with half the octahedral holes and one eighth of the tetrahedral holes occupied by cations.

thixotropy The behaviour shown by some *quick clays* on *deformation* in which there is a partial *recovery* of *residual strength* at some time after *failure*.

tholeiite (tholeiitic basalt) A *silica-oversaturated basalt* comprising calcium-rich *plagioclase*, *pyroxene* (in which *orthopyroxene* or *pigeonite* exceeds calcium-rich *pyroxene*) and Fe-Ti oxide, possibly with *augite* ± *olivine* in a siliceous *groundmass*. Forms the most abundant *basalt* group and makes up layer 2 of the *oceanic crust*.

tholeiitic basalt See *tholeiite*.

thompsonite ($NaCa_2(Al_5Si_5O_{20}).6H_2O$) A *zeolite* found in *amygdales* in *basic igneous rocks*.

thorianite (ThO_2) An *ore mineral* of thorium found in *pegmatites* and *alluvial deposits*.

thorite ($ThSiO_4$) A thorium *mineral* isostructural with *zircon*, found in *syenites*.

thorium-lead dating A *radiometric dating* method based on the decay of ^{232}Th to ^{208}Pb with a *half-life* of 13.9 Ga.

threshold angle The angle of a *threshold slope*.

threshold slope A graded or equilibrium hillslope whose inclination is controlled by the resistance of its soil cover to a dominant *degradational* process.

threshold velocity The minimum velocity of wind or water required to move particles of material.

thrombolite A *stromatolite*-like organism of uncertain origin with a *macroscopic* clotted *fabric*.

throughflow (subsurface flow) The downslope flow of water through soil approximately parallel to the ground surface, generated where the *hydraulic conductivity* decreases with depth.

throw The vertical *displacement* of a *dip-slip fault*.

thrust See *thrust fault*.

thrust belt A *tectonic* zone in which most of the *shortening* occurs on *thrust faults*, usually making up the external parts of an *orogenic belt*.

thrust fault (thrust) A *reverse fault* with a low angle of *dip*.

thrust fault system A set of related *faults* on which individual *displacements* give a net *contraction* in the system as a whole.

thrust sheet A *structure* like a *nappe*, but with a lower intensity of internal *deformation*.

thrust zone A zone of *thrusting* between two undeformed blocks.

thufur A small (10–50 cm high, 0.3–1 m diameter) earth hummock in a *periglacial* area above the treeline, resulting from differential freezing of the ground.

thulite A pink variety of *zoisite* containing manganese.

thuringite ($(Fe^{2+},Fe^{3+},Mg)_6(Al,Si)_4O_{10}(OH)_8$) A *chlorite* group *mineral* found in *veins* in *metamorphic rocks*.

tidal bar A *coastal bar* intermittently exposed by *tides*.

tidal bore (bore) A large *wave* which progresses up a funnel-shaped river or *estuary* with the rising *tide*.

tidal bundle A sediment unit deposited during a spring/neap tidal cycle comprising *sand* units transported by ebb and/or flood *tides*, separated by *non-conformities* or *pause planes*, commonly draped by fine-grained sediment which reflects fall-out during slack water.

tidal correction The correction applied to *gravity survey* data during *reduction* for the variation in *gravity* resulting from solid *Earth tides*, whose effect is to cause a change of ~3 g.u. over a 12 hour period.

tidal current The horizontal water movement caused by the gravitational attractions of the Sun and Moon on the Earth linked to the vertical movements of *tides*.

tidal delta A fan-shaped accumulation of sediment at the mouth of a *tidal inlet*, formed as tidal flow expands and decelerates seawards.

tidal friction The phenomenon which causes a time lag between the oceanic *tides* and solid *Earth tides* generated by frictional forces within the Earth.

tidal inlet A subaqueous channel between segments of a *barrier island* through which *tidal currents* flow, whose morphology is controlled by the relative magnitudes of the *wave* and *tidal* energy.

tidal palaeomorph An oversized, frequently meandering valley believed to represent a former tidal channel originating at a time of higher sea level.

tidal range The vertical height between consecutive high and low *tides*.

tidal wave A term often applied to a *tsunami*, but which gives a false impression of its origin.

tidalite A sediment deposited by *tidal currents*.

tide The vertical water movement arising from the gravitational attraction of the Sun and Moon, the effect of the latter being predominant.

tiger-eye A fibrous, golden brown *quartz pseudomorphous* after *crocidolite* and showing *chatoyancy*, used as an ornamental stone.

tight fold A *fold* with a small angle between its *limbs*.

tight gas sand A *reservoir* rock with a *permeability* so low as to allow production of no more than the equivalent of five barrels of oil per day without costly *well* stimulation.

tight sand A *reservoir* rock with a *permeability* so low as to allow production of no more than five barrels of oil per day without costly *well* stimulation.

tile clay A *clay* similar to *brick clay*, suitable for the manufacture of tiles.

till See *boulder clay*.

till fabric analysis The measurement of the orientations of *clasts* in *till*, whose plotting may indicate the direction of former *glacier* advance.

tillite A *conglomerate* of *glacial* origin, formed from *boulder clay*.

tilloid A chaotic mixture of large blocks in a *clay*-rich *matrix*, formed by *mudflows*, *landslides* and *glaciers*.

Tilodonta An order of infraclass *Eutheria*, subclass *Theria*, class *Mammalia*; primitive, carnivore-like *mammals*. Range U. *Palaeocene*–L. *Eocene*.

tilt-angle The angle between the horizontal primary electromagnetic field and the resultant field, used to quantify the disturbance in certain simple *electromagnetic induction methods*.

tilted fold See *inclined fold*.

tiltmeter An instrument which measures ultra-long period, angular displacements of the Earth's surface, usually for the purposes of *earthquake prediction*.

time correction The *static* or *dynamic correction* applied to seismic data.

time-distance curve A plot of *arrival time* of *seismic waves* against *shot-seismometer* separation, used in the interpretation of seismic data.

time-term method A method of interpretation of *seismic refraction* data in terms of an undulating refractor which makes use of the difference in travel time of a *head wave* to and from a refractor and the time a wave would take to travel the same horizontal distance at the *seismic velocity* of the lower medium.

tin (Sn) A very rare *native metal*.

tin girdle The region of SE Asia in which important deposits of *tin* are found.

tin pyrites See *stannite*.

tincalconite ($Na_2B_4O_5(OH).3H_2O$) A *mineral* formed by the *alteration* of borate *minerals* by *hydration* or dehydration.

tinguaite An *undersaturated*, medium- to coarse-grained *igneous rock* comprising *essential alkali feldspar*, *nepheline* and *aegirine* ± sodic *amphibole* or *biotite*. The *hypabyssal* equivalent of *phonolite*.

tinstone A popular name for *cassiterite*.

tip See *tip line*.

tip fold A *fold* which maintains the *strain* continuity between the end of a *fault* of finite *displacement* and unfaulted rock.

tip line (tip) The termination of a *thrust* surface.

Tirek An *Ordovician/Silurian* succession in the Mirnyy Creek area of NE Siberia equivalent to the *Hirnantian* and *Rhuddanian*.

tirodite (manganoan cummingtonite) ($(Mg,Mn,Fe)_7Si_8O_{22}(OH)_2$) A rare, yellow, manganese-rich, magnesium-bearing end-member of the *amphiboles*.

titanaugite Titanium-bearing *augite*.

titaniferous iron ore See *ilmenite*.

titanite (sphene) ($CaTiO(SiO_4)$) A common *accessory mineral*, sometimes exploited as an *ore* of titanium.

titanium hornblende See *kaersutite*.

titanomaghemite Titanium-bearing *maghemite*.

titanomagnetite Titanium-bearing *magnetite*.

Tithonian The highest *stage* of the *Jurassic*, 152.1–145.6 Ma.

titration A method of wet *chemical analysis* in which a solution of the unknown is gradually mixed with a suitable reagent of known concentration until some form of indicator, such as a colour change, is activated. The quantity of the reagent used can then be converted into the concentration of the unknown.

Tiverton A *Permian* succession in Queensland, Australia, covering part of the *Asselian*, the *Sakmarian* and part of the *Artinskian*.

toad's-eye tin A *botryoidal* or *reniform* variety of *cassiterite* which shows an internal concentric and fibrous structure.

toadstone A mining term for various types of altered *igneous rock*.

Toarcian A *stage* of the *Jurassic*, 187.0–178.0 Ma

todorokite ($(Mn,Ca,Mg)Mn_3O_7 .H_2O$) An *ore mineral* of manganese found as a *secondary mineral* formed from the *alteration* of other manganese-bearing *minerals*.

toe 1. The base of the working face of a mine or quarry. 2. The lowest, most *distal* part of a *fan, delta, scree, dune* etc.

toe erosion *Erosion* at the base of a steep slope, such as a cliff.

toe method A mapping technique used in a mine or quarry in which the *toe* is charted as a line and all features in a face are recorded and projected down to the reference level. The final map represents a single horizontal plane with all contacts shown in their true *strike* direction.

toeset bed The basal, asymptotic part of *cross-laminated* beds, usually of fine-grained sediment. cf. *foreset bed, topset bed*.

Tolbonskaya A *Triassic* succession in Siberia equivalent to the *Anisian* and *Ladinian*.

Tolen An *Ordovician* succession in Kazakhstan covering part of the *Pusgillian*, the *Cautleyan*, *Rawtheyan* and *Hirnantian*.

tombolo A *bar* or *spit* of *sand* linking an island to the mainland or another island, usually forming on the sheltered side of the island.

Tommotian The lowest *stage* of the *Cambrian*, 570–560 Ma.

tommotiid A small, conical or tubular *fossil* composed of phosphate, of uncertain taxonomic status, found in the *Tommotian*.

tomography See *seismic tomography*.

tonalite A *phaneritic* rock comprising *plagioclase*, which makes up $>90\%$ of the *feldspar*, and 20–30% *quartz*, with *alkali feldspar* absent or making up $<10\%$ of the rock. Associated with *plutons* and *batholiths* of the *diorite-granodiorite-granite* association.

Tonawandian A *Silurian* succession in North America covering parts of the *Telychian* and *Sheinwoodian*.

Tongaporutian A *Miocene* succession in New Zealand covering parts of the *Tortonian* and *Messian*.

tonstein A *kaolin*-rich *seat earth* possibly formed by the *alteration* of volcanic *ash* or *illite*-rich *clay*.

tool mark A mark on a sediment surface made by a mobile object.

tool track (groove, scratch) An incongruous *lineation* on a *fault* surface caused by *asperity ploughing*.

topaz ($Al_2SiO_4(F,OH)_2$) A *nesosilicate* found in *granite*, used as a *gem*.

topazolite A honey-yellow, *transparent* variety of *andradite*.

topographic correction (terrain correction) The correction applied to *gravity survey* data for the gravitational attraction of topography around an observation point, often calculated with a *Hammer chart*.

toponomy The mode of preservation of a *trace fossil*.

toposequence See *catena*.

topotype A specimen from the same locality as the *holotype*.

topset bed The upper, near-horizontal layers of a *cross-laminated bed*, usually of coarse-grained sediment. cf. *foreset bed*, *toeset bed*.

tor An exposure of bare rock with free faces on all sides resulting from differential *weathering*, *mass wasting* and stripping, whose form is often controlled by *joints*.

torbanite See *boghead coal*.

torbernite (cuprouranite) ($Cu(UO_2)_2(PO_4)_2.8–12H_2O$) A green *secondary mineral* found where uranium *ore* deposits have been *oxidized* and *weathered*.

Torian A *Cambrian* succession in Siberia covering part of the *Maentwrogian*.

toroidal field A *magnetic field* confined within a sphere with no radial components. cf. *poloidal field*.

Torridonian The upper division of the *Precambrian* in NW Scotland, generally of *fluvial* origin.

torsion balance An early type of *gravimeter* based on the oscillation of a beam suspended by a torsion fibre.

Tortonian A *stage* of the *Miocene*, 10.4–6.7 Ma.

total strain See *finite strain*.

touchstone A piece of stone scratched across a *precious metal* or alloy to assess its purity from the *streak*.

tourmaline ($(Na,K,Ca)(Li,Mg,Fe,Mn,Al)_3(Al,Fe,Cr,V)_6(BO_3)_3(Si_6O_{18})(O,OH,F)_4$) A complex *cyclosilicate* found in *granite pegmatite* and as an *accessory mineral* in *igneous* and *metamorphic rocks*.

tourmalinization A type of *wall rock alteration* in which significant amounts of new *tourmaline* are formed. Usually associated with medium- to high-temperature mineralization.

Tournaisian The oldest *epoch* of the *Carboniferous*, 362.5–358.3 Ma.

Tournaisien A *Devonian/Carboniferous* succession in France and Belgium covering part of the *Famennian* and the *Hastarian*.

tower karst (turmkarst) A *karst* landscape characterized by steep, flat-topped towers of *limestone* rising from a flat plain.

town gas See *coal gas*.

trace The *seismogram* for one *seismometer*.

trace d'activité animale (lebens-

spur) A *trace fossil* in unconsolidated sediment.

trace element An element present in a rock in such low concentration, usually in the parts per million range, that it does not control the presence of *minerals* such as *feldspar* and *pyroxene* but occurs within them by ionic substitution for *major elements*. Often indicative of the source of the rock. cf. *major element*.

trace fossil (biogenic sedimentary structure, ichnofossil) A *structure* in sediment produced by the activity of an ancient organism. See also *agrichnia, cubichnia, domichnia, endichnia, epichnia, exichnia, fodichnia, hypichnia, pascichnia, repichnia*.

Trachaeophyta A division of kingdom *Plantae*; the *vascular plants*. Range *Carboniferous–Recent*.

trachyandesite A *volcanic rock* intermediate in composition between *andesite* and *trachyte*.

trachybasalt A *volcanic rock* intermediate in composition between *basalt* and *trachyte*.

trachydacite A *volcanic rock* with a lower $Na_2O + K_2O$ content than *trachyte* and with no *plagioclase phenocrysts*.

trachyte A *silica-saturated* or *oversaturated*, weakly *porphyritic*, *alkaline volcanic rock* of *intermediate* composition containing mainly *sanidine* or *orthoclase phenocrysts*, but also *alkali feldspar, clinopyroxene*, Fe–Ti oxides, *amphibole, fayalite, aenigmatite* and *biotite* in a *groundmass* often showing *trachytic texture*.

trachytic texture A *texture* in a *volcanic rock* in which there is strong flow alignment of small *acicular alkali feldspar* crystals.

trachytoidal texture The *texture* of a *phaneritic igneous rock* in which *feldspars* are in parallel to subparallel alignment and thus resembles a *trachytic texture*.

trailing continental margin See *passive continental margin*.

transcurrent fault (wrench fault) A vertical *strike-slip fault*.

transfer fault A steep *fault* with a *strike-slip displacement* which transfers *displacement* from one *dip-slip* plane to another; a small-scale equivalent of an oceanic *transform fault*.

transform fault A special form of *strike-slip fault* which joins the ends of *constructive* and *destructive plate margins*.

transformation twinning A type of *twinning* comprising two *polymorphs*, caused by a change in *crystal structure* under different temperature-pressure conditions during *twin* formation.

transformation-enhanced deformation *Deformation* whose mechanism is enhanced by chemical reaction or phase changes.

transformational faulting (anticrack faulting) A sudden phase change in a *mineral* which takes place by rapid shearing of the *crystal lattice* along planes on which minute crystals of the new phase have grown. For example, the change from *olivine* to *spinel* in the downgoing slab probably contributes to *subduction zone seismicity*.

transgranular displacement A *displacement* across a grain boundary during *deformation*. cf. *intergranular displacement, intragranular displacement*.

transgression An incursion of the sea over a land area or over a shallow sea. cf. *regression*.

transient creep A slow *deformation* in rocks when they are subjected to *stress* at the temperatures and pressures found at or near the Earth's surface.

transition zone Any large-scale, diffuse boundary within the Earth, but usually the *mantle transition zone*.

translation The *displacement* of the centre of a body to a new position after *deformation*.

translation gliding A crystal *deformation* mechanism whereby one or more rows of atoms are displaced laterally along a glide plane.

translucent mineral A *mineral* in which the transmission of light is poorer than for a *transparent mineral*, generally displayed by *minerals* with a *non-metallic lustre*.

transmission capacity The state of a soil which can accommodate the flow of no more water.

transmission coefficient The ratio of the amplitude of a *seismic wave* transmitted across an *acoustic impedance* contrast to the amplitude of the incident wave, described by *Zoeppritz' equations*.

transmission electron microscopy (TEM) A technique of *electron microscopy* used in the investigation of the *microstructure* of crystalline specimens from the transmission of electrons through them. The electron beam is diffracted according to the *Bragg law*, with a very small angle between incident and diffracted rays. This allows magnification of up to 10^6, providing resolution of ~ 0.3 nm.

transmissivity The product of *hydraulic conductivity* and the thickness of a *bed* acting as an *aquifer*.

transmissometer See *nephelometer*.

transparent mineral A *mineral* in which the transmission of light is more effective than for a *translucent mineral*, generally displayed by *minerals* with a *non-metallic lustre*.

transportation slope A slope which suffers neither *erosion* nor deposition because the amount of material arriving from upslope is the same as that leaving downslope. Usually concave in form and situated at the base of a more extensive slope.

transposed foliation A *foliation* developed by the *deformation* of a primary surface, commonly by folding accompanied by *diffusive mass transfer*.

transposition The creation of a new *fabric*, commonly a *foliation*, by *deformation* of an existing one.

transpression The compression associated with movement along a curved *strike-slip fault*. cf. *transtension*.

transtension The tension associated with movement along a curved *strike-slip fault*. cf. *transpression*.

Transvaal jade (South African jade) A massive, light green variety of *hydrogrossular*, used as a substitute for *jade*.

transverse bar A *coastal bar* in a nearshore or intertidal zone normal to the shore. Its formation requires a restricted *fetch* length and low energy wave conditions.

transverse coast See *Atlantic-type coast*.

transverse dune A *dune* whose long axis is normal to the main *sand*-moving wind direction.

transverse wave See *S wave*.

trap 1. A *structure* in which *hydrocarbons* are trapped. 2. An obsolete term for a fine-grained, compact *igneous rock*.

traverse mapping A method of *geological mapping* for scales of 1:250 000–1:50 000 in which traverses are systematically spaced or follow primary topographic features.

travertine Calcareous material formed by *precipitation* from flowing fresh water at a *hot spring* or *waterfall* after passage through calcareous rock or sediment, aided by biochemical activity.

Tremadoc The oldest *epoch* of the *Ordovician*, 510.0–493.0 Ma.

tremolite (grammatite) ($Ca_2Mg_5Si_8O_{22}(OH)_2$) An *inosilicate* of the *amphibole* group, found mainly in *contact* and *regionally metamorphosed dolomitic* and low-*grade ultrabasic rocks*.

Trempealeauan A *Cambrian/Ordovician* succession in the E USA, covering parts of the *Dolgellian* and *Tremadoc*.

trench See *ocean trench*.

Tresca's failure criterion '*Failure* occurs when the maximum *shear stress* reaches a critical constant value for the material.'

trevorite ($NiFe_2O_4$) An *oxide mineral* of the *magnetite* series.

Trias See *Triassic*.

Triassic (Trias) The oldest *period* of the *Palaeozoic*, 245.0–208.0 Ma.

triaxial compression *Compression* with components along three orthogonal axes.

triaxial stress A *stress* system with components along three orthogonal axes.

triclinic system A *crystal system* whose members have three unequal, mutually oblique axes; i.e. no symmetry other than a possible centre.

tridymite (SiO_2) A high-temperature form of *silica*, stable in the range 870°–1470°C.

trigonal system A *crystal system* whose members have three lateral axes of equal length intersecting at an angle of 60° to each other and reaching the centres of the three vertical faces, and a vertical axis of different length perpendicular to the other three.

Trilobita/trilobites A subphylum of phylum *Arthropoda*; *arthropods* characterized by trilobation of the exoskeleton into cephalon (head), thorax (body) and pygidium (tail). Range L. *Cambrian*–U. *Permian*.

trimacerite A lithotype of *coal* with >5% of *macerals* from all three groups.

Trimerphytopsida A class of division *Trachaeophyta*, kingdom *Plantae*; the ancestors of megaphyllous *plants*. Range L. *Devonian*–U. *Devonian*.

Trinity A *Cretaceous* succession on the Gulf Coast of the USA covering parts of the *Aptian* and *Albian*.

triphylite ($Li(Fe,Mn)PO_4$) A lithium *mineral* found in *pegmatites*, *isomorphous* with *lithiophilite*.

triple junction The point where three *plate margins* intersect. Only certain configurations are possible, determined by the relative velocity vectors of the *plates* at the intersection. Quadruple junctions are inherently unstable.

triple point The point on a phase diagram at which the boundary curves of three phases meet and they can coexist.

tripoli A *microcrystalline*, soft, friable, porous siliceous material exploited as an *abrasive*.

tripolite (infusorial earth) A variety of *opal*, a *diatomite* with an earthy appearance.

trittkaren *Karren* in the form of heel-shaped pits.

Trituberculata An infraclass of subclass *Theria*, class *Mammalia* comprising the *Pantotheria* and *Symmetrodonta*. Range U. *Jurassic*–L. *Cretaceous*.

TRM See *thermoremanent magnetization*.

trochiform Descriptive of a coiled shell in which the sides are evenly conical and the base flat.

troctolite (troutstone) A *gabbro* with *olivine* and no *pyroxene*.

troilite (FeS) An iron *sulphide mineral* mainly found in *meteorites*.

trona ($Na_3H(CO_3)_2.2H_2O$) A *mineral* found in *saline lake* deposits.

trondhjemite A *phaneritic*, *quartz-albite*-rich *plagioclase* rock, including *plagiogranite* within *ophiolite* complexes, *tonalite* within *calc-alkaline batholiths* and forming part of a distinctive *Archaean* rock assemblage (*bimodal trondhjemite* and *metamorphosed basaltic* rocks with no *intermediate rocks*).

troostite A manganiferous variety of *willemite*.

trottoir A narrow organic *reef* in the intertidal zone.

trough cross-bedding See *trough cross-stratification*.

trough cross-stratification (trough cross-bedding) *Cross-stratification* in which the lower bounding surface is highly curved and concave-up so that *cross-sets* are scoop- or spoon-shaped, probably resulting from the migration of irregular, sinuous or *linguoid dunes*.

trough line A line joining the topographically lowest points on a folded surface, normally parallel to the *fold hinge line*.

troutstone See *troctolite*.

true dip The angle between the horizontal and the surface of a layer, measured in the vertical plane perpendicular to the *strike* of the layer and thus the maximum *dip* measurable. cf. *apparent dip*.

true north The direction towards the Earth's axis of rotation. cf. *magnetic north*.

true polar wander The postulated movement of the Earth's surface relative to the plane of the ecliptic.

tschermakite ($Ca_2Mg_3(Al,Fe)_2Al_2Si_6O_{22}(OH)_2$) An *amphibole* found in *igneous* and *metamorphic rocks*.

Tschermak's component The charge-coupled substitution of Al for divalent ions at *B* and for Si at *T* in *pyroxenes*.

Tselinograd An *Ordovician* succession in Kazakhstan covering part of the Late *Llanvirn*, the *Llandeilo* and part of the *Costonian*.

tsunami A long-wavelength water *wave* caused by sudden movement of the seafloor, such as caused by an *earthquake*, sediment *slump* or *volcanic eruption*, which travels at tens of m s^{-1}. Of low amplitude at sea, but growing to up to 20 m on reaching shallow water.

Tubulidentata An order of infraclass *Eutheria*, subclass *Theria*, class *Mammalia*; the aardvarks. Range *Miocene–Recent*.

tufa (calc-tufa) A spongy, porous variety of *travertine* associated with *algal* colonies.

tuff A volcanic sediment.

tuff cone A *structure* similar to a *tuff ring*, but generally larger and steeper sided, with a maximum thickness of 100–350 m and a maximum *dip* of 24°–30°, and with a higher proportion of *mudflow* deposits.

tuff ring A monogenetic *volcano*, of any composition but often of *basalt*, formed in a *phreatomagmatic eruption*, with a low rim and a broad, flat *crater*. Comprises *phreatic* air-fall *breccias* and finely-bedded air-fall and *pyroclastic surge* beds, the maximum *dip* of the *tephra* being 3°–12°. See also *tuff cone*.

tuffisite (tuffite) A fine-grained *matrix* in a volcanic *breccia*, comprising finely-divided fragments of *magma* and grains of sediment, which appear to have been *fluidized* and intruded along small cracks in blocks in the *breccia*.

tuffite See *tuffisite*.

tugtupite ($Na_4BeAlSi_4O_{12}Cl$) A rare, intense pink, beryllium-bearing form of *sodalite*.

Tumen A *Jurassic* succession in W Siberia covering the *Hettangian*, *Sinemurian*, *Pliensbachian* and part of the *Toarcian*.

tundra A vast, level, treeless, marshy region, usually with a permanently frozen subsoil.

tungstates *Minerals* containing tungsten in the form $(WO_4)^{2-}$, similar to the *molybdates* with which there is free substitution of molybdenum. Form two series, the *wolframite* group and the *scheelite* group.

tungstenite (WS_2) A *sulphide mineral* of the *layered sulphides group*.

tungstic ochre See *tungstite*.

tungstite (tungstic ochre) ($WO_3.H_2O$) An earthy, yellow or green, hydrated tungsten oxide, usually found as a *secondary mineral* associated with *wolframite*.

Tunicata *Sessile*, filter-feeding animals of phylum *Chordata* with free-swimming larvae which may be ancestral to the *Pisces*.

tunnel erosion A form of *pipe erosion* initiated by an abnormally large *hydraulic gradient*.

tunnel valley (rinnentaler, tunneldale) A long, deep channel cut by a subglacial meltwater river.

tunneldale See *tunnel valley*.

Turam EM method An *electromagnetic induction method* for *mineral* exploration using a large fixed source and a pair of receiving coils about 10 m apart. EM components are measured at two frequencies, providing information on features with anomalous *electrical conductivities* in the subsurface.

turbidite A deposit from a waning *turbidity current*.

turbidity current (turbidity flow) A generally turbulent, subaqueous *density current* of suspended sediment driven by *gravity*, with the potential of flowing down slopes of $< 1°$ and even up-slope. Occurs in non-marine to marine, shallow to deep water environments, generated by storms, the entry of sediment laden river *plumes*, shelf currents or slope *failure*. Flow velocities are up to 10 m s^{-1} and the flow length may reach thousands of kilometres.

turbidity flow See *turbidity current.*

turbidity meter An instrument using a light source which provides continuous measurement of *suspended load.*

turbinate Descriptive of a coiled shell shaped like a top, but with a rounded base.

turbulence A series of quasi-random motions at different scales in a fluid; the root mean square of the instantaneous fluctuating velocity component in *turbulent flow*.

turbulence intensity The ratio of *turbulence* to mean flow velocity in *turbulent flow*.

turbulent flow Flow with three-dimensional motion in which the velocity at any point fluctuates with time and *streamlines* may not be parallel. Occurs at *Reynolds number* > 200.

turgite See *hydrohaematite.*

turlough A seasonal lake, up to 5 km^2 in area, found in a glacially-influenced *karst* area, which fills and empties through *springs* and *sinkholes.*

turmkarst See *tower karst.*

Turonian A *stage* of the *Cretaceous*, 90.4–88.5 Ma.

turquoise $(CuAl_6(PO_4)_4(OH)_8.4H_2O)$ A blue-green *secondary mineral* of copper valued as a *gem.*

twin A composite crystal produced by *twinning.*

twin axis The axis of symmetry about which an individual in a *twin* can be rotated to produce the orientation of the other.

Twin Creek A *Jurassic* succession in Utah/Idaho, USA, covering part of the *Bajocian*, the *Bathonian* and part of the *Callovian.*

twin gliding A crystal *deformation* mechanism whereby a *displacement* is taken up by each row in a *crystal lattice.*

twin law The fundamental elements along or about which a crystal is *twinned* in terms of a *twin plane* or *twin axis.*

twin plane The plane of symmetry in which an individual in a *twin* can be reflected to produce the orientation of the other.

twinning The formation of composite crystals (*twins*) in which the two individuals have different orientations in simple *crystallographic* relationship to each other. Described by the symmetry operation necessary to convert one orientation to the other. See also *normal twin, parallel twin.*

two-way travel-time (TWT) The time taken for a *seismic wave* to return to the surface after undergoing a *seismic reflection.*

TWT See *two-way travel time.*

type locality 1. **(type section)** A locality selected as a standard for a *stratigraphic* unit. 2. The locality of a *type specimen.*

type section See *type locality.*

type specimen (lectotype, neotype) A specimen (*holotype*) on which the full description of a species is based.

typological method The *stratigraphic* description of a unit in terms of its *lithology* or *fossils*.

tyuyamunite ($Ca(UO_2)_2(VO_4)_2.5$–$8.5H_2O$) A green, *secondary ore mineral* of uranium and vanadium.

U

Ubendian orogeny An *orogeny* affecting central Africa at ~1800–1700 Ma.

Udden-Wentworth scale A scale for the subdivision of *clastic rocks* on the basis of *clast* size, based on a ratio of two between successive class boundaries. See also *phi unit.*

Udoteaceae A group of *calcareous algae* characterized by internal medullary filaments which branch and curve outwards to form an exterior cortical layer. Range M. *Ordovician–Recent.*

Ufimian A *stage* of the *Permian,* 256.1–255.0 Ma.

ugrandite An acronym for the *garnets* ***u****varovite,* ***gr****ossular* and ***and****radite.* cf. *pyralspite.*

uintaite (gilsonite) A brilliant black variety of *asphalt* occurring in rounded masses.

Uivakian orogeny An *orogeny* affecting the NE Canadian *shield* at ~3000 Ma.

Ulatizian An *Eocene* succession on the west coast of the USA covering parts of the *Ypresian* and *Lutetian.*

ulexite ($NaCaB_5O_6(OH)_6.5H_2O$) A borate *mineral* found associated with *borax,* deposited from *brine* formed in enclosed *basins* in arid areas.

ullmannite (nickel antimony glance) (NiSbS) A *sulphide mineral* found in *hydrothermal veins.*

Ulsterian A *Silurian/Devonian* succession in E North America covering the upper *Silurian, Lochkovian, Pragian* and part of the *Emsian.*

ultimate strength The *strength* shown by a material which fails after *ductility.* cf. *brittle strength.*

ultrabasic rock An *igneous rock* comprising essential *ferromagnesian minerals* and *feldspathoids* to the virtual exclusion of *quartz.*

ultracataclasite A cohesive *fault rock* with a random *fabric* and with >90% *matrix.* cf. *cataclasite, protocataclasite.*

ultramafic rock A rock comprising >90% *ferromagnesian minerals,* composed of *olivine, orthopyroxene* and *clinopyroxene* ± *amphibole.* Found in continents in association with *gabbro* and in *ophiolite* complexes.

ultramarine A pigment made from *lapis lazuli.*

ultramylonite A foliated *fault rock* with 90–100% *matrix.*

ulvöspinel (Fe_2TiO_4) An *oxide mineral* with *spinel* structure found as *exsolution* bodies within *magnetite* and also in lunar rocks.

umber A naturally-occurring brown pigment composed of iron and manganese oxides with minor *silica,* alumina and *lime,* formed by the *weathering* of rocks rich in *ferromagnesian minerals.*

unconfined aquifer An *aquifer* lacking an upper *aquiclude*, so that the upper limit of saturation is the *water table*.

unconformable Not following the underlying sequence of rock in *structure* or age.

unconformity A break in the *stratigraphic* record which represents a period of no sediment deposition. See also *diastem*, *disconformity*, *non-sequence*, *paraconformity*.

unconformity-related uranium deposit An *epigenetic* uranium deposit in which most of the mineralization lies at or just below an *unconformity*.

unconformity trap A *hydrocarbon trap* in which the *seal* is provided by impermeable material deposited on an *unconformity* above permeable *strata*.

underclay A fine-grained sediment immediately beneath a *coal seam*, generally with no *bedding* and strongly *leached*, which causes a high *kaolinite* content.

underconsolidated sediment A sediment as it was deposited, before the commencement of *diagenesis*, etc. cf. *normally consolidated sediment*, *overconsolidated sediment*.

underfit stream A large valley channel, often meandering, which contains smaller alluvial channels, i.e. the current channel is too small to account for the valley channel. Possibly formed in response to climate change, *river capture* etc.

underflow *Groundwater* flow in *alluvial* sediments parallel to and beneath a river channel.

underground mapping Mapping executed in tunnel drivage, working mines and abandoned mines, generally at a large scale.

underplating (offplating, plastering, subcretion) The addition of material to the base of a *structure*, such as occurs in the growth of an *accretionary prism*.

undersaturated Descriptive of an *igneous rock* deficient in *silica* so that *minerals* develop which are unstable in the presence of *silica*. See also *silica saturation*.

undersaturated oil pool An oil pool undersaturated with respect to gas, commonly arising from an increase in the depth of burial.

underthrusting The movement of an oceanic *plate* under the leading edge of a second *plate* at a *subduction zone*.

Undillian A *Cambrian* succession in Australia covering the middle part of the *Menevian*.

undulose extinction The *extinction* of a *mineral* which takes place over a range of angles, which may be caused by the presence of *microscopic kink bands*.

unequal slopes A theory which considered the *erosional* development of a *badland* divide between two slopes of unequal *declivity*, which acquire their smooth curves according to the *Law of Divides*. See also *Law of Equal Declivities*.

uneven fracture (irregular fracture) A *fracture* producing a rough and irregular surface.

uniaxial compressive strength The *strength* under uniaxial *compression*.

uniaxial stress A *stress* system in which the *principal stress* is compressive and the *intermediate* and *minimum principal stresses* are zero.

uniclinal shifting The migration of a river in the direction of *strata*.

uniformitarianism (actualism) The present viewed as the key to the past. Essentially the converse of *catastrophism*.

unimodal Possessing one mode, i.e. a frequency distribution in which there is only one most common value.

Uniramia A phylum of mainly terrestrial *arthropods* characterized by single-branched limbs. The head bears a pair of antennae, mandibles and two feeding maxillae. Range ?*Cambrian–Recent*.

unit cell The fundamental building block of a crystal comprising an ordered arrangement of atoms repeated exactly in all directions.

universal gravitational constant See *gravitational constant*.

universal soil loss equation (USLE) The most widely used method of *soil loss prediction*; $A = RKLSCP$, where A = *soil loss*, R = *rainfall erosivity factor*, K = *soil erodibility factor*, S = *slope gradient factor*, C = *crop management factor* and P = *erosion control practice factor*.

unloading The removal of *stress*, commonly gravitational, by *erosion*, melting of ice etc., followed by *pressure release* effects.

unsaturated Descriptive of a *mineral* which cannot develop in stable equilibrium in the presence of free *silica*.

unsaturated zone See *vadose zone*.

uphole shooting A seismic technique in which *shots* are detonated down a borehole and their arrivals monitored by a surface *seismometer* in order to determine *seismic velocities* and assist in *static corrections*.

uplift 1. The upward movement of part of the Earth's surface, typically 500–1000 km in width. 2. A *structure* caused by the upward movement of part of the Earth's surface.

upper continental crust The upper 10–12 km of the *continental crust*, previously termed the 'granitic layer'. cf. *lower continental crust*.

upper flow regime The *flow regime* commencing when *shear stress* on a bed abruptly decreases and *bedforms* are washed out and replaced by a *plane bed*. *Flow separation* no longer occurs and *shear stresses* start to increase as *standing waves* and then *antidunes* form as velocity increases. cf. *lower flow regime*.

Upper Greensand A *Cretaceous* succession in England covering the upper part of the *Albian*.

upper mantle 1. The *mantle* above ~700 km depth, including the *mantle transition zone*. 2. The *mantle* above ~400 km depth, i.e. not including the *mantle transition zone*. cf. *lower mantle*.

Upper Sandugan A *Silurian* succession in the Mirnyy Creek area of NE Siberia equivalent to the *Wenlock*.

upper-stage plane bed A flat *bed* in *sand*-grade sediment produced at *Froude numbers* approaching unity, over which there is intense sediment transport, characterized by extensive parallel *laminae* and *primary current lineations.*

upright fold A *fold* with a near horizontal *hinge line* and a vertical *axial surface.*

upthrown Descriptive of the side of a *fault* with relative upwards movement. cf. *downthrown.*

upthrust A *reverse fault* flattening upwards from a steep attitude to become a *thrust* at a shallow level.

upward continuation The computation of how a *potential field* would appear at an elevation above the level of observation, used to enhance the anomalies of deep sources. cf. *downward continuation.*

upwelling The transport of subsurface waters to the surface at a wide variety of scales in oceans, *estuaries* and lakes, usually in response to wind *stress* on the surface.

Uralian emerald A green variety of *andradite* used as a semi-precious *gem*, found in *nodules* in *ultrabasic rocks.*

Uralianeny An *orogeny* in the former USSR coeval with the *Variscan orogeny.*

uralite A blue-green *amphibole* of generally *actinolite* composition formed by the *alteration* of *pyroxene.*

uralitization (amphibolitization) A form of *alteration* affecting *pyroxene*-bearing rocks in which the *pyroxene* is replaced by fibrous *amphiboles.*

uraniferous calcrete A *calcrete* mineralized with uranium.

uraninite (UO_2) The major *ore mineral* of uranium, found in *pegmatites*, high-temperature *hydrothermal veins* and *placer deposits.*

uranium-lead age dating A dating method based on the radioactive decay of ^{238}U and ^{235}U to ^{206}Pb and ^{207}Pb respectively with *half-lives* of 4498 and 713 Ma. *Zircon*, which contains trace amounts of uranium, is the preferred *mineral* as its primary lead concentration is low and it is resistant to uranium *leaching.*

uranophane (uranotile) ($Ca(UO_2)_2Si_2O_7.6H_2O$) Hydrated calcium-uranium silicate, found in *pegmatites* as *pseudomorphs* after *uraninite*, as coatings on *fractures* and in the oxidized zone of *uraninite vein* deposits.

uranotile See *uranophane.*

Urgonian A massive, landscape-dominating, U. *Barremian* to L. *Aptian limestone* facies in S Europe.

Urodela An order of subclass *Lissamphibia*, class *Amphibia*; the newts and salamanders. Range late *Jurassic–Recent.*

urtite An *ijolite* with 0–30% *mafic minerals.*

'uruq See *draa.*

Ururoan A *Jurassic* succession in New Zealand equivalent to the *Pliensbachian, Toarcian* and *Aalenian.*

Urutawan A *Cretaceous* succession in New Zealand covering part of the middle *Albian.*

USLE See *universal soil loss equation.*

Ust'kel'Terskaya (Induan) A *Triassic* succession in Siberia covering the *Griesbachian* and part of the *Nammalian.*

uvala A series of complex closed depressions in a *karst* area, often caused by coalescing *dolines.*

uvarovite ($Ca_3Cr_2Si_3O_{12}$) A rare green *garnet* found in association with *chromite* in *serpentine*, in *skarns* and in *metamorphosed limestones.*

V

vacancy crystal defect (Schottky crystal defect) The simplest type of *point crystal defect* in which an atom is missing from its usual position in the regular arrangement of atoms.

vadose zone (unsaturated zone, zone of aeration) A shallow subsurface zone of *infiltration* and percolation of rainfall, *runoff* or melt water between the surface and the *water table*. Up to tens of metres in thickness with intergranular pores and fissures unsaturated with water and containing air at atmospheric pressure.

Valanginian A *stage* of the *Cretaceous*, 140.7–135.0 Ma.

valence bond theory An approach to chemical bonding based on pairs of electrons considered as shared between atoms, localized on atoms or on separate atoms with opposed spins that are arranged among the atoms of a molecule to form so-called 'structures'. These structures have characteristic energy and interact to form a stable representation of a molecule.

valentinite (Sb_2O_3) A white antimony trioxide formed by the decomposition of other antimony *ores*.

valley-bottom bulge See *valley bulge*.

valley bulge (valley-bottom bulge) *Strata* bulged up in a valley floor by *erosion*'s acting on rocks of different character, e.g. fine-grained, water-charged sediments extruded through *competent* rock.

valley meander A meander with a greater wavelength than the present river system, produced under conditions of higher *runoff*.

vallon de gélivation A small valley formed by the widening of *joints* in the *bedrock* by frost action rather than *fluvial* processes.

valve An exoskeletal unit which makes up a shell.

van Krevelen diagram A plot of *kerogen* composition in terms of oxygen : carbon ratio against hydrogen : carbon ratio.

vanadinite ($Pb_5(VO_4)_3Cl$) A rare *secondary mineral* found in the oxidized parts of lead *veins*.

vapour phase crystallization The *crystallization* of *minerals* in cavities or pores in *ignimbrites* from hot *volcanic gases*.

vapour transport A mechanism of *magmatic differentiation* in which constituents such as sodium, potassium and silicon become concentrated in a vapour phase which can migrate within, or escape from, a *magma* body.

Varanger The lower *epoch* of the *Vendian*, 610–590 Ma.

variable source area model A model for *runoff* processes in which the surface *runoff* is produced by saturation-excess *overland flow*. cf. *partial area model*.

variation diagram A graph showing the chemical variation in a suite of rocks to reveal genetic and other relationships between them.

variation method A method of determining the energies of stable *molecular orbitals* by minimizing their energy.

variegated copper ore A popular name for *bornite*.

variolitic structure A form of *spherulitic texture* comprising radiating fibres in a *glassy rock* at the chilled margin of an intrusion.

variometer See *magnetic variometer*.

Variscan orogeny (Armorican orogeny, Hercynian orogeny) An *orogeny* in late *Carboniferous*–early *Permian* in Europe and eastern North America.

varisicite ($Al(PO_4).2H_2O$) A blue-green, massive *mineral* similar to *turquoise* and used as a *gem*, found mainly in deposits formed by the action of phosphatic *meteoric waters* on aluminous rocks.

varve A thin layer of a *laminite* and *rhythmite* which represents an annual cycle of deposition.

varve dating The measurement and counting of *varves* in lake deposits so that a chronology can be established, often linked to *radiocarbon dating*.

vascular plant A plant with a fluid-conducting vascular system.

vasques A series of wide (up to several decimetres), shallow, flat-bottomed pools forming a network of tiered, terrace-like steps on *limestone* coastal platforms in the intertidal zone, separated by narrow, lobed ridges 10–200 mm in height.

vaterite A *metastable* form of *calcite* or *aragonite*.

vector 1. A line which represents the direction and magnitude of a motion. 2. The relative *displacement* direction across a *fault*.

vegetation arc A band of dense vegetation in *brousse tigrée*, generally convex downslope on *interfluves* and convex upslope in shallow drainage ways.

vein A sheet-like or *tabular*, *discordant*, mineralized body formed by complete or partial infilling of a *fracture* within a rock.

vein system A group of *veins* with a regular orientation.

velocity defect law A relationship describing the shape of the outer region of a *turbulent boundary layer* where the *Karman-Prandtl velocity law* does not apply. $(U_m - U)/U^* = -5.75 \log_{10}(y/\delta)$, where U_m = freestream velocity, U = mean velocity at a point at distance y from the wall to the *shear velocity* U^*, δ = *boundary layer* thickness or flow depth.

velocity log See *continuous velocity log*.

Vendian The upper *period* of the *Sinian*, 610–570 Ma.

ventifact An object modified in shape by wind action, primarily *sand abrasion*. See also *einkanter*, *dreikanter*, *zweikanter*, *yardang*.

Venturian A *Pliocene* succession on the west coast of the USA covering the upper part of the *Piacenzian*.

verdite A green rock mainly formed of green *mica* and *clay* used as an ornamental stone.

Vereiskian A *stage* of the *Carboniferous*, 311.3–309.2 Ma.

vergence The sense of *shear* deduced from the asymmetry of minor *structures* (*folds*, *faults* etc.) in an *orogenic belt*.

vermiculite ($(Mg,Ca)_{0.3}(Mg,Fe,Al)_3(Al,Si)_4O_{10}(OH)_2.nH_2O$) A *phyllosilicate* formed mainly by the *alteration* of *biotite*.

vernadite (δ-MnO_2) A manganese *mineral* found in *manganese nodules*.

vertical electrical sounding (VES, electric drilling) A *resistivity method* in which the current electrodes are expanded symmetrically so that the current penetrates progressively deeper into the subsurface. A *geoelectric section* can be obtained by inverting the sounding data.

vertical fold A *fold* in which the *hinge line* is vertical.

vertical seismic profile (VSP) A seismic profile obtained by recording a surface *seismic source* at a series of depths down a borehole.

very long baseline interferometry (VLBI) A method for determining the distance between two stations from the phase difference of the same signal received from a quasar, at an accuracy sufficient to monitor *continental drift*.

very low frequency method See *VLF method*.

VES See *vertical electrical sounding*.

vesicle A gas-filled cavity in a *magma* or *volcanic rock*.

vesiculation The growth of *vesicles* in a *magma*, caused by the *exsolution* of *volcanic gases*.

Veslyanskiy A *Permian* succession in the Timan area of the former USSR covering part of the *Wordian*.

vesuvian eruption A *volcanic eruption* characterized by very violent activity with the ejection of large quantities of material.

vesuvianite (idocrase) ($Ca_{10}(Mg,Fe)_2Al_4(SiO_4)_5(Si_2O_7)_2(OH)_4$) A *cyclosilicate* formed by the *contact metamorphism* of impure *limestone*.

VHN See *Vickers hardness number*.

vibration magnetometer A *magnetometer* in which a sample is vibrated to produce an oscillating voltage in a detector of the same frequency as the vibration.

Vibroseis® A mechanical *seismic source* in which a vibrator generates a wavetrain whose frequencies vary with time. Of low energy and thus environmentally advantageous. A conventional *seismogram* is obtained by *correlation* of the wavetrain with the received signal.

Vickers hardness number (VHN) A measure of the *hardness* of an *ore mineral* used in identification.

Vicksberg An *Oligocene* succession on the Gulf Coast of the USA covering the lower part of the *Rupelian*.

Vine-Matthews hypothesis A model which explains the origin of linear *oceanic magnetic anomalies* in terms of a *thermal remanent magnetization* acquired as newly-generated oceanic *lithosphere* cools in the *geomagnetic field*. As *seafloor spreading* continues, the *TRM* changes direction as the *geomagnetic field* reverses polarity.

Vinice An *Ordovician* succession in

Bohemia covering the lower part of the *Soudleyan*.

violane A massive, violet-blue variety of *diopside*.

violarite ($FeNi_2S_4$) A *sulphide mineral* of the *thiospinel group*.

Virgilian A *Carboniferous* succession in the USA covering parts of the *Kasimovian* and *Gzelian*.

viridine A green iron- and manganese-bearing variety of *andalusite*.

viscoelasticity *Deformation* in which *elastic* and *viscous* components act in parallel. Applied *stress* produces an initial rapid *strain* which decreases exponentially with time. On removal of the *stress*, the *strain* is recovered exponentially towards the initial value.

viscosity A *rheological* physical property of a material that deforms so that the ratio of *shear stress* to *shear strain rate* is constant, this ratio being the definition of the viscosity. Units Pa s.

viscosity coefficients The coefficients of proportionality in the tensor representation of the linear *stress–strain rate* relationship for a pure *Newtonian fluid*.

viscous flow (viscous strain) Behaviour in which there is a positive relationship between *shear stress* and *strain rate*.

viscous remanent magnetization (VRM) A low stability *remanent magnetization* acquired when a sample is held in a *magnetic field*, such as the *geomagnetic field*.

viscous strain See *viscous flow*.

viscous sublayer The region of flow closest to the bed in a *turbulent boundary layer* dominated by viscous forces and in which velocity increases linearly with height above the bed. $\delta = 11.5\ \upsilon/U^\star$, where δ = thickness of viscous sublayer, $U^\star$ = *shear velocity* of fluid of *kinematic viscosity* υ.

Viséan An *epoch* of the *Carboniferous*, 349.5–332.9 Ma.

vishnevite $(Na,Ca,K)_{6-7}Al_6Si_6O_{24}(SO_4,CO_3,Cl)_{1.0-1.5}.1\text{-}5H_2O$ A sulphate-bearing variety of *cancrinite*, found in *nepheline syenite*.

visual roundness scale A method of estimating the *roundness* of a *clast* by comparison with a template of twelve standard *clasts* of different *roundness* and *sphericity*.

vital effects Non-equilibrium fractionations of stable isotopes of carbon and oxygen in the rapid *precipitation* of biogenic carbonate.

vitrain A *lithotype of banded coal* comprising black, *vitreous*, *brittle* material with a *conchoidal fracture* that occurs in thin (6–8 mm) bands and forms the major constituent of *bituminous coal*.

vitreous lustre A *lustre* similar to broken glass.

vitric tuff A *tuff* with more *glassy* fragments than *lithic* or crystal fragments.

vitrification The formation of a *glass*.

vitrified fort An ancient Celtic fort that had been burned so intensely, possibly deliberately, that the *building stone* (*sandstone*, *conglomerate*, *gneiss* or *igneous rock*) is fused.

vitrinertite A bi-*maceral microlithotype of coal* comprising *vitrinite* and *inertinite*, characteristic of high *rank coals*.

vitrinite A *coal maceral* group mainly formed of the remains of trunks, branches, stems, roots and leaves of *land plants*, the principal group in most *coals*.

vitrite A *microlithotype of coal* composed of *vitrinite macerals*.

vitroclastic structure A structure formed in volcanic *ash* by the disruption of *vesicular glassy rocks*, so that most fragments have a concave outline.

vitrodurain A *lithotype of banded coal* intermediate between *vitrain* and *durain*.

vitrophyre A *porphyritic igneous rock* with a *glassy matrix*.

vitrophyric texture An *igneous rock texture* in which *phenocrysts* are embedded in a *glassy matrix*.

vivianite ($Fe_3(PO_4)_2.8H_2O$) A rare *secondary mineral* formed by the *weathering* of phosphates in *pegmatites*.

VLBI See *very long baseline interferometry*.

VLF method (very low frequency method) An *electromagnetic induction method* utilizing, as the primary field, low frequency signals in the waveband 15–25 kHz transmitted by large-scale navigational networks or military communication systems. Only a receiver is required and the surveying can be performed by a single operator.

vogesite A *calc-alkaline lamprophyre* with *phenocrysts* of *hornblende*, *pyroxene* ± *olivine* in a *matrix* in which *alkali feldspar* is in excess of *plagioclase*.

Voight model See *viscoelasticity*.

volcanic arenite A *lithic arenite* with *clasts* dominantly of volcanic origin.

volcanic-associated massive oxide deposit (Kiruna-type deposit) A *stratiform* oxide *ore* similar to a *massive sulphide deposit*, often characterized by *magnetite-haematite-apatite* in volcanic or volcanic-sedimentary terrains.

volcanic-associated massive sulphide deposit See *massive sulphide deposit*.

volcanic bomb (bomb) A mass of liquid *lava* thrown through the air which rotates and takes on a characteristic shape and structure.

volcanic breccia A non-sedimentary fragmental rock of volcanic origin.

volcanic earthquake An *earthquake* in a volcanically active area with an origin different from a *tectonic earthquake*. Possible origins include the modification of regional *stress* by *magma*, *volcanic eruption*, the oscillation of liquid ± gas in a *magma chamber* and *magma* flow in conduits compressed by *hydrostatic pressure*.

volcanic eruption An expulsion of *magma* ± broken rock from a *volcanic vent*, excluding the emission of gas from a *fumarole* and hot water from a *geyser*. See also *Hawaiian eruption*, *peléean eruption*, *plinian eruption*, *sub-plinian eruption*, *strombolian eruption*, *surtseyan activity* and *vulcanian eruption*.

volcanic-exhalative deposit (sedimentary exhalative deposit) A lenticular to sheet-like, *massive sulphide* *deposit* often found at the interface between volcanic or volcanic and sedimentary *beds*. Believed to be formed by *magmatic* fluids or the circulation of seawater.

volcanic gas The gas discharged into the atmosphere by *magmatic* activity, either *exsolved* from *magma* or through *fumaroles*. Common gases are H_2, N_2, HCl, HF, CH_4 and NH_4.

volcanic hazard The risk to life or property from volcanic activity, assessed for an area from historical records, the mapping of prehistoric deposits and comparison with the types of activity of *volcanoes* with similar evolution.

volcanic neck A positive topographic feature formed when a *volcanic plug* is exhumed by *erosion*.

volcanic pipe An approximately cylindrical conduit underlying a *volcanic vent*, 50 m–1 km in diameter and sometimes flaring towards the surface.

volcanic plug A mass of solidified *magma* or *breccia* filling a *volcanic pipe*.

volcanic rock A rock formed by the solidification of *lava* or *pyroclastic* material.

volcanic vent An orifice through which a *volcanic eruption* has occurred or is occurring.

volcanicity (vulcanicity, vulcanism) Volcanic activity.

volcaniclastic Descriptive of a *clastic rock* containing volcanic material.

volcano A location where *magma* and volatiles issue through the *crust* and accumulate.

Volgian A *stage* of the *Jurassic* in the *boreal* realm approximately equivalent to the *Tithonian*.

Von Karman's constant (κ) A constant, commonly taken as 0.4184, relating the mean size of eddies (l) in a flow to the depth of the flow (y): $l = \kappa y$.

Von Mises criterion The condition whereby a general *strain* at constant volume, excluding rotation, requires the simultaneous operation of five independent *slip systems*.

vortex ripple A steep variety of *wave ripple* in which the wavelength of the ripple-mark is 65% of the orbital diameter of the water motion near the bed, although this relationship breaks down above a critical *Reynolds number* and the *ripple* flattens.

vorticity The average *angular velocity* of lines between known points in a rock undergoing *strain*, used to describe *rotational strain*.

Vrzhumskiy A *Permian* succession on the eastern Russian Platform covering parts of the *Capitanian* and *Longtanian*.

VSP See *vertical seismic profile*.

vug A small, irregular cavity in an intrusive rock or carbonate sediment.

vulcanian eruption A type of *volcanic eruption* comprising a discrete, powerful, cannon-like blast with the expulsion of *lithic* and *scoria clasts*, *ash* and gas at high velocity from a *volcanic vent*, generating a short-lived *ash* plume. Possibly occurs after a build-up of *volcanic gas* pressure in a vent trapped beneath viscous *magma*.

vulcanicity See *volcanicity*.

vulcanism See *volcanicity*.

vulcanite A general name for a fine-grained *igneous rock*, usually forming a *lava flow*.

Vyatskiy A *Permian* succession on the eastern Russian Platform covering part of the *Changxinian*.

Vychegodskiy A *Permian* succession in the Timan area of the former USSR covering the lower part of the *Ufimian*.

wacke A *sandstone* with 15–75% *mud matrix* by volume.

wackestone A *limestone* comprising *matrix*-supported carbonate particles in *mud*.

wad A fine-grained mixture of manganese oxide and hydrated *oxide minerals*.

wadeite ($K_2CaZrSi_4O_{12}$) A colourless to lilac, transparent *mineral* found in potassium-rich rocks.

wadi (ouady, oued, wady) A steep-sided watercourse with sporadic flow in an arid region.

wady See *wadi*.

Waiauan A *Miocene* succession in New Zealand covering the lower part of the *Tortonian*.

Waipawan A *Palaeocene/Eocene* succession in New Zealand covering parts of the *Thanetian* and *Ypresian*.

Waipipian A *Pliocene* succession in New Zealand covering the lower part of the *Piacenzian*.

wairakite ($Ca_8(Al_{16}Si_{32}O_{96}).16H_2O$) A *zeolite* found in the deep parts of zones of *hydrothermal alteration*.

wairauite (CoFe) A rare alloy found in *peridotites*.

Waitakian An *Oligocene* succession in New Zealand covering the upper part of the *Chattian*.

Wall A *Permian* succession in Queensland, Australia, covering the lower part of the *Artinskian*.

wallrock The rock adjacent to a *mineral* deposit or igneous intrusion.

wallrock alteration The reaction of *hydrothermal fluids* with *wall rock*, decreasing in intensity with distance.

Walther's law A concept which suggests that a *conformable* vertical succession of *sedimentary facies* with no *unconformities* reflects their former lateral juxtaposition.

Wanganui A *Neogene* succession in New Zealand covering the *Opoitian*, *Waipipian*, *Mangapanian* and part of the *Nukumaruan*.

waning slope A slope on which weathered *scree* is eroded to form a concave *footslope*, which develops, and extends uphill, when stream incision ceases to be the dominant slope-forming process. cf. *waxing slope*.

want The missing part of a deposit resulting from *extensional faulting*.

Warendian An *Ordovician* succession in Australia covering the upper part of the *Tremadoc*.

Warepan A *Triassic* succession in New Zealand equivalent to the *Norian*.

Warthe Glaciation The third of the major *glaciations* affecting northern Europe in the *Quaternary*.

wash 1. See *arroyo*. 2. Loose debris.

wash load A fine-grained sediment that can be *suspended* at the lowest *discharges*.

Washita A *Cretaceous* succession on the Gulf Coast of the USA covering parts of the *Albian* and *Cenomanian*.

washout A lenticular body of *clastic sediment*, commonly *sandstone*, projecting from the roof of a *coal seam* and replacing all or part of it. Formed by scour and fill by a stream.

washover (fan) Sediment deposited by overwash of a *barrier island*, characteristic of microtidal *barrier islands* where *storm surge*-enhanced seas cannot pass through poorly-developed *tidal inlets* and break through the island.

washplain A nearly flat surface of *alluvium* covering deeply *weathered bedrock* in an environment where seasonal floods cannot incise due to the lack of abrasive *bedload* and large sediment volume.

waste The unwanted material which is removed during mining and quarrying.

water drive The buoyant effect of underlying water which drives oil through permeable rocks during *secondary migration*.

water gun A mechanical marine *seismic source* in which a pulse of high pressure water is released into the sea, creating a vacuum into which the surrounding water implodes.

water lane A tract of concentrated *sheetflow* in *brousse tigrée* with a uniformly dense tree cover.

water sapphire See *sapphir d'eau*.

water table The level of free-standing water within intergranular pores or fissures at the top of the *phreatic zone*, below which the pores of the host are saturated with water.

waterfall A stream falling over a precipice.

watershed (divide) The boundary of a drainage *basin*.

watten An area of tidal marshland and sand flats between the mainland and an *offshore bar* or island, exposed only at low *tide*.

Waucoban A *Cambrian* succession in the E USA covering the *Tommotian*, *Atdabanian* and part of the *Lenian*.

Waukesha Dolomite An *exceptional fossil deposit* of L. *Silurian* age at Brandon Bridge, Wisconsin with a marine biota of *arthropods*, worms and *conodonts* whose soft parts are preserved in phosphate.

wave-cut bench See *shore platform*.

wave-cut platform See *shore platform*.

wave diffraction The fanning out of a *wave* as it passes through an opening, for example in a breakwater.

wave rays (orthogonals) Lines drawn normal to wave crests, between which wave energy is constant and which converge and diverge at the coast to concentrate energy on headlands and spread it in bays.

wave refraction The process by which water wave crests are bent until they become parallel to submarine contours according to *Snell's law*, caused by the wave velocity decreasing with decreasing depth.

wave-ripple A periodic *bedform* generated by progressive *gravity waves*, usually symmetrical to slightly asymmetrical and trochoidal in section with wavelengths of 10 mm–1 m. See also *decaying ripple*, *growing ripple*, *orbital ripple*, *rolling grain ripple*, *vortex ripple*, *post-vortex ripple*.

wave-ripple cross-lamination A *cross-lamination* formed by the *migration* and/or *aggradation* of *wave*-generated *ripples*, characterized by unidirectional cross-laminae, lensoid and complexly interwoven *cross-sets*, irregular, undulating *cross-set* bases and *laminae discordant* with the external *ripple* form.

wavefront The surface over which the phase of a wave is the same and which propagates according to *Huygen's principle*.

wavelength dispersive analysis An *electron microprobe* technique, using a crystal *spectrometer* and gas-filled proportional counters, which provides low detection limits and a high resolution between elements.

wavellite $(Al_3(PO_4)_2(OH)_3.5H_2O)$ A *secondary mineral* found in low *grade metamorphic rocks*.

wavenumber The number of cycles per unit distance for a spatial waveform.

waves Regular oscillations of the water surface of oceans and lakes resulting mainly from wind-generated differences in pressure.

wavy bedding A type of *heterolithic bedding* comprising rippled *sand* with continuous *mud* draping the *ripples*.

waxing slope A convex slope at the crest of a cliff. cf. *waning slope*.

way-up See *younging*.

way-up criteria (way-up indicators) Phenomena used to determine the *younging* direction of *strata*, such as *sole marks*, *cross-stratification* and *graded bedding*.

way-up indicators See *way-up criteria*.

Weald Clay A *Cretaceous* succession in England equivalent to the *Hauterivian* and *Barremian*.

wear groove A feature on a *fault* surface caused by scratching by a hard particle.

weathered layer A heterogeneous surface layer up to tens of metres thick with an anomalously low *seismic velocity* caused by the presence of open *joints* and *microfractures* and its unsaturated state. Corrections are required for its effects on *seismic wave* travel times.

weathering The disintegration and decomposition of rock and sediment by near-surface mechanical and chemical processes.

weathering front A transition zone between unweathered *bedrock* and the overlying *saprolite* of a *deep weathering* profile, which may be at depths of up to 100 m in the tropics.

weathering index A measure of the intensity of chemical *weathering* in which comparison is made between a material which is stable with respect to the material considered.

weathering rind A layer of partly *weathered* rock a few centimetres thick which forms the surface of a *boulder* or *outcrop*. Often yellow, orange or red due to the *oxidation* of iron *minerals*.

weathering series A sequence of

common silicate *minerals* arranged in order of their susceptibility to chemical *weathering*.

websterite A *pyroxenite* with >95% *orthopyroxene* and *clinopyroxene* (each >10%). Occurs within *ultramafic* intrusions and as *ultramafic xenoliths* in *basalt*.

wehrlite 1. A *peridotite* with >95% *clinopyroxene* and *olivine* (>40%) and minor (<5%) *orthopyroxene*. Occurs within *ultramafic* intrusions and as *ultramafic xenoliths* in *basalt*. 2. See *telluric bismuth*.

Weichert-Herglov inversion A method of deriving the *seismic velocity*-depth distribution within the Earth from the variation of *arrival times* with *range* for *diving seismic waves*.

Weichsel Glaciation The fourth of the major *glaciations* affecting northern Europe in the *Quaternary*.

Weiss zone law (addition rule) A relationship defining the condition that a crystal plane is parallel to, or a face lies in, a given direction.

Weissenberg camera A technique of *X-ray diffraction analysis* in which the X-ray beam is incident normal to the axis of rotation of a crystal and *goniometer*. A cylindrical film concentric with the rotation axis oscillates through a large angle and screens limit the diffracted beams to those from the zero layer of the reciprocal axis. Different beam inclinations allow study of other lattice layers.

Weissliegendes A *Permian* succession in NW Europe covering the lower part of the *Kungurian*.

welded tuff A *volcanic rock* made up of explosively fragmented *tephra* which retained sufficient heat on deposition for the *glass* shards and *pumice clasts* to deform and adhere to produce a compact, *lava*-like rock.

well A dug or drilled hole which yields a fluid.

well logging See *geophysical borehole logging*.

well loss The part of the *drawdown* which results from water flowing into a *well* across the *well* face.

well shooting A method used to determine the *seismic velocity* of *strata* penetrated by a *borehole* by exploding a *shot* adjacent to it and measuring *arrival times* at a string of *geophones* down the *well*.

Wengen A *Triassic* succession in the Alps covering the middle part of the *Ladinian*.

Wenlock An *epoch* of the *Silurian*, 430.4–424.0 Ma.

Wenner configuration A common electrode arrangement in the *resistivity method* in which the distance between adjacent electrodes is constant.

Wentworth-Udden scale A scale used for *clast* size classification.

Werfen A *Triassic* succession in the Alps equivalent to the *Scythian*.

wernerite Members of the *scapolite series* intermediate in composition between *marialite* and *meionite*.

Werra A *Permian* succession in NW Europe covering the upper part of the *Ufimian*.

western states-type deposit See *sandstone-uranium-vanadium base metal deposit*.

Westphalian A *Carboniferous stage* in NW Europe covering part of the *Bashkirian* and the *Moscovian.*

westward drift The westwards motion, at ~0.2° a^{-1}, of most components of the *geomagnetic non-dipole field*, probably originating in the passage of magnetic lines of force through the outer *core*.

wet chemical analysis A classical method of chemical analysis in which each element is determined by a distinct, appropriate method, usually involving *colorimetry* or *titration.*

wet gas *Natural gas* containing significant amounts of gases other than *methane*. cf. *dry gas.*

wet-sediment deformation (soft-sediment deformation) The post- or syn-depositional *liquefaction* and/or *fluidization* of sediment, identified where *strain* has altered the geometry of primary sediment *lamination, bedding, fabric* or *texture.*

Wettersteinkkalk A *Triassic* succession in the Alps equivalent to the *Ladinian.*

Whaingaroan An *Oligocene* succession in New Zealand covering the *Rupelian* and part of the *Chattian.*

whetstone (honestone) A sharpening stone for a metal blade, typically rectangular in shape. *Schist* and *sandstone* were used in Britain from the Bronze Age; other utilized rocks include *basalt* and some *limestones.*

whin stone A popular term for any dark, fine-grained *igneous rock.*

white copperas See *goslarite.*

white corundum See *white sapphire.*

white iron pyrites See *marcasite.*

white lead ore See *cerussite.*

white mica See *muscovite.*

white nickel A popular name for *chloanthite.*

white sapphire (leucosapphire, white corundum) A pure, colourless variety of *corundum* used as a *gem.*

white smoker A plume of *hydrothermal fluid* from an ocean floor vent producing a white cloud of sulphate *minerals*. cf. *black smoker.*

white vitriol A popular name for *goslarite.*

Whiterockian Series A *series* of the *Ordovician* in North America.

whitings A suspension of *aragonite* needles forming milky-white patches tens to hundreds of metres long in subtropical oceans. Of uncertain origin.

whitlockite ($(Ca,Mg)_3(PO_4)_2$) Calcium phosphate, found as a *secondary mineral* in *granite pegmatites* and in phosphate deposits.

Whitwellian A *stage* of the *Silurian*, 426.1–425.4 Ma.

Widmanstätten pattern A pattern of intersecting *lamellae* on the polished, acid-etched surface of an *iron meteorite*, formed by the *exsolution* of *taenite* and *kamacite* from a nickel-iron alloy during very slow cooling.

Wiener filter A filter that converts the input signal into an output which comes closest, by a least squares criterion, to some desired form.

Wilcox An *Eocene* succession on the Gulf Coast of the USA covering the *Ypresian* and part of the *Lutetian.*

wildcat well A speculative exploratory *well.*

wildflysch A *turbidite*-like, *mass flow mictite* containing numerous exotic *clasts.*

willemite ($Zn2SiO_4$) A *nesosilicate* sometimes occurring as an *ore mineral.*

Wilson cycle A term referring to the sequence in the life of an ocean, from *continental splitting* through ocean growth by *seafloor spreading* to its destruction following *subduction* in a *continent–continent collision.*

wind ripple A small-scale, *aeolian bedform* created by wind acting on dry *sand* or *granules.* Wavelengths may reach 20 m and heights of 1 m, and are significantly smaller than the dimensions of a *dune.* See also *impact ripple, aerodynamic ripple.*

wind ripple lamination A low-angle *lamination* in a *sand sheet* deposit punctuated by subhorizontal *erosion* surfaces.

window An area where *erosion* has cut through a *thrust* or *recumbent fold* to expose the rocks beneath.

winnowing The removal of sedimentary particles by the wind or water currents.

wire-line logging See *geophysical borehole logging.*

Wisconsin Glaciation The fifth of the major *glaciations* affecting North America in the *Quaternary.*

withamite A variety of *piemontite.*

witherite ($BaCO_3$) A rare *mineral* found associated with *galena* in *hydrothermal veins.*

within-plate basalt (WPB) A *basalt* formed within a continental or oceanic *plate.*

Wolfcampian A *Carboniferous/Permian* succession in the Delaware Basin, USA covering part of the *Gzelian,* the *Asselian* and part of the *Sakmarian.*

wolframite ($(Fe,Mn)WO_4$) The major *ore mineral* of tungsten, found mainly in high-temperature *ore* and *quartz veins* in or near *granitic* rocks, in medium-temperature *ore* deposits and in *placer deposits.*

wollastonite ($CaSiO_3$) A *pyroxenoid* mainly formed by the *contact metamorphism* of *limestones.*

Wonokan A *stage* of the *Vendian,* 590–580 Ma.

wood opal Wood fossilized by *petrifaction* with *opal.*

wood tin A variety of *cassiterite* with a fibrous appearance.

Woodbine A *Cretaceous* succession on the Gulf Coast of the USA covering part of the *Cenomanian.*

Word A *Permian* succession in the Delaware Basin, USA equivalent to the *Wordian.*

Wordian A *stage* of the *Permian,* 255.0–252.5 Ma.

work hardening See *strain hardening.*

work softening See *strain softening.*

World-Wide Standarized Seismograph Network (WWSSN) A network designed and installed in the early 1960s to provide the means of monitoring underground nuclear explosions. The first global network utilizing common *seismometer* design and also used in the monitoring of *earthquakes*. Now superseded by the *Federation of Digital Seismic Networks*.

WPB See ***w**ithin-**p**late **b**asalt*.

wrench fault See *transcurrent fault*.

wrench fault system See *strike-slip fault system*.

wrinkle mark See *runzelmarken*.

Wujiapping A *Permian* succession in China equivalent to the *Longtanian*.

wulfenite ($PbMoO_4$) An orange-red molybdate *mineral* found in the oxidized parts of *lead veins*.

Wulff net (equal angle net) A type of *stereographic net* in which equal angles on the surface of a sphere project as equal distances on the net. Used in calculating angular relationships between planes and lines. cf. *Schmidt net*.

Würm Glaciation The fourth of the major *glaciations* affecting the Alps in the *Quaternary*.

wurtzite (ZnS) A *polymorph* of *sphalerite* found mainly in *hydrothermal vein* deposits.

wurtzite group A group of *sulphide minerals* with a structure similar to the *sphalerite group*, but with the arrangement of tetrahedra such that a lattice with hexagonal rather than cubic symmetry is formed.

WWSSN See ***W**orld-**W**ide **S**tandarized **S**eismograph **N**etwork*.

Wyandot A *Cretaceous* succession on the Scotian shelf of Canada covering part of the *Santonian*, the *Campanian* and part of the *Maastrichtian*.

wyomingite An alkaline *volcanic rock* comprising *leucite*, *phlogopite* and *diopside*.

X

X-ray crystallography The study of *crystal structure* using X-ray techniques.

X-ray diffraction analysis (XRD) A method of instrumental analysis making use of the *diffraction* of X-rays by crystalline materials. X-ray wavelengths are effective as they are of the same order of size as the atom spacings in crystals (~ 0.1 nm). X-rays are scattered by the electronic charges surrounding atomic nuclei and diffracted according to *Bragg's law* where they are detected by sensors at appropriate angles to provide information on the *crystal structure*.

X-ray fluorescence analysis (XRF) A method of instrumental analysis making use of the *fluorescence* of crystalline materials by X-rays. Widely used in archaeology to characterize artefacts and their sources.

xalostocite A pale rose-pink variety of *grossular*.

xanthophyllite See *clintonite*.

xeno- Foreign, strange or different.

xenoblastic texture A *texture* in a *metamorphic rock* in which *mineral* grains have developed without showing crystal faces.

xenocryst A foreign rock inclusion in an *igneous rock*. cf. *autolith*.

xenolith A foreign crystal inclusion in an *igneous rock*. cf. *autolith*.

xenomorphic With no crystal form.

xenotime (YPO_4) Yttrium phosphate resembling *zircon*, often with small amounts of cerium, erbium and thorium, found in *granites* and *pegmatites* as an *accessory mineral*.

xenotopic fabric A *fabric* of a crystalline carbonate rock or cement, or an *evaporite*, in which most of the crystals are *anhedral*.

Xinchang An *Ordovician* succession in China covering part of the *Tremadoc*.

Xiphosura A class of subphylum *Chelicerata*, phylum *Arthropoda*; diverse *arthropods* with a large prosoma, a variable number of opisthosomal somites and a long pointed telson. Range *Silurian–Recent*.

xonotlite ($Ca_6Si_6O_{17}(OH)_2$) A hydrous calcium silicate found mainly as small *veins* in *serpentine* or *contact metamorphic* zones.

XRD See *X-ray diffraction analysis*.

XRF See *X-ray fluorescence analysis*.

Xuzhuang A *Cambrian* succession in China covering parts of the *Solvan* and *Menevian*.

Y

Y shear A subsidiary *fault* in a *shear zone* parallel to the *shear* direction.

Yangyuan A *Carboniferous* succession in China equivalent to the *Tournaisian*.

Yapeenian An *Ordovician* succession in Australia covering part of the Late *Arenig*.

yardang An elongate landform resembling the hull of an inverted boat sculpted by wind *erosion* from weakly consolidated rocks.

yazoo A tributary stream running parallel to the main stream for some distance.

Yeadonian A *stage* of the *Carboniferous*, 320.6–318.3 Ma.

yellow ground Weathered and oxidized *kimberlite* at the surface. cf. *blue ground*.

yellow quartz See *citrine*.

yellow tellurium See *sylvanite*.

yellowcake Concentrated, precipitated and dried uranium oxide.

Yerkebidaik An *Ordovician* succession in Kazakhstan covering part of *Costonian*, the *Harnagian* and part of the *Soudleyan*.

Yichang (Ichang) An *Ordovician* succession in China comprising the *Xinchang* and *Ningguo*.

yield criterion The relation between the *principal stresses* at *yield stress* in the *deformation* of a *plastic* material.

yield depression curve A graph of water extraction rate from a *well* against *drawdown* for different pumping rates. Used to assess the rate for minimum *drawdown* which uses least energy in overcoming *head loss*.

yield point See *elastic limit*.

yield strength See *strength*.

yield stress The critical value of *stress* above which continuous, permanent *strain* results in a *plastic* material.

yield surface A representation of the *yield criterion* on a diagram showing the three *principal stresses* which define the *stress* conditions at which yield occurs.

Ynezian A *Palaeocene* succession on the west coast of the USA covering parts of the *Danian* and *Thanetian*.

yoderite ($(Al,Mg,Fe)_2Si(O,OH)_5$) A rare iron-magnesium silicate found in *schist*.

Yongningzhen A *Triassic* succession in China covering part of the *Nammalian* and the *Spathian*.

Yoredale facies A British name for the *cyclothem* sequence *limestone*, *shale*, *sandstone*, *coal* produced in a sinking *delta* environment during the *Carboniferous*.

Young's modulus An *elastic modulus*

defined as the ratio of applied *stress* to the resultant *extension* along the axis of a plane-ended cylinder.

younging The property of *strata* by which they become *stratigraphically* younger in a certain direction, which can be determined from *way-up criteria*. See also *facing*.

Ypresian A *stage* of the *Eocene*, 56.5–50.0 Ma.

yttrocerite A rare, massive, *granular* or earthy variety of cerian fluorite.

yu-stone A Chinese name for *jade* of *gem* quality.

yugawaralite ($Ca_2(Al_4Si_{12}O_{32}).8H_2O$) A *zeolite* found in networks and *veins* and as crystals in cavities in *andesite tuffs* which have been altered by waters from *hot springs*.

Yumatin A *period* of the *Riphean*, 1350–1050 Ma.

Z

z fold An *asymmetrical fold* with one *limb* shorter than the other whose profile defines a 'Z' shape. cf. *s fold.*

Zahorany An *Ordovician* succession in Bohemia covering the upper part of the *Soudleyan.*

Zanclian The lower *stage* of the *Pliocene*, 5.2–3.40 Ma.

Zechstein The younger *epoch* of the *Permian*, 256.1–245.0 Ma.

Zechstein Sea A sea of restricted circulation covering parts of E England, the Low Countries, Germany and North Sea in U. *Permian* times, whose evaporation gave rise to thick *evaporite* deposits.

Zechsteinkalk A *Permian* succession in NW Europe covering the lower part of the *Ufimian.*

Zemorrian A *Miocene* succession on the west coast of the USA covering the *Oligocene* and part of the *Aquitanian.*

zeolite facies A *metamorphic facies* characterized by the development of *smectite zeolites* in *basic igneous rocks* at low temperatures and pressures.

zeolites Hydrated aluminosilicate *minerals* with a framework structure which encloses cavities occupied by large ions and water molecules, which both have considerable mobility, allowing ion exchange and reversible dehydration. Generally white to colourless in the pure state but often coloured by the presence of *iron oxides* or other impurities. Occur in *amygdales* and fissures in *basic volcanic rocks*, as *authigenic minerals* in *sedimentary rocks*, in *tuffs* and in low *grade metamorphic rocks* as a result of *hydrothermal alteration.*

zeolitization A type of *wall rock alteration* in which *stilbite, natrolite, heulandite* and other *zeolites* are developed.

zero length spring A special type of spring used in modern *gravimeters* which is pretensioned during manufacture so that the restoring force is proportional to its length, thus increasing the range and sensitivity of the instrument.

zeuge A *tabular* rock mass perched on a pinnacle of softer rock as the result of differential *erosion* of the underlying rock.

Zhangxia A *Cambrian* succession in China covering part of the *Menevian.*

Zharyk An *Ordovician* succession in Kazakhstan covering the upper part of the *Onnian.*

Zi-Liu-Jing A *Jurassic* succession in Sichuan, China, equivalent to the *Sinemurian, Pliensbachian, Toarcian* and *Aalenian.*

zibar A coarse-grained, low relief (< 10 m high), *aeolian* depositional feature with no *slip-faces*, found in regular spacings of up to 400 m and forming undulating surfaces on *sand sheets* and in *interdunes*.

zig-zag fold See *chevron fold*.

zinc bloom See *hydrozincite*.

zinc spinel See *gahnite*.

zincblende See *sphalerite*.

zincite (red zinc ore) (ZnO) An *ore mineral* of zinc found associated with *franklinite*, *willemite* and *calcite* in zinc deposits.

Zingg diagram A diagram used to plot the relative dimensions of a particle in terms of its maximum, intermediate and minimum dimensions.

zinkenite ($PbSb_2S_4$) A steel-grey, lead-antimony sulphide occurring as columnar, often very thin, hexagonal crystals forming a fibrous mass in low- to medium-temperature *vein* deposits.

zinnwaldite ($K(Li,Al,Fe)_3(Al,Si)_4O_{10}(OH,F)_2$) A grey to brown *mica* similar to *lepidolite* containing appreciable iron, found in *tin veins* and *granite pegmatites*.

zircon ($ZrSiO_4$) A *nesosilicate*, a common *accessory mineral* in *igneous rocks* and used as a *gem* when transparent.

Zlichovian A *Devonian* succession in Czechoslovakia covering the lower part of the *Emsian*.

Zoantharia A subclass of class *Anthozoa*; solitary or *colonial corals* with polyps bearing multiples of six tentacles. Range M. *Triassic–Recent*.

Zoeppritz' equations Formulae which describe how the *seismic wave* energy of a *P* or *S wave* is partitioned between reflected, transmitted and mode-converted waves incident on a planar discontinuity.

zoisite ($Ca_2Al_3O(SiO_4)(Si_2O_7OH)$) A *sorosilicate mineral*, an orthorhombic *polymorph* of *clinozoisite*, found in *metamorphic rocks*.

zonal index fossil A *fossil* with a wide geographic range and short temporal range used in *biostratigraphy*.

zonation The subdivision of a *stratigraphic* unit by means of *fossils*.

zone 1. A *biostratigraphical* subdivision, several of which comprise a *stage*. 2. A set of crystal faces all parallel to the *zone axis*.

zone axis The direction to which the crystal faces of a *zone* are parallel.

zone fossil A *fossil* species characterizing a *zone*.

zone of aeration See *vadose zone*.

zone of fluctuation The zone between the highest and lowest levels reached by the *water table*.

zone of oxidation 1. The zone in the upper part of a sulphide *orebody* in which copper, zinc and silver sulphides are soluble and the *orebody* is *oxidized* and *leached* of many of its valuable elements down to the *water table*, leaving a *gossan*. 2. Any zone in which *oxidation* occurs.

zone of permanent saturation See *phreatic zone*.

zone refining A process of *magma diversification* during *partial melting* in which a succession of zones of molten or partially molten *magma* pass through a rock by causing melting at the front of each zone followed by *crystallization* and deposition behind. Certain elements are preferentially included in the melt and carried forward by each zone.

zoned crystal A crystal with concentric zones of varying composition, reflecting a complex *crystallization* history.

zoning of ore deposits The zoning of the *mineralogy* or chemistry of *ore*, *industrial* or *gangue minerals* in a *mineral* deposit or district.

Zooxanthellae Symbiotic *algae* found in the endoderm of *reef corals*.

Zosterophyllopsida A class of division *Trachaeophyta*, kingdom *Plantae*; the ancestors of microphyllous plants. Range U. *Silurian*–U. *Devonian*.

zunyite ($Al_{13}Si_5O_{20}(OH,F)_{18}Cl$) A colourless to grey to pink *mineral* found in *veins*, disseminated in *porphyries* and as an *alteration* product of *feldspar*.

zussmanite ($K(Fe,Mg,Mn)_{13}(Si,Al)_{18}O_{42}(OH)_{14}$) A pale green, *tabular*, hydrated iron-magnesium-potassium silicate found in *metamorphosed shale*, siliceous *ironstone* and impure *limestone*.

zweikanter A *ventifact* with two curved surfaces intersecting at a sharp edge. cf. *einkanter*, *dreikanter*.

BIBLIOGRAPHY

General

Allaby, A. and Allaby, M. (eds) (1990) *The Concise Oxford Dictionary of Earth Sciences*. Oxford University Press, Oxford.

Bott, M. H. P. (1982) *The Interior of the Earth*. Arnold, London.

Brown, G. C. and Mussett, A. E. (1993) *The Inaccessible Earth*. Allen & Unwin, London.

Brown, G. C., Hawkesworth, C. J. and Wilson, R. C. L. (1992) *Understanding the Earth*. Cambridge University Press, Cambridge.

Compton, R. R. (1985) *Geology in the Field*. John Wiley & Sons, New York.

Emiliani, C. (1992) *Planet Earth*. Cambridge University Press, Cambridge.

Finkl, C. W. (ed.) (1984) *The Encyclopedia of Applied Geology*. Van Nostrand Reinhold, New York.

Finkl, C. W. (ed.) (1988) *The Encyclopedia of Field and General Geology*. Van Nostrand Reinhold, New York.

Harben, P. W. and Bates, R. L. (1984) *Geology of the Nonmetallics*. Metal Bulletin Inc., New York.

Jacobs, J. A. (1987) *The Earth's Core*. Academic Press, London.

James, D. E. (ed.) (1989) *Encyclopedia of Solid-Earth Geophysics*. Van Nostrand, New York.

Kearey, P. (ed.) (1993) *The Encyclopedia of the Solid Earth Sciences*. Blackwell Scientific Publications, Oxford.

Kennett, J. (1982) *Marine Geology*. Prentice-Hall, Englewood Cliffs, NJ.

Le Roy, L. W. and Le Roy, D. O. (eds) (1977) *Subsurface Geology*. Colorado School of Mines, Golden.

Moseley, F. (1981) *Methods in Field Geology*. W. H. Freeman, San Francisco.

Press, F. and Siever, R. (1982) *Earth*. W. H. Freeman, San Francisco.

Scrutton, R. A. and Talwani, M. (eds) (1982) *The Ocean Floor*. John Wiley & Sons, Chichester.

Taylor, S. R. and McClennan, S. M. (1985) *The Continental Crust: its Composition and Evolution*. Blackwell Scientific Publications, Oxford.

Economic Geology

Armstead, H. C. H. and Tester, J. W. (1987) *Heat Mining*. Spon, London.

Armstead, H. C. H. (1978) *Geothermal Energy*. Spon, London.

Barnes, H. L. (ed.) (1979) *Geochemistry of Hydrothermal Ore Deposits*. John Wiley & Sons, New York.

Berger, B. R. and Bethke, P. M. (eds) (1986) *Geology and Geochemistry of Epithermal Systems*. Society of Economic Geologists, El Paso, TX.

Brooks, J. and Fleet, A. J. (eds) (1987) *Marine Petroleum Source Rocks*. Blackwell Scientific Publications, Oxford.

Brooks, J. (ed.) (1983) *Petroleum Geology and Exploration of Europe*. Geological Society Special Publication 12. Blackwell Scientific Publications, Oxford.

Butler, E. W. and Pick, J. B. (1982) *Geothermal Energy Development*. Plenum, New York.

Collins, A. G. (1975) *Geochemistry of Oilfield Waters*. Developments in Petroleum Science 1. Elsevier, Amsterdam.

Downing, R. A. and Gray, D. A. (1986) *Geothermal Energy – The Potential in the United Kingdom*. British Geological Survey, HMSO, London.

Edwards, R. and Atkinson, K. (1986) *Ore Deposit Geology*. Chapman & Hall, London.

Erickson, A. J. Jr (ed.) (1984) *Applied Mining Geology*. American Institute of Mining, Metallurgical and Petroleum Engineers, New York.

Evans, A. M. (1993) *Ore Geology and Industrial Minerals*. Blackwell Scientific Publications, Oxford.

Freidrich, G. H., Genkin, A. J., Naldrett, A. J. *et al.* (eds) (1986) *Geology and Metallogeny of Copper Deposits*. Springer-Verlag, Berlin.

Gray, P. H. J., Bowyer, G. J., Castle, J. F., Vaughan, D. J. and Warner, N. A. (eds) (1990) *Sulphide Deposits – Their Origin and Processing*. Institute of Mineralogy and Metallurgy, London.

Guilbert, J. M. and Park, C. F. Jr (1986) *The Geology of Ore Deposits*. W. H. Freeman, New York.

Hobson, G. D. (ed.) (1980) *Developments in Petroleum Geology*. Applied Science Publishers, London.

Hutchison, C. S. (1983) *Economic Deposits and Their Tectonic Setting*. Macmillan, London.

Jenson, M. L. and Bateman, A. M. (1979) *Economic Mineral Deposits*. John Wiley & Sons, New York.

Macdonald, E. H. (1983) *Alluvial Mining*. Chapman & Hall, London.

Magara, K. (1978) *Compaction and Fluid Migration – Practical Petroleum Geology*. Developments in Petroleum Science. Elsevier, Amsterdam.

Maynard, J. B. (1983) *Geochemistry of Sedimentary Ore Deposits*. Springer-Verlag, New York.

North, F. K. (1985) *Petroleum Geology*. Allen & Unwin, Boston.

Peters, W. C. (1987) *Exploration and Mining Geology*. John Wiley & Sons, New York.

Roberts, W. H. and Cordell, R. J. (eds) (1980) *Problems of Petroleum Migration*. Studies in Geology 10. American Association of Petroleum Geologists, Tulsa, OK.

Stach, E., Mackowsky, M.-Th., Teichmuller, M. *et al.* (1982) *Stach's Textbook of Coal Petrology*. Gebrüder Borntraeger, Berlin.

Tarling, D. H. (ed.) (1981) *Economic Geology and Geotectonics*. Blackwell Scientific Publications, Oxford.

Titley, S. R. (1982) *Advances in Geology of Porphyry Copper Deposits*. University of Arizona Press, Tucson, AZ.

Ward, C. R. (1984) *Coal Geology and Coal Technology*. Blackwell Scientific Publications, Melbourne.

Wolf, K. H. (ed.) (1976) *Handbook of Strata-bound and Stratiform Ore Deposits*. Elsevier, Amsterdam.

Engineering Geology and Hydrogeology

Anderson, J. G. C. and Trigg, C. F. (1976) *Case Histories in Engineering Geology*. Elek Sciences, London.

Atkinson, B. K. (ed.) (1987) *Fracture Mechanics of Rock*. Academic Press, New York.

Chow, V. T. (ed.) (1964) *Handbook of Applied Hydrogeology*. McGraw-Hill, New York.

Clayton, C. R. I., Symonds, N. E. and Matthews, M. C. (1982) *Site Investigation – A Handbook for Engineers*. Granada, London.

Francis, J. R. D. (1975) *Fluid Mechanics for Engineering Students*. Arnold, London.

Jaeger, J. C. and Cook, N. G. (1979) *Fundamentals of Rock Mechanics*. Chapman & Hall, London.

Price, M. (1985) *Introducing Groundwater*. Allen & Unwin, London.

Geochemistry

Henderson, P. (1982) *Inorganic Geochemistry*. Pergamon, Oxford.

Iler, R. K. (1979) *Chemistry of Silica*. Wiley-Interscience, New York.

Krauskopf, K. B. (1979) *Introduction to Geochemistry*. Dowden, Hutchinson & Ross, Stroudsburg, PA.

Lasaga, A. C. and Kirkpatrick, R. J. (eds) (1981) *Kinetics of Geochemical Processes*. Reviews in Mineralogy 8. Mineralogical Society of America, Washington, DC.

Lerman, A. (1979) *Geochemical Processes. Water and Sediment Environments*. John Wiley & Sons, New York.

Newman, A. C. D. (ed.) (1987) *Chemistry of Clays and Clay Minerals*. Mineralogical Society Monograph 6.

Saxena, S. K. (1973) *Thermodynamics of Rock-forming Crystalline Solutions*. Springer-Verlag, Berlin.

Sosman, R. B. (1965) *The Phases of Silica*. Rutgers University Press, New Brunswick.

Stumm, W. and Morgan, J. J. (1981) *Aquatic Chemistry*. John Wiley & Sons, New York.

Vaughan, D. J. and Craig, J. R. (1978) *Mineral Chemistry of Metal Sulphides*. Cambridge University Press, Cambridge.

Geomorphology

Bird, E. C. F. (1984) *Coasts*. Basil Blackwell, Oxford.

Bowden, K. F. (1983) *Physical Oceanography of Coastal Waters*. Ellis Horwood, Chichester.

Büdel, J. (1982) *Climatic Geomorphology*. Princeton University Press, Princeton, NJ.

Coates, D. R. (1981) *Environmental Geology*. John Wiley & Sons, New York.

Cooke, R. U. and Doornkamp, J. C. (1974) *Geomorphology and Environmental Management*. Clarendon Press, Oxford.

Drewry, D. (1986) *Glacial Geologic Processes*. Arnold, London.

Eyles, N. (ed.) (1983) *Glacial Geology*. Pergamon Press, Oxford.

Gardner, R. and Scoging, H. (eds) (1983) *Mega-geomorphology*. Oxford University Press, Oxford.

Goudie, A. S. (1973) *Duricrusts in Tropical and Subtropical Landscapes*. Clarendon Press, Oxford.

Goudie, A. S. (1981) *Geomorphological Techniques*. Allen & Unwin, London.

Goudie, A. S. (1986) *The Human Impact on the Environment*. Basil Blackwell, Oxford.

Goudie, A. S. (ed.) (1985) *The Encyclopaedic Dictionary of Physical Geography*. Blackwell Scientific Publications, Oxford.

Goudie, A. S. and Pye, K. (eds) (1983) *Chemical Sediments and Geomorphology*. Academic Press, London.

Greeley, R. and Iverson, J. D. (1985) *Wind as a Geological Process*. Cambridge University Press, Cambridge.

Jennings, J. M. (1985) *Karst Geomorphology*. Basil Blackwell, Oxford.

Johnson, A. M. (1970) *Physical Processes in Geology*. Freeman Cooper, San Francisco.

King, C. A. M. (1972) *Beaches and Coasts*. Arnold, London.

La Fleur, R. G. (ed.) (1984) *Groundwater as a Geomorphic Agent*. Allen & Unwin, Boston.

Lerman, A. (ed.) (1978) *Chemistry, Geology and Physics of Lakes*. Springer-Verlag, New York.

Nickling, W. G. (ed.) (1986) *Aeolian Geomorphology*. Unwin Hyman, Boston.

Paterson, K. and Sweeting, M. M. (eds) (1986) *New Directions in Karst*. Geo Books, Norwich.

Pethick, J. S. (1984) *An Introduction to Coastal Geomorphology*. Arnold, London.

Pond, S. and Pickard, G. L. (1983) *Introductory Dynamical Oceanography*. Pergamon, Oxford.

Richards, K. S. (1982) *Rivers: Form and Process in Alluvial Channels*. Methuen, London.

Richards, K. S., Arnett, R. R. and Ellis, S. (eds) (1984) *Geomorphology and Soils*. Allen & Unwin, London.

Schrumm, S. A. (1977) *The Fluvial System*. Wiley-Interscience, New York.

Selby, M. J. (1985) *The Earth's Changing Surface*. Clarendon Press, Oxford.

Shepard, F. P. (1978) *Geological Oceanography*. Heinemann, London.
Syvitski, J. P. M., Burrell, D. C. and Skei, J. M. (1987) *Fjords: Processes and Products*. Springer-Verlag, New York.
Trenhaile, A. S. (1987) The *Geomorphology of Rock Coasts*. Clarendon Press, London.
Trudgill, S. (1985) *Limestone Geomorphology*. Longman, London.
Trudgill, S. T. (ed.) (1986) *Solute Processes*. John Wiley & Sons, Chichester.
Verstappen, H. T. (1983) *Applied Geomorphology*. Elsevier, Amsterdam.
Washburn, A. L. (1979) *Geocryology: A Survey of Periglacial Processes and Environments*. Arnold, London.

Geophysics

Aki, K. and Richards, P. G. (1980) *Quantitative Seismology; Theory and Methods*. W. H. Freeman, San Francisco.
Anstey, N. A. (1977) *Seismic Interpretation: the Physical Aspects*. IHRDC Publications, Boston.
Anstey, N. A. (1981) *Seismic Prospecting Instruments*. Gebrüder Borntraeger, Berlin.
Bolt, B. A. (1988) *Earthquakes – A Primer*. W. H. Freeman, San Francisco.
Bullen, K. E. and Bolt, B. A. (1985) *An Introduction to the Theory of Seismology*. Cambridge University Press, Cambridge.
Butler, R. F. (1992) *Paleomagnetism*. Blackwell Scientific Publications, Oxford.
Ellis, D. V. (1987) *Well Logging for Earth Scientists*. Elsevier, Amsterdam.
Jacobs, J. A. (1984) *Reversals of the Earth's Magnetic Field*. Adam Hilger, Bristol.
Kearey, P. and Brooks, M. (1991) *An Introduction to Geophysical Exploration*. Blackwell Scientific Publications, Oxford.
Khramov, A. N. (1987) *Paleomagnetology*. Springer-Verlag, Berlin.
Labo, J. (1986) *A Practical Introduction to Borehole Geophysics*. Society of Exploration Geophysicists, Tulsa, OK.
Lambeck, K. (1980) *The Earth's Variable Rotation*. Cambridge University Press, Cambridge.
Mörner, N. (ed.) (1980) *Earth Rheology, Isostasy and Eustasy*. John Wiley & Sons, Chichester.
O'Reilly, W. (1984) *Rock Magnetism*. Blackie, Glasgow.
Piper, J. D. A. (1987) *Palaeomagnetism and the Continental Crust*. Halsted Press, New York.
Rikitake, T. (1976) *Earthquake Prediction*. Elsevier, Amsterdam.
Serra, O. (1984) *Fundamentals of Well Log Interpretation*. 1 & 2. Developments in Petroleum Science 15A & 15B. Elsevier, Amsterdam.
Sheriff, R. E. (1980) *Seismic Stratigraphy*. IHRDC Publications, Boston.
Simon, R. B. (1981) *Earthquake Interpretation: A Manual for Reading Seismograms*. Kauffmann, Los Altos, CA.
Tarling, D. H. (1983) *Palaeomagnetism*. Chapman & Hall, London.
Telford, W. M., Geldart, L. P. and Sheriff, R. E. (1990) *Applied Geophysics*. Cambridge University Press, Cambridge.

Mineralogy and Crystallography

Bailey, S. W. (ed.) (1984) *Micas*. Reviews in Mineralogy 13. Mineralogical Society of America, Washington, DC.

Bailey, S. W. (ed.) (1988) *Hydrous Phyllosilicates Exclusive of Micas*. Reviews in Mineralogy 19. Mineralogical Society of America, Washington, DC.

Ballhausen, C. J. and Gray, H. B. (1965) *Molecular Orbital Theory*. Benjamin, New York.

Ballhausen, C. J. (1962) *Introduction to Ligand Field Theory*. McGraw-Hill, New York.

Berry, L. G. and Mason, B. (1983) *Mineralogy: Concepts, Descriptions, Determinations*. W. H. Freeman, San Francisco.

Bishop, A. C. (1967) *An Outline of Crystal Morphology*. Hutchinson, London.

Brindley, G. W. and Brown, G. (eds) (1984) *Crystal Structures of Clay Minerals and their X-ray Identification*. Mineralogical Society Monograph 5.

Brown, W. L. (ed.) (1984) *Feldspars and Feldspathoids. Structures, Properties and Occurrences*. NATO ASI Series, Series C, 137. Riedel, Dordrecht.

Clark, A. H. (1993) *Hey's Mineral Index*, 3rd edition. British Museum of Natural History and Chapman & Hall, London.

Cox, P. A. (1987) *The Electronic Structure and Chemistry of Solids*. Oxford University Press, Oxford.

Cuilty, B. D. (1978) *Elements of X-ray Diffraction*. Addison Wesley, Reading, MA.

Deer, W. A., Howie, R. A. and Zussman, J. (1978) *Rock Forming Minerals*. Longman, Harlow.

Deer, W. A., Howie, R. A. and Zussman, J. (1992) *An Introduction to the Rock Forming Minerals*, 2nd edition. Longman, London.

Dent Glasser, L. S. (1977) *Crystallography and its Applications*. Van Nostrand Reinhold, New York.

Ernst, W. G. (1968) *Amphiboles*. Springer-Verlag, New York.

Farmer, V. C. (ed.) (1974) *The Infrared Spectra of Minerals*. Mineralogical Society Monograph 4.

Faure, G. (1977) *Principles of Isotope Geology*. John Wiley & Sons, New York.

Fleischer, M. and Mandarino, J. A. (1991) *Glossary of Mineral Species 1991*. Mineralogical Record, Tucson, AZ.

Goldstein, J. I., Newbury, D. E., Echin, P. *et al.* (1981) *Scanning Electron Microscopy and X-ray Microanalysis*. Plenum, London.

Gottardi, G. and Galli, E. (1985) *Natural Zeolites*. Springer-Verlag, Berlin.

Gribble, C. D. and Hall, A. J. (1985) *A Practical Introduction to Optical Mineralogy*. Allen & Unwin, London.

Grim, R. F. (1968) *Clay Mineralogy*. McGraw-Hill, New York.

Hawthorne, F. C. (ed.) (1988) *Spectroscopic Methods in Mineralogy and Geology*. Reviews in Mineralogy 18. Mineralogical Society of America, Washington, DC.

Hurbut, C. S. Jr and Klein, C. (1985) *Manual of Mineralogy*. John Wiley & Sons, New York.

Kalló, D. and Sherry, H. S. (eds) (1988) *Occurrence, Properties and Utilization of Natural Zeolites.* Akadémiai Kiadó, Budapest.

Klug, H. P. and Alexander, L. E. (1974) *X-ray Diffraction Procedures.* John Wiley & Sons, London.

Ladd, M. F. C. and Palmer, R. A. (1985) *Structure Determination by X-ray Crystallography.* Plenum, New York.

Leake, B. E. (1978) 'The nomenclature of amphiboles.' *Mineralogical Magazine* 42, 533–65.

McKie, D. and McKie, C. (1986) *Essentials of Crystallography.* Blackwell Scientific Publications, Oxford.

Morimoto, N. (1988) 'Nomenclature of pyroxenes.' *Mineralogical Magazine* 52, 535–50.

Nickel, E. H. and Nichols, M. C. (1991) *Mineral Reference Manual.* Chapman & Hall, London and Van Nostrand Reinhold, New York.

Papike, J. J. (ed.) (1969) *Pyroxenes and Amphiboles: Crystal Chemistry.* Special Paper 2. Mineralogical Society of America.

Phillips, F. C. (1971) *An Introduction to Crystallography.* Longman, London.

Poirier, J. P. (1986) *Creep of Crystals.* Cambridge University Press, Cambridge.

Putnis, A. and McConnell, J. D. C. (1980) *Principles of Mineral Behaviour.* Blackwell Scientific Publications, Oxford.

Ribbe, P. H. (ed.) (1982) *Orthosilicates.* Reviews in Mineralogy 5. Mineralogical Society of America, Washington, DC.

Roberts, W. L., Rapp, G. P. Jr and Weber, J. (1974) *Encyclopedia of Minerals.* Van Nostrand Reinhold, New York.

Rumble III, D. (ed.) (1976) *Oxide Minerals.* Reviews in Mineralogy 3. Mineralogical Society of America, Washington, DC.

Sand, L. B. and Mumpton, F. H. (eds) *Natural Zeolites: Occurrence, Properties, Use.* Pergamon, Oxford.

Smith, J. V. and Brown, W. L. (1988) *Feldspar Minerals* Vol. 1: *Crystal Structure, Physical, Chemical and Microtextural Properties.* Springer-Verlag, Berlin.

Urch, D. S. (1979) *Orbitals and Symmetry.* Macmillan, London.

Veblen, D. R. and Ribbe, P. H. (eds) (1983) *Amphiboles* – Phase Relations. Reviews in Mineralogy 9B. Mineralogical Society of America, Washington, DC.

Veblen, D. R. (ed.) (1983) *Amphiboles and Other Hydrous Pyriboles* – Mineralogy. Reviews in Mineralogy 9A. Mineralogical Society of America, Washington, DC.

Velde, B. (1977) *Clays and Clay Minerals in Natural and Synthetic Systems.* Elsevier, Amsterdam.

Verma, A. R. and Krishna, P. (1966) *Polymorphism and Polytypism in Crystals.* John Wiley & Sons, New York.

Woodgate, G. K. (1980) *Elementary Atomic Structure.* Oxford University Press, Oxford.

Miscellaneous

Allum, J. A. E. (1986) *Photogeology and Regional Mapping*. Pergamon, Oxford.

Bauer, J. and Bouska, V. (1983) *A Guide in Colour to Precious and Semiprecious Stones*. Octopus Books, London.

Carter, D. J. (1986) *The Remote Sensing Sourcebook*. Kogan Page, London and McCarta Ltd, London.

Drury, S. A. (1987) *Image Interpretation in Geology*. Allen & Unwin, London.

Durrance, E. M. (1986) *Radioactivity in Geology*. Ellis Horwood, Chichester.

Ford, T. E. and Cullingford, C. H. D. (1976) *The Science of Speleology*. Academic Press, London.

Kempe, D. R. C. and Harvey, A. P. (eds) (1983) *The Petrology of Archaeological Artefacts*. Clarendon Press, Oxford.

Menard, H. W. (1986) *Islands*. Scientific American Library, New York.

Moore, C. A. (1973) *Handbook of Subsurface Geology*. Harper & Row, New York.

Ollier, C. D. (1969) *Weathering*. Oliver & Boyd, Edinburgh.

Roberts, J. L. (1982) *Introduction to Geological Maps and Structures*. Pergamon, Oxford.

Tite, M. S. (1972) *Methods of Physical Examination in Archaeology*. Seminar Press, London.

Tritton, D. J. (1977) *Physical Fluid Dynamics*. Van Nostrand Reinhold, Wokingham.

Webster, R. A. (1979) *Practical Gemmology*. NAG Press, London.

Palaeontology

Aldridge, R. J. (ed.) (1987) *Palaeobiology of Conodonts*. Ellis Horwood, Chichester.

Alexander, R. McN. (1981) *The Chordates*. Cambridge University Press, Cambridge.

Anderson, O. R. (1983) *Radiolaria*. Springer-Verlag, New York.

Benton, M. J. (1990) *Vertebrate Palaeontology*. Unwin Hyman, London.

Bolli, H. M., Saunders, J. B. and Perch-Nielsen, K. (eds) (1985) *Plankton Stratigraphy*. Cambridge University Press, Cambridge.

Boucot, A. (1975) *Evolution and Extinction Rate*. Developments in Palaeontology and Stratigraphy. Elsevier, Amsterdam.

Brasier, M. D. (1980) *Microfossils*. Allen & Unwin, London.

Carroll, R. L. (1987) *Vertebrate Palaeontology and Evolution*. W. H. Freeman, Chicago.

Clarkson, E. N. K. (1986) *Invertebrate Palaeontology and Evolution*. Allen & Unwin, London.

Dodd, J. R. and Stanton, R. J. (1981) *Paleoecology, Concepts and Applications*. John Wiley & Sons, Chichester.

Flügel, E. (ed.) (1977) *Fossil Algae, Recent Results and Developments*. Springer-Verlag, Berlin.

Frey, R. W. (ed.) (1975) *The Study of Trace Fossils*. Springer-Verlag, Berlin.

Glaessner, M. F. (1984) *The Dawn of Animal Life – A Biohistorical Study*. Cambridge University Press, Cambridge.
Hamilton, G. B. and Lord, A. R. (eds) (1982) *A Stratigraphical Index of Calcareous Nannofossils*. Ellis Horwood, Chichester.
Haynes, J. R. (1981) *Foraminifera*. Macmillan, London.
House, M. R. (ed.) (1979) *The Origin of Major Invertebrate Groups*. Academic Press, London.
Laport, L. F. (ed.) (1974) *Reefs in Time and Space*. Special Publication 18. Society of Economic Paleontologists and Mineralogists, Tulsa, OK.
Moore, R. C. (ed.) (1955, 1969) *Treatise on Invertebrate Paleontology*. Geological Society of America and University of Kansas Press, Boulder and Lawrence.
Murray, J. W. (ed.) (1985) *Atlas of Invertebrate Macrofossils*. Longman, Harlow.
Nielson, C. and Larwood, G. P. (eds) (1985) *Bryozoa: Ordovician to Recent*. Olsen & Olsen, Fredensborg.
Nitecki, M. H. (ed.) (1984) *Extinctions*. University of Chicago Press, Chicago.
Purchon, R. D. (1968) *The Biology of the Mollusca*. Pergamon, Oxford.
Raup, D. M. and Jablonski, D. (eds) (1986) *Patterns and Processes in the History of Life*. Springer-Verlag, Berlin.
Romer, A. S. (1966) *Vertebrate Paleontology*. Chicago University Press, Chicago.
Rudwick, M. J. S. (1970) *Living and Fossil Brachiopods*. Hutchinson, London.
Sarjeant, W. S. (1974) *Fossil and Living Dinoflagellates*. Academic Press, London.
Schopf, T. J. M. (ed.) (1972) *Models in Paleobiology*. W. H. Freeman, San Francisco.
Schram, F. R. (1986) *Crustacea*. Oxford University Press, Oxford.
Smith, A. B. (1984) *Echinoid Palaeobiology*. Allen & Unwin, Hemel Hempstead.
Stanley, S. M. (1979) *Macroevolution, Pattern and Process*. W. H. Freeman, San Francisco.
Stanley, S. M. (1987) *Extinctions*. Scientific American Books, New York.
Tappan, H. (1980) *The Paleobiology of Plant Protists*. W. H. Freeman, San Francisco.
Tevesz, M. J. S. and McCall, P. L. (eds) (1983) *Biotic Interactions in Recent and Fossil Communities*. Plenum, London.
Thomas, B. A. and Spicer, R. A. (1987) *The Evolution and Palaeobiology of Land Plants*. Croom Helm, London.
Tiffney, B. H. (ed.) (1986) *Geological Factors in the Evolution of Plants*. Yale University Press, New Haven, CT.
Walter, M. R. (ed.) (1976) *Stromatolites*. Developments in Sedimentology 20. Elsevier, Amsterdam.
Wray, J. L. (1977) *Calcareous Algae*. Developments in Paleontology and Stratigraphy 4. Elsevier, Amsterdam.

Petrology, Igneous and Metamorphic

Arndt, N. T. and Nisbet, E. G. (eds) (1982) *Komatiites*. Allen & Unwin, London.
Atherton, M. P. and Tarney, J. (1981) *The Origin of Granite Batholiths*. Geochemistry Group of the Mineralogical Society. Shiva, Kent.

Barker, F. (ed.) (1979) *Trondhjemites, Dacites and Related Rocks*. Elsevier, Amsterdam.

Best, M. G. (1982) *Igneous and Metamorphic Petrology*. W. H. Freeman, New York.

Bowes, D. R. (ed) (1990) *The Encyclopedia of Igneous and Metamorphic Petrology*. Van Nostrand Reinhold, New York.

Coleman, R. G. (1977) *Ophiolites*. Springer-Verlag, Heidelberg.

Cox, K. G., Bell, J. D. and Pankhurst, R. J. (1979) *The Interpretation of Igneous Rocks*. Allen & Unwin, London.

Crawford, A. J. (ed.) (1989) *Boninites*. Unwin Hyman, London.

Fitton, J. G. and Upton, B. G. J. (eds) (1987) *Alkaline Igneous Rocks*. Geological Society Special Publication 30. Blackwell Scientific Publications, Oxford.

Gupta, A. K. and Yaki, K. (1979) *Petrology and Genesis of Leucite-bearing Rocks*. Springer-Verlag, Berlin.

Kornprobst, J. (ed.) (1984) *Kimberlite 1. Kimberlites and Related Rocks*. Elsevier, Amsterdam.

Morse, S. A. (1980) *Basalts and Phase Diagrams*. Springer-Verlag, Berlin.

Nixon, P. H. (ed.) (1987) *Mantle Xenoliths*. John Wiley & Sons, Chichester.

Park, R. G. and Tarney, J. (eds) (1987) *Evolution of Lewisian and Comparable Precambrian High Grade Terrains*. Geological Society Special Publication 27. Blackwell Scientific Publications, Oxford.

Ringwood, A. E. (1975) *Composition and Petrology of the Earth's Mantle*. McGraw-Hill, New York.

Streckeisen, A. L. (1967) 'Classification and nomenclature of igneous rocks (Final report of an enquiry).' *Neues Jahrbuch für Mineralogie, Abhandlung*, 107, 144–240.

Thorpe, R. S. (1982) *Andesites: Orogenic Andesites and Related Rocks*. Wiley & Sons, Chichester.

Wager, L. R. and Brown, G. M. (1968) *Layered Igneous Rocks*. Oliver & Boyd, Edinburgh.

Wimmaenauer, W. (ed.) (1974) *The Alkaline Rocks*. John Wiley & Sons, New York.

Yoder, H. S. (ed.) (1979) *The Evolution of the Igneous Rocks*. Princeton University Press, Princeton, NJ.

Plate Tectonics

Carey, S. W. (1976) *The Expanding Earth*. Elsevier, Amsterdam.

Condie, K. C. (1989) *Plate Tectonics and Crustal Evolution*. Pergamon, Oxford.

Cox, A. and Hart, R. B. (1986) *Plate Tectonics. How it Works*. Blackwell Scientific Publications, Palo Alto, CA.

Cox, A. (1973) *Plate Tectonics and Geomagnetic Reversals*. W. H. Freeman, San Francisco.

Davies, P. A. and Runcorn, S. K. (eds) (1980) *Mechanisms of Continental Drift and Plate Tectonics*. Academic Press, London.

Gill, J. B. (1981) *Orogenic Andesites and Plate Tectonics*. Springer-Verlag, Berlin.

Kearey, P. and Vine, F. J. (1990) *Global Tectonics*. Blackwell Scientific Publications, Oxford.

Le Grand, H. E. (1988) *Drifting Continents and Shifting Theories*. Cambridge University Press, Cambridge.

Park, R. G. (1988) *Geological Structures and Moving Plates*. Blackie, Glasgow.

Toksöz, M. N., Uyeda, S. and Francheteau, J. (eds) (1980) *Ocean Ridges and Arcs*. Elsevier, Amsterdam.

Windley, B. F. (1984) *The Evolving Continents*. John Wiley & Sons, New York.

Sedimentology

Allen, J. R. L. (1982) *Sedimentary Structures, Their Character and Physical Basis*. Developments in Sedimentology 30. Elsevier, New York.

Allen, J. R. L. (1985) *Principles of Physical Sedimentology*. Allen & Unwin, London.

Bathurst, R. G. C. (1971) *Carbonate Sediments and their Diagenesis*. Developments in Sedimentology 12. Elsevier, Amsterdam.

Berner, R. A. (1980) *Early Diagenesis: A Theoretical Approach*. Princeton University Press, Princeton, NJ.

Blatt, H., Middleton, G. V. and Murray, R. C. (1980) *Origin of Sedimentary Rocks*. Prentice-Hall, Englewood Cliffs, NJ.

Boggs, S. H. Jr (1987) *Principles of Sedimentology and Stratigraphy*. Merrill, Columbus, OH.

Bradshaw, P. (1971) *An Introduction to Turbulence and its Measurement*. Pergamon, Oxford.

Brenchley, P. J. and Williams, B. P. J. (1985) *Sedimentology: Recent Developments and Applied Aspects*. Geological Society Special Publication 18. Blackwell Scientific Publications, Oxford.

Brodozikowski, K. and Van Loon, A. J. (1991) *Glacigenic Sediments*. Elsevier, Amsterdam.

Brookfield, M. E. and Ahlbrandt, T. S. (eds) (1983) *Aeolian Sediments and Processes*. Developments in Sedimentology 38. Elsevier, Amsterdam.

Collinson, J. D. and Lewin, J. (eds) (1983) *Modern and Ancient Fluvial Systems*. Special Publications of the International Association of Sedimentologists 6. Blackwell Scientific Publications, Oxford.

Dans, R. A. (ed.) (1987) *Coastal Sedimentary Environments*. Springer-Verlag, New York.

Dowdeswell, J. A. and Scourse, J. D. (1990) *Glacimarine Environments*. Geological Society, London.

Duchafour, P. (1982) *Pedology, Pedogenesis and Classification*. Allen & Unwin, London.

Ethridge, F. C. and Flores, R. M. (eds) (1987) *Recent Developments in Fluvial Sedimentology*. Special Publication 39. Society of Economic Paleontologists and Mineralogists, Tulsa, OK.

Ethridge, F. C. and Flores, R. M. (eds) (1981) *Recent and Ancient Nonmarine*

Depositional Environments: Models for Exploration. Special Publication 31. Society of Economic Paleontologists and Mineralogists, Tulsa, OK.

Fitzpatrick, E. A. (1983) *Soils and their Formation, Classification and Distribution*. Longman, London.

Flügel, E. (1982) *Microfacies Analysis of Limestones*. Springer-Verlag, Berlin.

Frostick, L. E. and Reid, I. (eds) (1987) *Desert Sediments: Ancient and Modern*. Geological Society Special Publication 35. Blackwell Scientific Publications, Oxford.

Ham, W. E. (ed.) (1962) *Classification of Carbonate Rocks*. Memoir 1. American Association of Petroleum Geologists, Tulsa, OK.

Head, K. H. (1982) *Manual of Soil Laboratory Testing*. Pentech Press, London.

Kaplan, J. R. (ed.) (1974) *Natural Gases in Marine Sediments*. Plenum, New York.

Kirkland, D. W. and Evans, R. (1973) *Marine Evaporites: Origin, Diagenesis and Geochemistry*. Benchmark Papers in Geology. Hutchinson & Ross, Stroudsburg, PA.

Komar, P. D. (1976) *Beach Processes and Sedimentation*. Prentice-Hall, Englewood Cliffs, NJ.

Larsen, G. and Chilingar, G. V. (eds) (1967) *Diagenesis in Sediments*. Developments in Sedimentology 8. Elsevier, Amsterdam.

Leeder, M. R. (1982) *Sedimentology: Process and Product*. Allen & Unwin, London.

Leggett, J. K. and Zuffa, G. G. (eds) (1987) *Marine Clastic Sedimentology*. Graham and Trotman, London.

Lerche, I. and O'Brien, J. J. (1986) *The Dynamical Geology of Salt Related Structures*. Academic Press, London.

Matter, A. and Tucker, M. (eds) (1978) *Modern and Ancient Lake Sediments*. Special Publication of the International Association of Sedimentologists 2. Blackwell Scientific Publications, Oxford.

Miall, A. D. (ed.) (1978) *Fluvial Sedimentology*. Canadian Society of Petroleum Geologists Memoir 5.

Parker, A. and Sellwood, B. (eds) (1983) *Sediment Diagenesis*. NATO ASI Series C, 115, Reidel, Dordrecht.

Peryt, T. (ed.) (1983) *Coated Grains*. Springer-Verlag, Berlin.

Peterson, J. A. and Osmond, J. C. (eds) (1961) *Geometry of Sand Bodies*. American Association of Petroleum Geologists, Tulsa, OK.

Pettijohn, F. J., Potter, P. E. and Siever, R. (1973) *Sand and Sandstone*. Springer-Verlag, Berlin.

Reading, H. G. (ed.) (1986) *Sedimentary Environments and Facies*. Blackwell Scientific Publications, Oxford.

Reeder, R. J. (ed.) (1983) *Carbonates: Mineralogy and Chemistry*. Reviews in Mineralogy 11. Mineralogical Society of America, Washington DC.

Rubin, D. M. (1987) *Cross-bedding, Bedforms and Paleocurrents*. Concepts in Sedimentology 1. Society of Economic Paleontologists and Mineralogists, Tulsa, OK.

Schlanger, S. O. and Cita, M. B. (eds) (1982) *Nature and Origins of Cretaceous Carbon-rich Facies*. Academic Press, New York.

Scholle, P. A. and Schluger, P. R. (eds) (1979) *Aspects of Diagenesis*. Special

Publication 26. Society of Economic Paleontologists and Mineralogists, Tulsa, OK.

Sieveking, G. de G. and Hart, M. B. (eds) (1986) *The Scientific Study of Flint and Chert*. Cambridge University Press, Cambridge.

Soil Survey Staff (1992) *Key to Soil Taxonomy*, 5th edition. Soil Management Support Services Technical Monograph No. 19. Pochahontas Press, Blacksburg.

Sonnenfeld, P. (1984) *Brines and Evaporites*. Academic Press, London.

Stride, A. H. (ed.) (1982) *Offshore Tidal Sands: Processes and Deposits*. Chapman & Hall, London.

Van der Meer, J. J. M. (ed.) (1989) *Tills and Glaciotectonics*. Balkema, Rotterdam.

Walker, R. G. (ed.) (1984) *Facies Models*. Geoscience Canada Reprint Series 1.

Wright, V. P. (ed.) (1986) *Paleosols: Their Recognition and Interpretation*. Blackwell Scientific Publications, Oxford.

Yalin, M. S. (1977) *Mechanics of Sediment Transport*. Pergamon, Oxford.

Stratigraphy

Audley-Charles, M. G. and Hallam, A. (1988) *Gondwana and Tethys*. Geological Society Special Publication 37. Oxford University Press, Oxford.

Berggren, W. A. and Van Couvering, J. A. (eds) (1984) *Catastrophes and Earth History*. Princeton University Press, Princeton, NJ.

Bruton, D. L. (ed.) (1983) *Aspects of the Ordovician System*. Palaeontological Contributions from the University of Oslo 295. Universitetsforlaget, Oslo.

Eicher, D. L. and McAlester, A. L. (1980) *History of the Earth*. Prentice-Hall, Englewood Cliffs, NJ.

Eicher, D. L. (1976) *Geologic Time*. Prentice-Hall, Englewood Cliffs, NJ.

Glover, J. E. and Groves, D. I. (eds) (1981) *Archaean Geology*. Special Publication of the Geological Society of Australia 7.

Hallam, A. (1987) *Jurassic Environments*. Cambridge University Press, Cambridge.

Hambrey, M. J. and Harland, W. B. (1981) *Earth's Pre-Pleistocene Glacial Record*. Cambridge University Press, Cambridge.

Harland, W. B., Armstrong, R. L., Cox, A. V., Craig, L. E., Smith, A. G. and Smith, D. G. (1990) *A Geologic Timescale 1990*. Cambridge University Press, Cambridge.

Harris, A. L., Holland, C. H. and Leake, B. E. (eds) (1979) *The Caledonides of the British Isles – Reviewed*. Geological Society Special Publication 8. Scottish Academic Press, Edinburgh.

Hoffman, A. and Nitecki, M. H. (eds) (1986) *Problematic Fossil Taxa*. Oxford University Press, New York.

Holland, C. H. (ed.) (1971) *Cambrian of the New World*. Wiley Interscience, New York.

Holland, C. H. (ed.) (1974) *Cambrian of the British Isles, Norden and Spitzbergen*. John Wiley & Sons, Chichester.

House, M. R., Scrutton, C. T. and Bassett, M. G. (eds) (1979) *The Devonian*

System. Special Paper in Palaeontology 23. The Palaeontological Association, London.

Imbrie, J. and Imbrie, K. P. (1979) *Ice-ages: Solving the Mystery*. Macmillan, London.

Kauffman, E. G. and Hazel, J. E. (eds) (1977) *Concepts and Methods of Biostratigraphy*. Dowden, Hutchinson & Ross, Stroudsburg, PA.

Nilsson, T. (1983) *The Pleistocene, Geology and Life of the Quaternary Ice Age*. Riedel, Dordrecht.

Nisbet, E. G. (1987) *The Young Earth*. Allen & Unwin, Boston.

Odin, G. S. (ed.) (1982) *Numerical Dating in Stratigraphy*. John Wiley & Sons, Chichester.

Pomerol, C. (1982) *The Cenozoic Era*. Ellis Horwood, Chichester.

Snelling, N. J. (ed.) *The Chronology of the Geological Record*. Memoirs of the Geological Society 10. Blackwell Scientific Publications, Oxford.

Zapfe, H. (ed.) (1974) *The Stratigraphy of The Alpine-Mediterranean Triassic*. Springer-Verlag, Vienna.

Structural Geology and Geodynamics

Coward, M. P., Dewey, J. F. and Hancock, P. L. (eds) (1987) *Continental Extensional Tectonics*. Geological Society Special Publication 28. Blackwell Scientific Publications, Oxford.

Friedel, J. (1964) *Dislocations*. Pergamon, London.

Griggs, D. and Handin, J. (eds) (1979) *Rock Deformation*. Geological Society of America Memoir 79.

Hobbs, B. E., Means, W. D. and Williams, P. F. (1976) *An Outline of Structural Geology*. John Wiley & Sons, New York.

Hsu, K. J. (ed.) (1982) *Mountain Building Processes*. Academic Press, London.

Hull, D. (1975) *Introduction to Dislocations*. Pergamon, Oxford.

Kroner, A. and Greiling, R. (eds) (1984) *Precambrian Tectonics Illustrated*. Schweizerbart, Stuttgart.

Lawn, B. R. and Wilshaw, T. R. (1975) *Fracture of Brittle Solids*. Cambridge University Press, Cambridge.

McClay, K. R. and Price, N. J. (eds) *Thrust and Nappe Tectonics*. Geological Society Special Publication 9. Blackwell Scientific Publications, Oxford.

Means, W. D. (1978) *Stress and Strain*. John Wiley & Sons, New York.

Miyashiro, A., Aki, K. and Sengör, A. M. C. (1982) *Orogeny*. John Wiley & Sons, New York.

Nicholas, R. and Poirier, J. P. (1976) *Crystalline Plasticity and Solid-state Flow*. Wiley Interscience, London.

Park, R. G. (1983) *Foundations of Structural Geology*. Blackie, Glasgow.

Paterson, M. S. (1978) *Experimental Rock Deformation – The Brittle Field*. Springer-Verlag, Berlin.

Phillips, F. C. (1971) *The Use of Stereographic Projection in Structural Geology*. Arnold, London.

Price, N. J. (1966) *Fault Development in Brittle and Semi-brittle Rock*. Pergamon, Oxford.

Ramsay, J. G. and Huber, M. I. (1987) *Techniques of Modern Structural Geology*. Academic Press, London.

Ramsay, J. G. (1967) *Folding and Fracturing of Rocks*. McGraw-Hill, New York.

Seyfert, C. K. (ed.) (1987) *The Encyclopedia of Structural Geology and Plate Tectonics*. Van Nostrand Reinhold, New York.

Suppe, J. (1985) *Principles of Structural Geology*. Prentice-Hall, Englewood Cliffs, NJ.

Turner, F. J. and Weiss, L. E. (1963) *Structural Analysis of Metamorphic Tectonites*. McGraw-Hill, New York.

Vita-Finzi, C. (1986) *Recent Earth Movements: An Introduction to Neotectonics*. Academic Press, London.

Volcanology

Cas, R. A. F. and Wright, J. G. (1987) *Volcanic Successions Modern and Ancient*. Allen & Unwin, London.

Chapin, C. E. and Elston, W. E. (eds) *Ash Flow Tuffs*. Special Paper of the Geological Society of America 180.

Fisher, R. V. and Schminke, H.-U. (1984) *Pyroclastic Rocks*. Springer-Verlag, Berlin.

Francis, P. (1976) *Volcanoes*. Penguin, Harmondsworth, Middlesex.

Le Bas, M. J. (1977) *Carbonatite – Nepheline Volcanism*. John Wiley & Sons, New York.

Macdonald, G. A. (1972) *Volcanoes*. Prentice-Hall, Englewood Cliffs, NJ.

Sheets, P. D. and Grayson, D. K. (1979) *Volcanic Activity and Human Ecology*. Academic Press, London.

Simkin, T., Siebert, L., McLelland, L. *et al.* (1981) *Volcanoes of the World*. Hutchinson Ross, Stroudsburg, PA.

Tazieff, H. and Sabroux, J. C. (1983) *Forecasting Volcanic Events*. Elsevier, Amsterdam.

Williams, H. and McBirney, A. R. (1979) *Volcanology*. Freeman, Cooper & Co., San Francisco.

READ MORE IN PENGUIN

POLITICS AND SOCIAL SCIENCES

National Identity Anthony D. Smith

In this stimulating new book, Anthony D. Smith asks why the first modern nation states developed in the West. He considers how ethnic origins, religion, language and shared symbols can provide a sense of nation and illuminates his argument with a wealth of detailed examples.

The Feminine Mystique Betty Friedan

'A brilliantly researched, passionately argued book – a time-bomb flung into the Mom-and-Apple-Pie image ... Out of the debris of that shattered ideal, the Women's Liberation Movement was born' – Ann Leslie

Faith and Credit Susan George and Fabrizio Sabelli

In its fifty years of existence, the World Bank has influenced more lives in the Third World than any other institution yet remains largely unknown, even enigmatic. This richly illuminating and lively overview examines the policies of the Bank, its internal culture and the interests it serves.

Political Ideas Edited by David Thomson

From Machiavelli to Marx – a stimulating and informative introduction to the last 500 years of European political thinkers and political thought.

Structural Anthropology Volumes 1–2 Claude Lévi-Strauss

'That the complex ensemble of Lévi-Strauss's achievement ... is one of the most original and intellectually exciting of the present age seems undeniable. No one seriously interested in language or literature, in sociology or psychology, can afford to ignore it' – George Steiner

Invitation to Sociology Peter L. Berger

Sociology is defined as 'the science of the development and nature and laws of human society'. But what is its purpose? Without belittling its scientific procedures Professor Berger stresses the humanistic affinity of sociology with history and philosophy. It is a discipline which encourages a fuller awareness of the human world ... with the purpose of bettering it.

READ MORE IN PENGUIN 

POLITICS AND SOCIAL SCIENCES

Conservatism Ted Honderich

'It offers a powerful critique of the major beliefs of modern conservatism, and shows how much a rigorous philosopher can contribute to understanding the fashionable but deeply ruinous absurdities of his times' – *New Statesman & Society*

The Battle for Scotland Andrew Marr

A nation without a parliament of its own, Scotland has been wrestling with its identity and status for a century. In this excellent and up-to-date account of the distinctive history of Scottish politics, Andrew Marr uses party and individual records, pamphlets, learned works, interviews and literature to tell a colourful and often surprising account.

Bricks of Shame: Britain's Prisons Vivien Stern

'Her well-researched book presents a chillingly realistic picture of the British sytstem and lucid argument for changes which could and should be made before a degrading and explosive situation deteriorates still further' – *Sunday Times*

Inside the Third World Paul Harrison

This comprehensive book brings home a wealth of facts and analysis on the often tragic realities of life for the poor people and communities of Asia, Africa and Latin America.

'Just like a Girl' Sue Sharpe
How Girls Learn to be Women

Sue Sharpe's unprecedented research and analysis of the attitudes and hopes of teenage girls from four London schools has become a classic of its kind. This new edition focuses on girls in the nineties – some of whom could even be the daughters of the teenagers she interviewed in the seventies – and represents their views and ideas on education, work, marriage, gender roles, feminism and women's rights.

READ MORE IN PENGUIN

REFERENCE

The Penguin Dictionary of Literary Terms and Literary Theory
J. A. Cuddon

'Scholarly, succinct, comprehensive and entertaining, this is an important book, an indispensable work of reference. It draws on the literature of many languages and quotes aptly and freshly from our own' – *The Times Educational Supplement*

The Penguin Spelling Dictionary

What are the plurals of *octopus* and *rhinoceros*? What is the difference between *stationery* and *stationary*? And how about *annex* and *annexe*, *agape* and *Agape*? This comprehensive new book, the fullest spelling dictionary now available, provides the answers.

Roget's Thesaurus of English Words and Phrases
Edited by Betty Kirkpatrick

This new edition of Roget's classic work, now brought up to date for the nineties, will increase anyone's command of the English language. Fully cross-referenced, it includes synonyms of every kind (formal or colloquial, idiomatic and figurative) for almost 900 headings. It is a must for writers and utterly fascinating for any English speaker.

The Penguin Dictionary of English Idioms
Daphne M. Gulland and David G. Hinds-Howell

The English language is full of pitfalls for the foreign student – but the most common problem lies in understanding and using the vast array of idioms. *The Penguin Dictionary of English Idioms* is uniquely designed to stimulate understanding and familiarity by explaining the meanings and origins of idioms and giving examples of typical usage.

The Penguin Wordmaster Dictionary
Martin H. Manser and Nigel D. Turton

This dictionary puts the pleasure back into word-seeking. Every time you look at a page you get a bonus – a panel telling you everything about a particular word or expression. It is, therefore, a dictionary to be read as well as used for its concise and up-to-date definitions.

READ MORE IN PENGUIN

REFERENCE

Medicines: A Guide for Everybody Peter Parish

Now in its seventh edition and completely revised and updated, this bestselling guide is written in ordinary language for the ordinary reader yet will prove indispensable to anyone involved in health care – nurses, pharmacists, opticians, social workers and doctors.

Media Law Geoffrey Robertson QC and Andrew Nichol

Crisp and authoritative surveys explain the up-to-date position on defamation, obscenity, official secrecy, copyright and confidentiality, contempt of court, the protection of privacy and much more.

The Slang Thesaurus Jonathon Green

Do you make the public bar sound like a gentleman's club? The miraculous *Slang Thesaurus* will liven up your language in no time. You won't Adam and Eve it! A mine of funny, witty, acid and vulgar synonyms for the words you use every day.

The Penguin Dictionary of Troublesome Words Bill Bryson

Why should you avoid discussing the *weather conditions*? Can a married woman be celibate? Why is it eccentric to talk about the aroma of a cowshed? A straightforward guide to the pitfalls and hotly disputed issues in standard written English.

The Penguin Dictionary of Musical Performers Arthur Jacobs

In this invaluable companion volume to *The Penguin Dictionary of Music* Arthur Jacobs has brought together the names of over 2,500 performers. Music is written by composers, yet it is the interpreters who bring it to life; in this comprehensive book they are at last given their due.

The Penguin Dictionary of Third World Terms Kofi Buenor Hadjor

Words associated with the Third World are rarely subject to analysis and definition. Yet many of these terms are loaded with assumptions and cultural attitudes. As a result, discussion of the Third World can be bogged down in bias, prejudice – and Western self-interest.